Statistical Abstract of Undergraduate Programs in the Mathematical Sciences in the United States

Fall 2005 CBMS Survey

Statistical Abstract of Undergraduate Programs in the Mathematical Sciences in the United States

Fall 2005 CBMS Survey

David J. Lutzer
The College of William and Mary

Stephen B. Rodi
Austin Community College

Ellen E. Kirkman
Wake Forest University

James W. Maxwell
American Mathematical Society

This survey was supported by the National Science Foundation under grant #DMS-0412843.

Any opinions, findings, and conclusions or recommendations expressed in this material are those of the authors and do not necessarily reflect the views of the National Science Foundation.

Library of Congress Cataloging-in-Publication Data

Statistical abstract of undergraduate programs in the mathematical sciences in the United States : fall 2005 CBMS Survey / David J. Lutzer . . .[et al.].
p. cm.
ISBN 978-0-8218-4332-1 (alk. paper)
1. Mathematics—Study and teaching (Higher)—United States—Statistics—Tables. I. Maxwell, James W., 1944–

QA13.S745 2007
510.71′1—dc22 2007060823

Visit the AMS home page at http://www.ams.org/

10 9 8 7 6 5 4 3 2 1 12 11 10 09 08 07

Contents

Acknowledgments

Many people and organizations played important roles in the CBMS2005 project, and we want to thank them. First is the CBMS Council, which authorized the project. Next is the National Science Foundation, which funded it. For the record, we note that the opinions expressed in this report do not necessarily reflect the views of the National Science Foundation. Third are the members of the CBMS2005 Steering Committee who contributed so much to the formulation of the questionnaire and to the follow-up efforts in fall 2005. In addition to the four authors of this report, the Steering Committee members were Ray Collings of Georgia Perimeter College, John Fulton of Clemson University, William Kalsbeek of the University of North Carolina, Darcy Mays of Virginia Commonwealth University, Emily Puckette of University of the South, Ron Rosier of CBMS, and Susan Wood of the Virginia Community College System. Fourth is Colleen Rose of the American Mathematical Society, whose technical assistance was crucial in the preparation of the CBMS questionnaires and in some of the data analysis. Fifth are the hundreds of department chairs and program directors who made sure that the very long CBMS2005 questionnaire was completed and returned. Sixth are Robert Agans and a group of graduate students at the University of North Carolina Survey Research Unit who devoted so many hours to cleaning data and writing the programs that carried out the statistical analyses. Seventh is a group of careful readers who studied drafts of the report and tables, identifying inconsistencies and pointing out findings of the survey that we had missed. In addition to members of the Steering Committee mentioned above, two others deserve special mention – Katherine Kulick of the College of William and Mary and Carolyn Neptune of Johnson County Community College. In the two-year college community, we are indebted to the Executive Board of the American Mathematical Association of Two-Year Colleges and to the Committee on Two-Year Colleges of the Mathematical Association of America for helping us develop the following team of volunteers who were active in obtaining survey response: Judy Ackerman, Geoffrey Akst, Steve Blasberg, Gary Britton, David Buchtal, Kevin Charlwood, Elizabeth Chuy, Cheryl Cleaves, Ruth Collins, Judy Devoe, Irene Doo, Anne Dudley, David Dudley, Jan Ford, Ben Fusaro, Wanda Garner, Larry Gilligan, Christie Gilliland, Margie Hobbs, Glenn Jacobs, Mary Ann Justinger, Robert Kimball, Stephen Krevisky, Reginald Luke, Shawna Mahan, Bob Malena, Jay Malmstrom, Abe Mantel, Lois Martin, Marilyn Mays, Kathy Mowers, Mary Robinson, Alfredo Rodriguez, Jim Roznowski, Jon Scott, Karen Sharp, Dale Siegel, Irene Starr, Jim Trefzger, Karen Walters, Ann Watkins, Pete Wildman, and Kathie Yoder. Finally we want to thank our spouses for their help and tolerance of the many weekends that we devoted to this project.

Ellen Kirkman
David Lutzer
James Maxwell
Stephen Rodi

Foreword

Every five years since 1965, the Conference Board of the Mathematical Sciences (CBMS) has sponsored a study of undergraduate mathematics and statistics in U.S. colleges and universities, and this is the ninth report in that series. With NSF support the CBMS2005 project surveyed a stratified random sample of three separate universes: two-year college mathematics programs, mathematics departments in four-year colleges and universities, and statistics departments in four-year colleges and universities.

As part of an ongoing cross-sectional study, the CBMS2005 project collected data on enrollments, bachelors degrees granted, and faculty demographics in each of the three universes mentioned above. Results of these studies appear in Chapters 1, 3, 4, 5, 6, and 7 of this report, with global data appearing in Chapter 1 and more fine-structured information in the other chapters. For example, data on the total number of bachelors degrees granted in the 2004–2005 academic year appear in Table S.4 of Chapter 1, and in Table E.2 of Chapter 3, where those data are broken out by the type of department through which the degrees were granted.

In addition, based on proposals from various professional society committees, the CBMS2005 project studied certain special topics that were judged to be especially timely. These were the mathematical education of pre-service teachers, academic resources available to undergraduates, dual-enrollments, mathematics in the general education curriculum, requirements of the national mathematics major, and assessment practices in college and university mathematics and statistics departments. Reports on these special projects appear in Chapter 2.

The CBMS2005 project differs from its predecessors in that the data in this report came from two separate surveys. Historically, CBMS surveys have not been the only source for faculty demographic data in the mathematics and statistics departments of four-year colleges and universities. A group of mathematical sciences professional societies have combined to sponsor a Joint Data Committee (JDC) that collects and publishes annual demographic data in the Notices of the American Mathematical Society, and in 1995 and 2000 there was considerable overlap between JDC and CBMS efforts to collect faculty demographic data. In response to complaints from department chairs about that overlap, the JDC and the CBMS2005 Steering Committee agreed to coordinate their efforts in fall 2005. See Chapter 4 for details.

To put the CBMS2005 data in context, this report sometimes refers to earlier CBMS reports (called CBMS2000, CBMS1995, etc.) and to other professional society reports. Publication data on the other reports cited appears in the CBMS2005 bibliography section.

Chapter 1

Summary of CBMS2005 Report

Highlights of Chapter 1

A. Enrollments

- Between fall 1995 and fall 2005, total enrollment in U.S. four-year colleges and universities grew by about 21%, while enrollment in those institutions' mathematics and statistics departments grew by only about 8%. See Table S.1.
- Between fall 1995 and fall 2005, mathematics and statistics enrollments in the nation's public two-year colleges grew by 18%, compared with the roughly 21% rise in overall public two-year college enrollment. See Table S.1.
- Between fall 2000 and fall 2005, enrollments in the mathematics and statistics departments of the nation's four-year colleges and universities declined slightly, and lagged far behind total enrollment growth. See Table S.1.
- Between fall 2000 and fall 2005, mathematics and statistics enrollments in the nation's public two-year colleges reached a new high, growing by about 26% and more than erasing a decline that occurred between 1995 and 2000. See Table S.1.
- Between fall 2000 and fall 2005, enrollments in pre-college-level courses (formerly called remedial courses) at four-year colleges and universities dropped slightly. Enrollments in pre-college-level courses in fall 2005 were about 10% below their levels in fall 1995. See Table S.2.
- Between fall 2000 and fall 2005, four-year college and university enrollments in introductory-level courses (including precalculus) dropped slightly, but fall 2005 introductory-level enrollments were still 15% above their levels in fall 1995. See Table S.2.
- In fall 2005, calculus-level course enrollments in four-year colleges and universities were about 3% higher than in fall 2000, and exceeded fall 1995 calculus-level enrollments by about 9%. See Table S.2.
- In fall 2005, advanced-level mathematics enrollments exceeded fall 2000 levels by about 10%, and surpassed fall 1995 levels by about 17%. See Table S.2.
- In four-year college and university mathematics departments, elementary-level statistics enrollments in fall 2005 exceeded the levels of fall 2000 by about 9% and were about a third larger than in fall 1995. Upper-level statistics enrollments declined slightly between 2000 and 2005 but still surpassed 1995 levels by about 20%. See Table S.2.
- In four-year college and university statistics departments, elementary-level enrollments in fall 2005 were essentially unchanged from fall 2000 levels and were 10% above 1995 levels. Upper-level statistics enrollments grew by about 20% between 2000 and 2005, after increasing by about 25% between 1995 and 2000. See Table S.2.
- In two-year colleges, statistics enrollments, which had increased by less than 3% between 1995 and 2000, increased by almost 60% between fall 2000 and fall 2005. See Table S.2.
- Computer science enrollments in mathematics departments of four-year colleges and universities, which had risen between fall 1995 and fall 2000, dropped by about 55% between fall 2000 and fall 2005, for a net decline of about 42% between 1995 and 2005. This decline occurred at all course levels, with upper-level computer science enrollments in mathematics departments dropping by nearly 70% between 2000 and 2005. See Table S.2.

B. Bachelors degrees granted

- The total number of bachelors degrees awarded through the nation's mathematics and statistics departments (including some computer science degrees) declined by about 5% between the 1999–2000 and 2004–2005 academic years, and about 6% fewer bachelors degrees were awarded in 2004–2005 than in 1994–1995 by mathematics and statistics departments. If computer science degrees are excluded from the count, then the five-year decline was only half as large, but the ten-year decline was slightly larger. See Table S.4.
- The number of bachelors degrees in computer science awarded through mathematics and statistics departments declined by about 21% between the 1999–2000 and 2004–2005 academic years. See Table S.4.
- The number of mathematics education bachelors degrees granted through mathematics departments dropped by about a third between 1999–2000 and 2004–2005 and by about 30% when 2004–2005 is compared with 1994–1995. See Table S.4.

- The percentage of bachelors degrees awarded to women through U.S. mathematics and statistics departments declined from 43.4% in 1999–2000 to 40.4% in the 2004–2005 academic year, a percentage that is below the 41.9% figure for 1994–1995. If computer science degrees are excluded, then the percentage of bachelors degrees awarded to women through mathematics and statistics departments declined from 46.7% in the 1999–2000 academic year to 43.4% in 2004–2005, which was also below the 45% figure from 1994–1995. See Table S.4.

C. Who taught undergraduate mathematics and statistics courses?

- The percentage of undergraduate mathematics and statistics sections in four-year colleges and universities taught by tenured and tenure-eligible (TTE) faculty declined between fall 2000 and fall 2005. In two-year colleges, the percentage of mathematics and statistics sections taught by permanent full-time faculty rose marginally from the levels of fall 2000. See Table S.6.

D. What pedagogical methods were used in undergraduate mathematics and statistics courses?

- Among four "reform pedagogies" studied by CBMS2005, four-year colleges and universities used graphing calculators in about half of their calculus courses, and computer assignments were used as a teaching tool in about a fifth of sections taught, while use of writing assignments and group projects in calculus courses fell to nearly single-digit levels. The four reform pedagogies were more widely used in two-year mathematics programs than in four-year departments, and were more widely used in Elementary Statistics courses than in calculus courses. See Tables S.11, S.12, and S.13.

E. The number of faculty

- Between 1995 and 2005, the number of full-time faculty members in four-year college and university mathematics departments grew by 12%, with the majority of the growth occurring after 2000. In doctoral statistics departments, the number of full-time faculty members reversed a decline that had occurred between 1995 to 2000, and in fall 2005 was about 13% larger than in fall 1995. In the mathematics programs of two-year colleges, the 21% growth in full-time faculty numbers matched the overall enrollment growth of two-year colleges and matched the increase in mathematics and statistics enrollments between 1995 and 2005. See Table S.14.
- Between fall 2000 and fall 2005, the number of part-time faculty in four-year mathematics departments declined by about 10% and increased by about 10% in doctoral statistics departments while the number of part-time faculty in two-year college mathematics programs increased by 22%. See Table S.14.
- The number of tenured and tenure-eligible faculty in four-year mathematics departments rose by 6% between fall 2000 and fall 2005. During that same five-year period, the number of TTE faculty in doctoral statistics departments grew by 10%, and the number of permanent full-time faculty members in mathematics programs at two-year colleges grew by 26%. See Table S.15.

F. Gender and ethnicity in the mathematical sciences faculty

- The percentage of women among the tenured faculty of mathematics departments grew from 15% to 18% between fall 2000 and fall 2005, with considerable variation in this percentage when departments are grouped by the highest degree that they offer. During that same period, the percentage of women among tenure-eligible faculty held steady at 29%. In doctoral statistics departments, the percentage of women among tenured faculty grew from 9% to 13% between fall 2000 and fall 2005, while the percentage of women among tenure-eligible faculty grew from 34% to 37%. The percentage of women in the permanent full-time faculty of two-year college mathematics programs rose slightly, reaching 50% in fall 2005. See Table S.17.
- The percentage of faculty classified as "White, not Hispanic" dropped from 84% to 80% in mathematics departments, and declined from 76% to 71% in doctoral statistics departments between fall 2000 and fall 2005. See Tables S.20 and S.21.

G. Changes in the mathematical sciences faculty due to deaths and retirements

The mathematics departments in two- and four-year colleges lost about three percent of their permanent full-time members (respectively, their TTE faculty) to deaths and retirements in the 1999–2000 and 2004–2005 academic years. In doctoral statistics departments, losses due to deaths and retirements were closer to 2% in each of those academic years. See Table S.22.

An overview of enrollments (Tables S.1, S.2, and S.3)

Total enrollment growth in four-year colleges and universities during the 1995–2005 decade outstripped mathematics and statistics enrollment growth, and in fall 2005 there were many more American college students taking substantially less mathematics and statistics courses than did their predecessors a decade earlier. Four-year colleges and universities saw fall-term enrollments in mathematics and statistics rise by about 8% between 1995 and 2005, at the same time that total enrollment in four-year colleges and universities grew by about 21%. The problem was even more pronounced in the decade's last five years, between fall 2000 and fall 2005, when mathematics and statistics enrollments in four-year colleges and universities actually declined, at the same time that total enrollment in four-year colleges and universities rose by about 13%.

Information about mathematics and statistics enrollments comes from CBMS surveys in 1995, 2000, and 2005, while estimates of total enrollment in four-year colleges and universities come from the National Center for Educational Statistics (NCES) and are based on data that post-secondary educational institutions must submit to the Integrated Post-secondary Education Data System (IPEDS). Most national data cited in this report are drawn from the NCES report *Projections of Education Statistics to 2015*, which is available at http://nces.ed.gov/programs/projections/tables/asp .

NCES data show that total enrollments in the nation's public two-year colleges (TYCs) also increased by about 21% between fall 1995 and fall 2005. CBMS survey data suggest that the same ten-year period saw a roughly 18% growth in the mathematics and statistics enrollments in the mathematics departments and programs of the nation's public TYCs.

That 18% estimate requires explanation because the TYC enrollment totals in Table S.1 (1,498,000 for fall 1995 and 1,697,000 for fall 2005) suggest a 13% increase. Two factors explain why the estimate is 18%. First, recall that the 1995 TYC total included some computer science course enrollments, as well as mathematics and statistics enrollments, while the data for 2005 included only mathematics and statistics enrollments. Table S.1 allows us to remove those computer science enrollments, and we see that there were approximately 1,455,000 mathematics and statistics enrollments in fall 1995. Second, as careful readers will already have noted, the TYC sample frames for CBMS1995 and CBMS2005 were different. The CBMS1995 sample frame included approximately

TABLE S.1 Enrollment (in 1000s) in undergraduate mathematics, statistics, and computer science courses taught in mathematics departments and statistics departments of four-year colleges and universities, and in mathematics programs of two-year colleges. Also NCES data on total fall enrollments in two-year colleges and four-year colleges and universities in fall 1990, 1995, 2000, and 2005. NCES data includes both public and private four-year colleges and universities, and includes only public two-year colleges.

	Four-Year College & University Mathematics & Statistics Departments						Two-Year College Mathematics Programs[4]			
	Fall				2005 by Dept		Fall			
	1990	1995	2000	2005	Math	Stat	1990	1995	2000	2005
Mathematics	1621[1]	1471[1]	1614	1607	1607	--	1241	1384	1273	1580
Statistics	169	208	245	260	182	78	54	72	74	117
Computer Science	180	100	124	59	57	2	98[2]	43[2]	39[2]	--[2]
Total	1970	1779	1984	1925	1845	80	1393	1498	1386	1697
NCES Total Fall Undergraduate Enrollments[3]	6719	6739	7207	8176			4996	5278	5697	6389

[1] These totals include approximately 2000 mathematics enrollments taught in statistics departments.

[2] Computer science totals in two-year colleges before 1995 included estimates of computer science courses taught outside of the mathematics program. In 1995 and 2000, only those computer science courses taught in the mathematics program were included. Starting in 2005, no computer science courses were included in the two-year mathematics survey.

[3] Data for 1990, 1995, and 2000, and middle alternative projection for 2005, are taken from Tables 16,18, and 19 of the NCES publication *Projections of Educational Statistics to 2015* at http://nces.ed.gov/programs/projections/tables.asp.

[4] Starting in 2005, data on mathematics, statistics, and computer sciences enrollments in two-year colleges include only public two-year colleges.

half of the nation's private, not-for-profit TYCs while the CBMS2005 frame consisted of public TYCs only. To estimate the impact of that sample-frame change, we note that NCES data from 2002 show that public TYC enrollment was just over 99% of the combined enrollment in private not-for-profit and public TYCs. If we assume that public TYCs also taught just over 99% of the mathematics and statistics enrollment in the combined public and private, not-for-profit TYCs, and that the 99% figure still applied in 2005, we estimate that the combined mathematics and statistics enrollment in public and private, not-for-profit TYCs grew from 1,455,000 in 1995 to 1,714,000 in 2005, which is roughly an 18% increase. Alternatively, assuming that the 99% figure applied in 1995 as well as in 2002, we get the same 18% growth estimate.

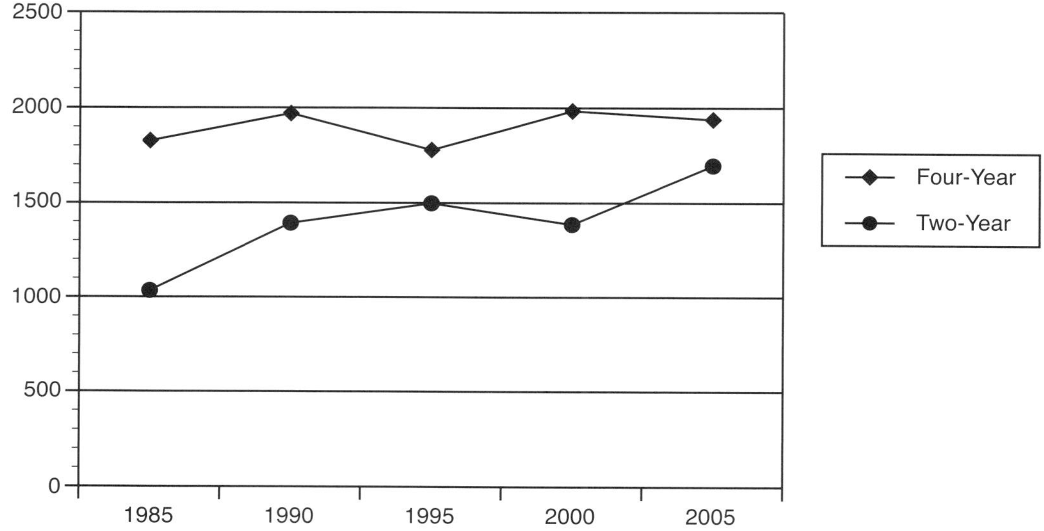

FIGURE S.1.1 Combined enrollment (in 1000s) in undergraduate mathematics, statistics, and computer science courses at four-year colleges and universities in mathematics departments and statistics departments, and in mathematics programs of two-year colleges: Fall 1985[1], 1990, 1995[2], 2000[2], and 2005[2]. Data for 2005 include only public two-year colleges.

[1] 1985 totals do not include computer science enrollments in mathematics and statistics departments.
[2] Before 1995, two-year enrollment totals included computer science enrollments taught outside of the mathematics program. In 1995 and 2000, only computer science courses taught within the mathematics program were counted. Starting in 2005, no computer science courses were included in the CBMS survey of two-year mathematics programs.

Table S.2 begins the process of breaking total mathematical sciences enrollment (shown in Table S.1) into its component parts. Among four-year mathematics and statistics departments, the course categories used in fall 2005 were pre-college courses, introductory-level courses, calculus-level courses, and advanced-level courses. The course category called "pre-college level" in CBMS2005 was called "remedial level" in previous CBMS studies, but the courses within the renamed category were essentially unchanged. Among four-year departments, the category of introductory-level courses was essentially unchanged from previous surveys, and included liberal arts mathematics courses, mathematics courses for elementary teachers, and a cluster of courses with names such as College Algebra, Precalculus, and Trigonometry. The category called "calculus-level courses" included all calculus courses and courses in linear algebra and differential equations. Appendix I shows that enrollments in various calculus courses accounted for about 82% of the 586,000 calculus-level enrollments reported in Table S.2. To see the complete listing of courses in each of the categories of Table S.2, see Appendix I or Section C of the questionnaires reproduced in Appendix IV.

Table S.2 also shows enrollments in various course categories in two-year mathematics programs. However, direct comparisons between course-category enrollments in four-year and two-year mathematics departments are problematic because the categories included different courses in the four-year and two-year mathematics questionnaires, as can be seen from Appendix 4 where the questionnaires are reproduced. In particular, the list of pre-college courses for two-year colleges is larger than the corresponding list for four-year colleges, and courses such as Linear Algebra and Differential Equations are not included in the two-year college calculus-level category.

In four-year mathematics departments, the sum of all mathematics course enrollments dropped marginally, from 1,614,000 in fall 2000 to 1,607,000 in fall 2005. Those totals mask more interesting changes. Between fall 2000 and fall 2005, the number of students in pre-college courses declined by about 8% (from 219,000 to 201,000) and introductory-level enrollments fell by about 2% (from 723,000 to 706,000). These declines were almost offset by other mathematics enrollment increases. Calculus-level enrollments, which, as noted above, include some sophomore-level courses as well as various calculus courses, increased by about 3% in four-year mathematics departments, and advanced-level mathematics enrollments increased by almost 10%.

When compared with the levels of fall 1995, pre-college-level enrollments in four-year mathematics departments were down by about 10%, while introductory-level and calculus-level enrollments were up by about 15% and 9% respectively, and advanced-level mathematics enrollments increased by about 17%. The total number of all mathematics enrollments in four-year mathematics departments increased by about 9% in the 1995–2005 decade.

Two-year college total mathematics enrollments rose by about 24%, from 1,273,000 in fall 2000 to 1,580,000 in fall 2005, with substantial increases in the pre-college, introductory, and "other" categories. These increases more than wiped out a moderate enrollment decline that occurred between 1995 and 2000 in two-year college mathematics programs.

Between fall 2000 and fall 2005, the nation's undergraduate statistics course enrollments continued their pattern of long-term growth. Enrollments in the elementary-level statistics category (which includes several courses in addition to Elementary Statistics) continued to rise, growing by about 9% in four-year mathematics departments and by 58% in two-year colleges between fall 2000 and fall 2005. The only exception to this growth pattern was in separate departments of statistics, where enrollment in elementary-level statistics held steady at about 54,000.

Ten-year growth for statistics enrollments between fall 1995 and fall 2005 was 62% in two-year colleges, 25% in four-year mathematics departments, and 20% in four-year statistics departments. As Table E.2 of Chapter 3 will show, almost all of the growth in statistics department enrollments occurred in masters-level departments—undergraduate enrollment in doctoral statistics departments began and ended the decade at about the 62,000 level.

The bottom row of Table S.2 shows that total course enrollments in four-year mathematics departments declined by about 3%, from 1,908,000 in fall 2000 to 1,845,000 in fall 2005. That decline is attributable primarily to a sharp decrease in computer science enrollments in mathematics departments, from 123,000 in fall 2000 to 57,000 in fall 2005. The decline in computer science enrollments in mathematics departments might be part of a broader national trend, but it might also be explained by the growth of computer science as a separate discipline with its own academic departments. If computer science enrollments are excluded, then the combination of mathematics and statistics course enrollments in four-year mathematics departments was essentially the same in fall 2005 as in fall 2000, and was about 11% larger in fall 2005 than in fall 1995.

In previous CBMS studies, computer science enrollments were included as a separate category in both the four-year and two-year CBMS questionnaires. In contrast, CBMS2005 did not collect data on computer science enrollments in two-year college mathematics programs, because anecdotal evidence suggested that these courses had moved into separate programs within the two-year-college system. It might have happened that some two-year mathematics programs included computer science enrollments in the "other mathematics courses" category in the two-year college questionnaire. In fact, the "other-courses" category in the two-year college total expanded from 130,000 enrollments in fall 2000 to 187,000 enrollments in fall 2005, a surprising 44% increase that happens to be close to the total number of computer science enrollments in two-year colleges in fall 2000. Alternatively, the 44% increase might be due to the creation of new courses that do not fit conveniently into any course description in the current two-year college questionnaire, e.g., a single course that combines high school algebra and college algebra (two separate courses in the CBMS2005 questionnaire) into a single course. The large number of "other course" enrollments in CBMS2005 suggests that a revision in the two-year course listing is in order for the CBMS2010 survey.

A frequently quoted number is the percentage of all undergraduate enrollments in the nation's mathematics and statistics departments and programs that occur in two-year colleges. The previous paragraph shows that there are two different ways to calculate that percentage; fortunately, the two methods give more or less the same answer. If a substantial number of two-year-college computer science enrollments were included under "Other mathematics courses," then two-year-college enrollments (1,697,000) should be compared with the sum of all enrollments in four-year mathematics and statistics departments (1,925,000). By that calculation, two-year colleges taught about 47% of all undergraduate enrollments in mathematical sciences departments and programs. Alternatively, if two-year college enrollments did not include a substantial number of computer science courses, then the two-year total (1,697,000) should be compared with the 1,867,000 mathematics and statistics enrollments in four-year mathematics and statistics departments, excluding computer science, which gives a percentage closer to 48%. For comparison, note that in fall 1995 the percentage of undergraduate mathematics and statistics enrollments (excluding computer science) taught in two-year colleges was 46%, and in 2000, it was 42%.

TABLE S.2 Total enrollment (in 1000s), including distance learning enrollment, by course level in undergraduate mathematics, statistics, and computer science courses taught in mathematics and statistics departments at four-year colleges and universities, and in mathematics programs at two-year colleges, in fall 1990,1995, 2000, and 2005. (Two-year college data for 2005 include only public two-year colleges and do not include any computer science.)

	Mathematics Departments				Statistics Departments				Two-year College Mathematics Programs			
Course level	1990	1995	2000	2005	1990	1995	2000	2005	1990	1995	2000	2005
Mathematics courses												
Precollege level	261	222	219	201	--	--	--	--	724	800	763	965
Introductory level (including Precalculus)	592	613	723	706	--	--	--	--	245	295	274	321
Calculus level	647	538	570	587	--	--	--	--	128	129	106	108
Advanced level	119	96	102	112	--	--	--	--	0	0	0	0
Other (2-year)									144	160	130	187
Total Mathematics courses	**1619**	**1469**	**1614**	**1607**	**--**	**--**	**--**	**--**	**1241**	**1384**	**1273**	**1580**
Statistics courses												
Elementary level	87	115	136	148	30	49	54	54	54	72	74	117
Upper level	38	28	35	34	14	16	20	24	0	0	0	0
Total Statistics courses	**125**	**143**	**171**	**182**	**44**[2]	**65**[2]	**74**	**78**	**54**	**72**	**74**	**117**
CS courses [1]												
Lower level	134	74	90	44	0	1	1	2	98	43	39	0
Middle level	12	13	17	8	0	0	0	0	0	0	0	0
Upper level	34	12	16	5	0	0	0	0	0	0	0	0
Total CS courses [1]	**180**	**99**	**123**	**57**	**0**	**1**	**1**	**2**	**98**	**43**	**39**	**0**
Grand Total	**1924**	**1711**	**1908**	**1845**	**44**[2]	**66**[2]	**75**	**80**	**1393**	**1499**	**1386**	**1697**

Note: Round-off may make column totals seem inaccurate.

[1] Computer science enrollment starting in 1995 and 2000 includes only courses taught in mathematics programs. For earlier years it also includes estimates of computer science courses taught outside of the mathematics program. Starting in 2005, computer science courses were no longer included in the two-year college survey.

[2] These totals were adjusted to remove certain mathematics enrollments included in statistics totals in 1990 and 1995.

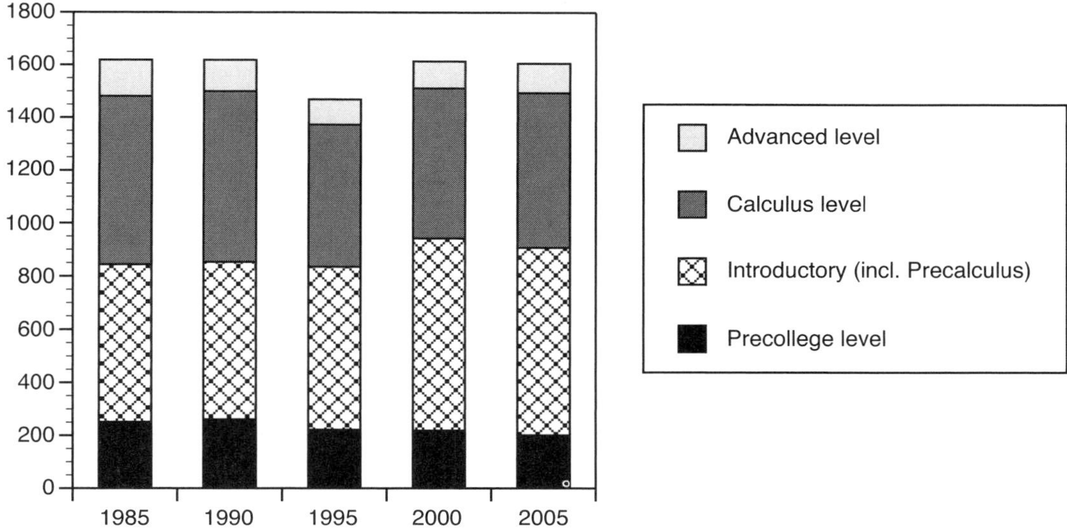

FIGURE S.2.1 Enrollments (in 1000s) in undergraduate mathematics courses in mathematics departments of four-year colleges and universities, by level of course: fall 1985, 1990, 1995, 2000, and 2005.

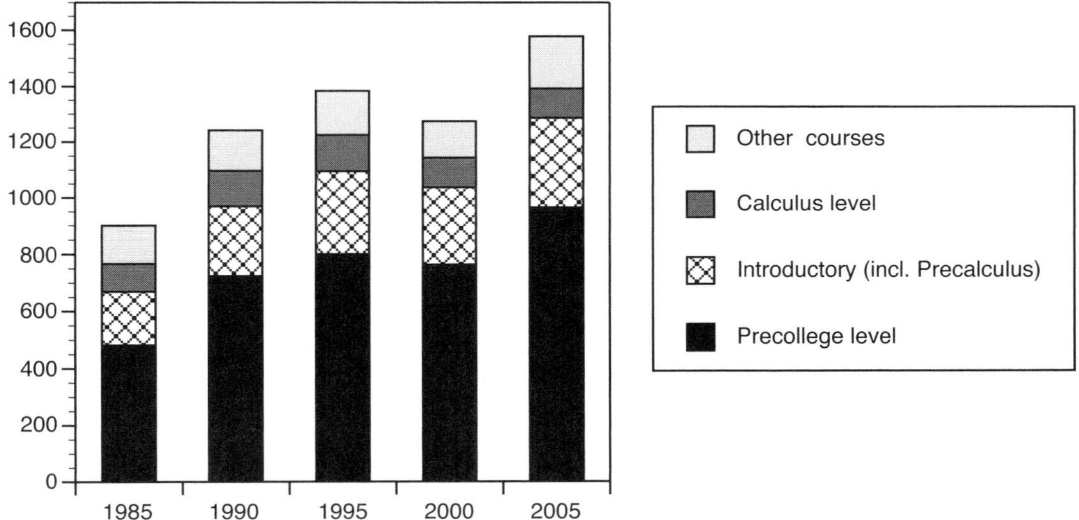

FIGURE S.2.2 Enrollments (in 1000s) in mathematics courses in two-year college mathematics programs by level of course in fall 1985, 1990, 1995, 2000, and 2005.

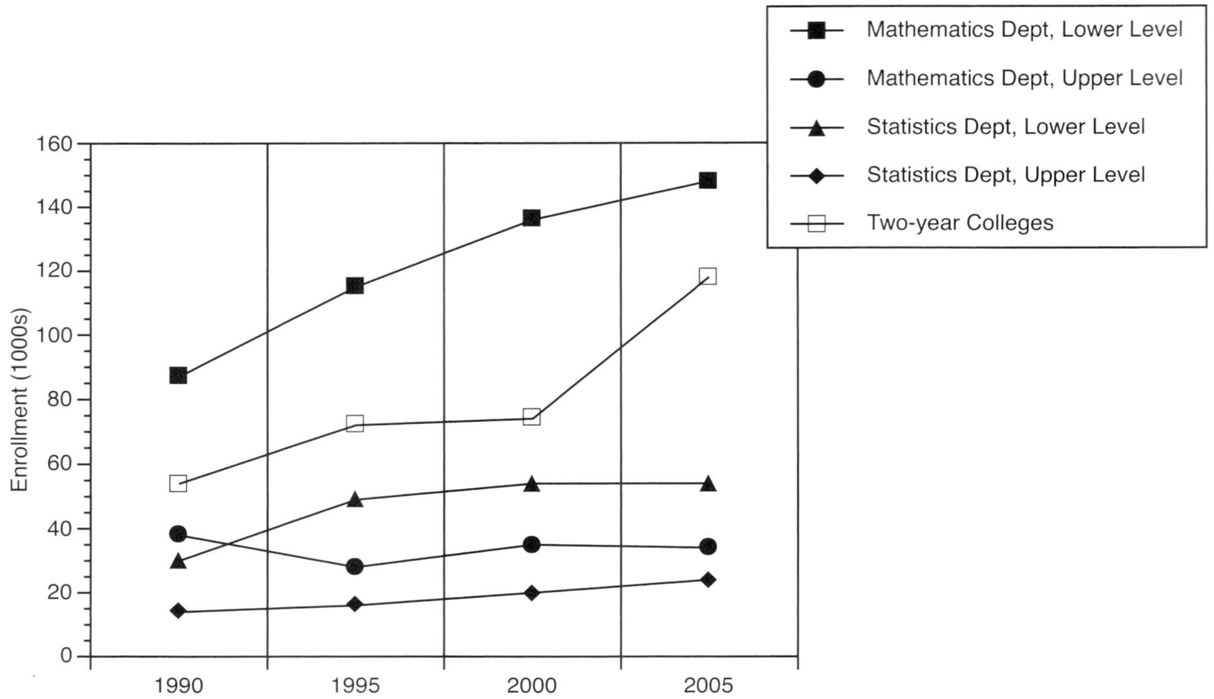

FIGURE S.2.3 Enrollments (in 1000s) in statistics courses in two year college mathematics programs, and in mathematics and statistics departments of four-year colleges and universities in fall 1990,1995, 2000, and 2005.

Academic year enrollments

CBMS surveys follow the NCES pattern and focus only on fall enrollments. However, CBMS data also make it possible to use fall enrollments to project full-year enrollments, and recent CBMS studies reveal an interesting trend among mathematics and statistics departments at four-year colleges and universities. In the surveys of fall 1990, 1995, 2000, and 2005, departments were asked to give their total enrollment for the previous academic year's fall term, and also their total enrollment for the entire previous academic year. Using this data one can estimate the national ratio of full-year enrollment to fall-term enrollment in the mathematical sciences programs of four-year colleges and universities. The ratios found in 1990, 1995, 2000, and 2005 were, respectively, 2, 2, 1.85 (SE = 0.03) and 1.75 (SE = 0.03), and those ratios can be used to project full-year enrollment from fall-term enrollment.

What is responsible for the change in that ratio from 2 to 1.85 to 1.75? Table S.3 provides one possible explanation, namely the widespread shift to the semester system. Why would the shift to the semester system cause the academic year to fall term ratio to decline? The authors of CBMS1995 (who found a ratio of 2) argued that "[t]he lesser Spring semester enrollment in those institutions with a two semester calendar is precisely balanced by those institutions on the term or quarter calendar, where the Fall enrollment is substantially less than half of the academic year enrollment." That argument, when combined with the substantial growth in the percentage of schools on the semester system (see Table S.3), probably explains the change in the academic-year-to-fall-term ratio noted above.

TABLE S.3 Percentages of four-year colleges and universities with various types of academic calendars in fall 1995, 2000 and 2005.

	Percentage of Four-year Colleges & Universities		
Type of calendar	1995 %	2000 %	2005 %
Semester	77	89	91
Trimester	0	1	1
Quarter	8	4	6
Other	15	6	2

Note: Zero means less than one-half of one percent.

Bachelors degrees in the mathematical sciences (Table S.4)

Table S.4 presents data on the total number of bachelors degrees awarded through the mathematics and statistics departments of four-year colleges and universities in the U.S. Because some mathematics departments also offer computer science programs, these totals include some degrees in computer science. In addition—see below—CBMS includes certain double majors and joint majors in its total of mathematics and statistics bachelors degrees.

The total number of degrees in the 2004–2005 academic year awarded through mathematics and statistics departments was down by more than 6% from the number awarded ten years earlier, in 1994–1995. Most of that decline occurred between 1999–2000 and 2004–2005. Women received 40.4% of all degrees awarded by mathematics and statistics departments in 2004–2005, down from the 41.8% figure in 1994–1995 and down from the 43.4% figure in 1999–2000.

Even if one excludes the number of computer science degrees granted through mathematics and statistics departments, a number that naturally declined as colleges and universities established separate computer science departments, the number of bachelors degrees in mathematics and statistics dropped by about 2% between 1999–2000 and 2004–2005, and by about 6% between 1994–1995 and 2004–2005. The number of mathematics education bachelors degrees granted through mathematics departments dropped by about a third over a five-year period, from 4991 in 1999–2000 to 3369 in 2004–2005. The number of bachelors degrees in mathematics increased between 1999–2000 and 2004–2005.

Table S.4 shows that the number of computer science bachelors degrees awarded through the nation's mathematics departments dropped from 3,315 in the 1999–2000 academic year to 2,603 in the 2004–2005 academic year. The annual Taulbee Surveys, published by the Computing Research Association, study the nation's doctoral computer science departments and include data on computer science bachelors degrees awarded through such departments. This can provide some context for the figures in Table S.4. Comparison of Table 9 of [BI] and Table 9 of [Z] shows that the number of computer science bachelors degrees granted through doctoral computer science departments rose from 12,660 in 1999–2000 to 15,137 in 2004–2005. Of the bachelors degrees awarded through doctoral computer science departments, 20% were awarded to women in 1999–2000, a percentage that dropped to 15% by 2004–2005. Table S.4 shows that in mathematics departments, the percentage of computer science degrees awarded to women in 1999–2000 was about 24% and declined to about 18% in 2004–2005.

As noted above, CBMS counts of bachelors degrees included double majors, i.e., students who completed two separate majors, one being mathematics or statistics. CBMS counts also included a separate category called "joint majors." What defines a joint major? In the CBMS questionnaire sent to mathematics departments, a joint major was defined as a student who "completes a single major in your department that integrates courses from mathematics and some other program or department and typically requires fewer credit hours than the sum of the credit hours required by the two separate majors". An analogous definition appeared in the questionnaire sent to statistics departments. Joint majors in mathematics and statistics, or in mathematics and computer science, are traditional joint majors. The number of mathematics and statistics joint majors rose slowly, from 188 in 1994–1995, to 196 in 1999–2000, to 203 in 2004–2005. The number of mathematics and computer science joint majors rose from 453 in 1994–1995 to 876 in 1999–2000 and fell back to 719 in 2004–2005, still registering a substantial increase over the decade 1994–1995 to 2004–2005. CBMS2005 Table S.4 contains a new category of joint major, one that combines upper-level mathematics with upper-level business or economics (or mixes statistics and business or economics). In 2004–2005, the number of bachelors degrees of this new type of joint major was somewhat larger than in the more traditional joint mathematics and statistics degree.

In Chapter 3, Table E.1 and its figures give more detail on the number of bachelors degrees awarded through mathematics and statistics departments of different types, classified by highest degree offered. There is considerable variation by type of department in terms of the number of bachelors degrees awarded and in the percentage of degrees awarded to women.

Bachelors-degree estimates from previous CBMS surveys have differed from NCES degree counts. This was in part because CBMS figures rely on departmental counts rather than on university-wide counts, with the result that any student who has a double major "Mathematics and X" is counted as a mathematics major by CBMS. How was such a student counted in the IPEDS reports that are the basis for NCES estimates? Before 2002, IPEDS data assigned each student one and only one major, so that a student who double majored in "Mathematics and X" might or might not be counted as a mathematics

TABLE S.4 Combined total of all bachelors degrees in mathematics and statistics departments at four-year colleges and universities between July 1 and June 30 in 1984-85, 1989-90, 1994-95, 1999-2000 and 2004-2005 by selected majors and gender.

Major	84-85	89-90	94-95	99-00	04-05
Mathematics (except as reported below)	13171	13303	12456	10759	12316
Mathematics Education	2567	3116	4829	4991	3369
Statistics (except Actuarial Science)	538	618	1031	502	527
Actuarial Mathematics	na	245	620	425	499
Operations Research	312	220	75	43	31
Joint Mathematics & Computer Science	2519	960	453	876	719
Joint Mathematics & Statistics	121	124	188	196	203
Joint Math/Stat & (Business or Economics)	na	na	na	na	214
Other	9	794	502	1507	954
Total Mathematics, Statistics, & joint degrees	**19237**	**19380**	**20154**	**19299**	**18833**
Number of women	na	8847	9061	9017	8192
Computer Science degrees	**8691**	**5075**	**2741**	**3315**	**2603**
Number of women	na	1584	532	808	465
Total degrees	**27928**	**24455**	**22895**	**22614**	**21437**
Number of women	na	10431	9593	9825	8656

Note: Round-off may make column totals seem inaccurate.

major. Since 2002, colleges and universities have the option of reporting double majors in "Mathematics and X" both under the mathematics disciplinary code and under the code for discipline X, but they are not required to do so. That would seem to introduce additional ambiguity into the IPEDS-based counts of

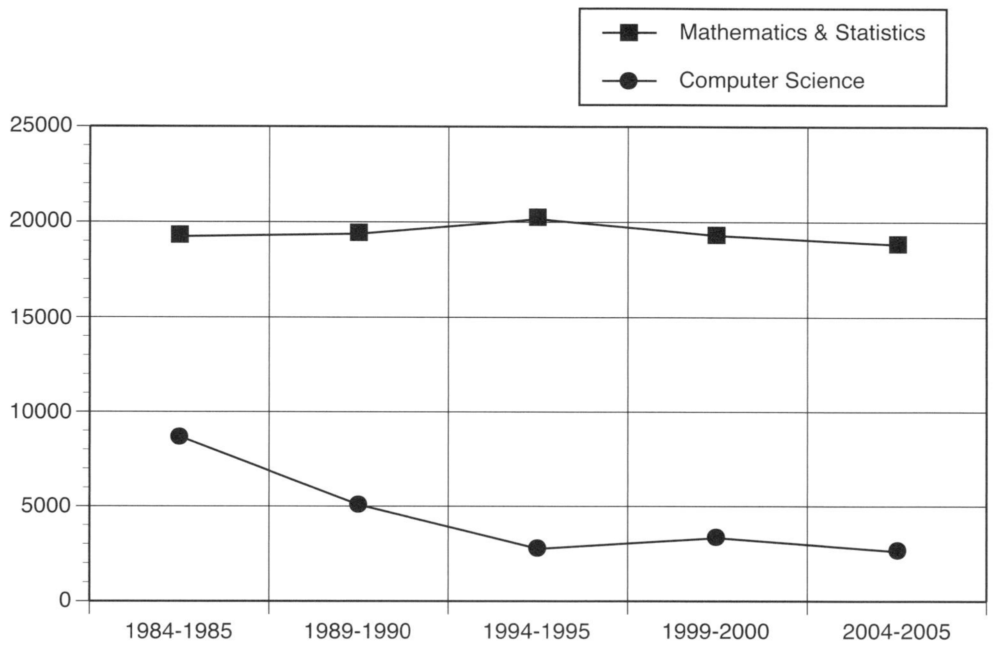

FIGURE S.4.1 Number of bachelors degrees in mathematics and statistics, and in computer science, granted through mathematics and statistics departments in academic years 1984-1985, 1989-1990, 1994-1995, 1999-2000, and 2004-2005.

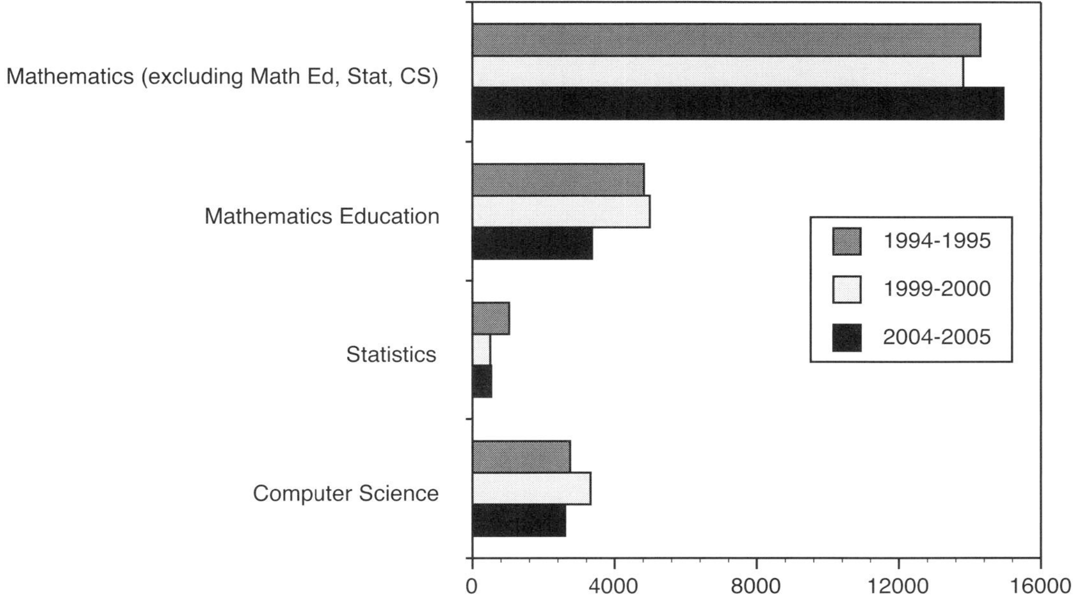

FIGURE S.4.2 Number of bachelors degrees awarded by mathematics and statistics departments (combined) at four-year colleges and universities between July 1 and June 30 in 1994-95, 1999-2000, and 2004-2005.

mathematics majors. Furthermore, CBMS estimates of mathematics majors include Mathematics Education majors so long as they receive their degrees through a mathematics or statistics department, and that is not necessarily the case in IPEDS reports. Finally, CBMS estimates of mathematical sciences majors include several thousands of computer science majors who received their bachelors degrees through mathematics departments, and these students would be reported in IPEDS data under a disciplinary code not included in the Mathematics and Statistics category used by NCES.

Who teaches undergraduates in mathematics and statistics departments? (Tables S.5 through S.10)

CBMS2005 Tables S.5 through S.10 study the kinds of instructors assigned to teach undergraduate mathematical science courses in two- and four-year colleges and universities. Faculty in four-year colleges and universities are broken into four broad categories: tenured and tenure-eligible (TTE) faculty, other full-time faculty who are not TTE (called OFT faculty), part-time faculty, and graduate teaching assistants (GTAs). For two-year colleges, which typically do not have a tenure-track system, CBMS2005 tables distinguish between courses taught by full-time faculty and part-time faculty.

The faculty categories used to study four-year college and university mathematics and statistics departments are self-explanatory, except the GTA category. Instructions in the CBMS questionnaires were very specific about GTA-taught courses; a course was to be reported as taught by a GTA if and only if the GTA was completely in charge of the course (i.e., was the "instructor of record" for the course). GTAs who ran discussion or recitation sections as part of a lecture/recitation course were not included in this special category.

The faculty-classification system described above for four-year colleges and universities is complicated by the fact that some colleges and universities do not recognize tenure. However, such schools typically distinguish between permanent and temporary full-time faculty. Departments in such schools were asked to report courses taught by permanent faculty in the column labeled TTE, while courses taught by temporary full-time faculty were to be reported as taught by OFT faculty. In addition, CBMS2005 found that the number of four-year college and university departments that do not recognize tenure was small; CBMS2005 projects that in fall 2005, only 5% of the nation's mathematics departments belonged to colleges and universities that did not recognize tenure. If departments are classified by the highest degree that they offer in the mathematical sciences, then CBMS2005 found that in fall 2005, 100% of the nation's doctorate- or masters-granting mathematics departments belonged to tenure-granting colleges or universities, as did 93% of all bachelors-granting departments. Among masters- and doctoral-level statistics departments, all belonged to tenure-granting universities.

Readers must take special precautions when comparing the findings of CBMS2000 and CBMS2005 because CBMS2000 sometimes presented its findings in terms of percentages of enrollment and sometimes in terms of percentages of sections offered. For statistical reasons, CBMS2005 presented most of its results in terms of percentage of sections offered.

Table S.5 presents a macroscopic view of faculty who taught undergraduate courses in the mathematics and statistics departments of four-year colleges and universities and in mathematics programs at two-year colleges in the fall of 2005. Less than half of mathematics sections in four-year colleges and universities were taught by tenured and tenure-eligible (TTE) faculty, and the same was true of statistics courses taught in statistics departments. If TTE and OFT faculty are combined, CBMS2005 shows that about 70% of all sections in mathematics and statistics departments were taught by full-time faculty in fall 2005. In mathematics programs of two-year colleges (which typically do not have tenure-track systems), 56% of sections were taught by full-time faculty.

No single table in CBMS2000 compares directly with CBMS2005 Table S.6. The historical data in Table S.6 present percentages of sections taught by various types of instructors and were derived from Tables E.12 to E.18 in Chapter 3 of the CBMS2000 report. Tables S.7 through S.10 contain some comparisons with data from the Chapter 1 tables (coded "SFY") in CBMS1995 and CBMS2000, and we ask the reader to notice that the historical data concern percentages of *enrollments*, while data from CBMS2005 involve percentages of *sections taught*.

CBMS2000 and independent American Mathematical Society surveys detected a trend toward using fewer tenured and tenure-eligible (TTE) faculty and markedly greater reliance on other full-time (OFT) faculty in teaching undergraduates between fall 1995 and fall 2000 [LM]. CBMS2005 found a continued decline in the percentage of TTE faculty teaching undergraduate mathematics courses between fall 2000 and fall 2005. The decrease in TTE-taught sections was most noticeable among pre-college-level courses, which were called "remedial courses" in previous CBMS studies.

CBMS2005 Table S.6 suggests that the percentage of sections in mathematics departments that were taught by part-time faculty in fall 2005 was not much different than in fall 2000. The same was true for two-year colleges. This is consistent with national data across all disciplines, but contrasts with data from Table S.14 of this report showing that the percentage of part-time faculty among all faculty in four-year mathematics and statistics departments declined between fall 2000 and fall 2005. See the discussion associated with S.14 for further details.

Table S.6 presents a new feature of CBMS2005—a study of those who taught upper-level mathematics courses. Previous CBMS surveys had made the assumption that essentially all upper-division courses were taught by TTE faculty, and once upon a time that may have been true. Anecdotal evidence suggested that such an assumption was problematic today, and to test that hypothesis CBMS2005 asked departments how many of their upper-division sections were taught by TTE faculty. In mathematics departments, CBMS2005 found that the percentage was 84% in fall 2005. The remaining 16% of sections—whose instructors might have been visiting scholars, postdocs, etc.—are listed as having unknown instructors.

It is perhaps interesting to note that between fall 2000 and fall 2005, the nation's mathematics departments actually increased the percentage of sections of statistics and of computer science that were taught by TTE faculty, at the same time they were decreasing the percentage of mathematics sections taught by TTE faculty.

In the nation's statistics departments, the percentage of sections taught by TTE faculty seemed to decrease slightly in elementary-level courses. Teaching by part-time faculty apparently fell by about a third between fall 2000 and fall 2005, as did teaching by GTAs. This appears to have been offset by a substantial increase in teaching by OFT faculty. These conclusions are somewhat tentative because data from statistics departments did not identify the type of instructors who taught 21% of statistics departments' elementary-level sections. Among upper-level sections in statistics departments, 74% were taught by TTE faculty, with the remaining 26% listed as taught by unknown instructors.

As noted above (see also Chapter 7), few two-year colleges have a tenure system, so CBMS2005 (and its predecessors) asked two-year college departments

TABLE S.5 Percentage of sections (excluding distance-learning sections) in various types of courses taught by different types of instructors in mathematics and statistics departments of four-year colleges and universities, and percentage of sections taught by full-time and part-time faculty in mathematics programs of public two-year colleges, in fall 2005. Also total enrollments (in 1000s), excluding distance-learning enrollments.

	Percentage of sections taught by					
Four-Year College & University	Tenured/ tenure-eligible %	Other full-time %	Part-time %	Graduate teaching assistants %	Unknown %	Total enrollment in 1000s
Mathematics Departments						
Mathematics courses 2005	46	21	20	8	5	1588
Statistics courses 2005	52	24	19	2	2	179
Computer Science courses 2005	70	11	11	0	7	56
All mathematics department courses 2005	48	21	19	7	5	1825
Statistics Departments						
All statistics department courses 2005	47	23	7	11	13	79
Two-Year College Mathematics Programs	Full-time		Part-time			Enrollment in 1000s
All TYC mathematics program courses 2005	56	--	44	--	--	1616

Note: zero means less than one-half of one percent.

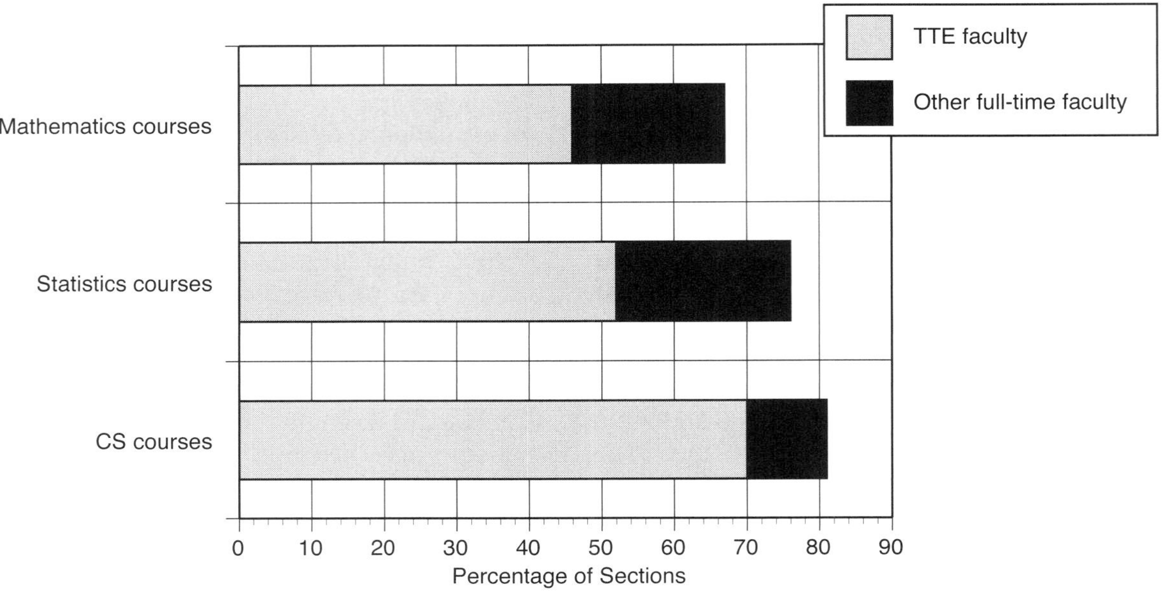

FIGURE S.5.1 Percentage of sections in four-year college and university mathematics departments taught by tenured/tenure-eligible (TTE) faculty and by other full-time (OFT) faculty in fall 2005, by type of course. Deficits from 100% represent courses taught by part-time faculty, graduate teaching assistants, and unknown faculty.

to report the number of sections of each course that were taught by full-time faculty. CBMS2005 found that in fall 2005, 56% of sections in the mathematics programs of two-year colleges were taught by full-time faculty, up two points from fall 2000.

Among first-year courses, calculus courses have long been of particular importance to mathematics departments, as well as to the client departments for which mathematics is a prerequisite (e.g., the sciences and engineering). Consequently, CBMS surveys pay special attention to calculus courses. Tables S.7 and S.8 present data on two types of calculus courses, traditionally called "mainstream" and "non-mainstream". The term "mainstream calculus" refers to courses that serve as prerequisites for upper-division mathematics courses and as prerequisites for physical science and engineering courses, while other calculus courses (often with names such as "Calculus for Business and Social Sciences" and "Calculus for the Life Sciences") are lumped together as "non-mainstream". Fall 2005 enrollments in Mainstream Calculus I were roughly double the fall 2005 enrollments in Non-mainstream Calculus I.

TABLE S.6 Percentage of fall 2005 sections (excluding distance-learning sections) in courses of various types taught in mathematics and statistics departments of colleges and universities by various types of instructors, and percentage of sections taught by full-time and part-time faculty in mathematics programs at public two-year colleges in fall 2005, with data from fall 2000 from CBMS2000 tables E12 to E18. Also total enrollments (in 1000s).

	Percentage of sections taught by					
Four-Year Colleges & Universities	Tenured/ tenure-eligible %	Other full- time %	Part-time %	Graduate teaching assistants %	Unknown %	Total enrollment in 1000s
Mathematics Department courses						
Mathematics courses						
Precollege level 2005	9	25	46	14	5	199
Precollege level 2000	20	18	43	10	10	219
Introductory level 2005	31	25	28	10	6	695
Introductory level 2000	35	21	28	10	6	723
Calculus level 2005	61	17	9	7	6	583
Calculus level 2000	64	14	10	6	5	570
Upper level 2005	84*				16*	112
Statistics courses						
Elementary level 2005	49	16	28	3	3	145
Elementary level 2000	47	16	24	5	8	136
Upper level 2005 sections	59*				41*	34
Computer Science courses						
Lower level 2005	63	12	17	1	8	43
Lower level 2000	42	19	28	0	11	90
Statistics Department Courses						
Elementary level 2005	25	21	13	20	21	53
Elementary level 2000	27	14	20	29	10	54
Upper level 2005	74*				26*	23
Two-Year College Mathematics Programs	Full-time		Part-time			
All 2005 sections	56		44			1739
All 2000 sections	54		46			1347

* CBMS2005 asked departments to specify the number of upper division sections and the number taught by tenured and tenure-eligible faculty. The deficit from 100% is reported as "unknown".

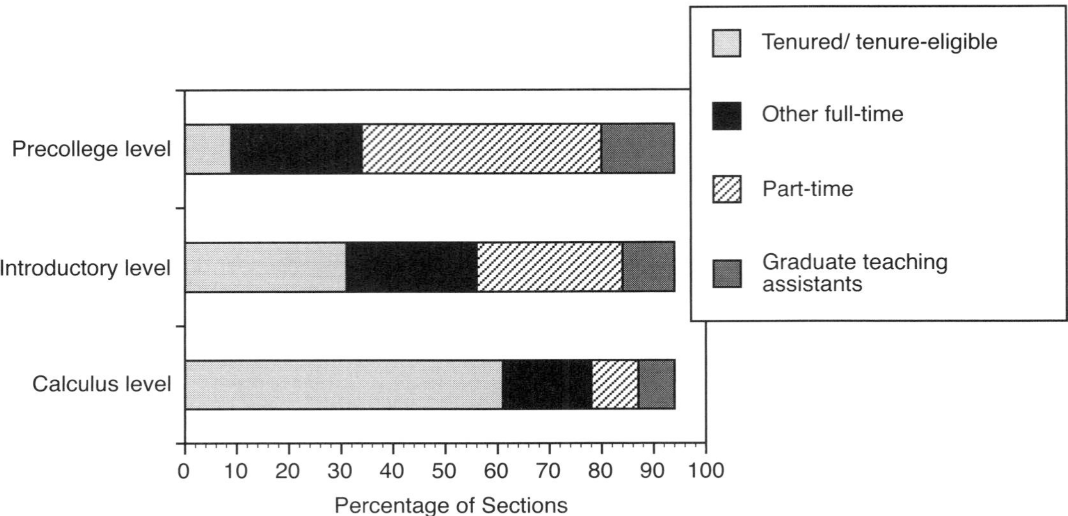

FIGURE S.6.1 Percentage of sections in lower-division undergraduate mathematics courses in mathematics departments at four-year colleges and universities by level of course and type of instructor in fall 2005. Deficits from 100% represent unknown instructors.

There are three major ways that mathematics departments organize their calculus teaching. The first, found primarily in larger universities, is based on the large lecture/small recitation model in which a large group of students meets with a faculty lecturer several times per week, and is broken into smaller recitation, discussion, problem, or laboratory sessions that typically meet just once per week, often with a graduate student. The second and third methods (called "regular sections" by CBMS studies) involve all enrolled students meeting in a single group throughout the week. Among these regular sections, CBMS2005 distinguished between sections of size thirty or less, and sections of size more than thirty. (The number thirty was chosen because it is the recommended maximum section size for mathematics courses in [MAA Guidelines].) Previous CBMS studies found that different types of faculty are typically used to teach the three different course models.

Tenure-track faculty (i.e., tenured and tenure-eligible faculty) taught almost two-thirds of Mainstream Calculus I sections in fall 2005, and only about a third of Non-mainstream Calculus I courses. Combining the TTE and OFT faculty categories shows that about 80% of Mainstream Calculus I sections were taught by full-time faculty, marginally higher than the percentage of enrollment taught by TTE faculty in fall 2000. (Recall the caveat about comparing CBMS2000 percentages, which are percentages of enrollments, with CBMS2005 percentages, which are percentages of sections taught.) Table S.9 shows an example of the different staffing patterns used to teach different types of sections. The differences are best understood in terms of the highest degree offered by the mathematics department, as can be seen in the tables in Chapter 5.

For Non-mainstream Calculus I, the percentages of sections taught by TTE faculty were substantially lower than for Mainstream Calculus I, and the percentage of

TABLE S.7 Percentage of fall 2005 sections in Mainstream Calculus I and II (not including distance-learning sections) taught by various kinds of instructors in mathematics departments at four-year colleges and universities by size of sections with historical data showing fall 2000 percentage of enrollments. Percentage of sections taught by full-time and part-time faculty in mathematics programs at two-year colleges in fall 2000 and 2005. Also total enrollments (in 1000s) and average section sizes. (Two-year college data for 2005 include only public two-year colleges.)

	Percentage of sections taught by						
Four-Year Colleges & Universities	Tenured/ tenure-eligible %	Other full-time %	Part-time %	Graduate teaching assistants %	Unknown %	Enrollment in 1000s	Average section size
Mainstream Calculus I							
Large lecture/recitation	52	27	9	5	7	80	46
Regular section <31	77	10	5	5	3	63	22
Regular section >30	49	17	10	16	8	58	36
Course total 2005	**63**	**17**	**7**	**8**	**5**	**201**	**32**
Course total 2000 (% of enrollment)	60	18	11	7	4	190	32
Mainstream Calculus II							
Large lecture/recitation	58	24	5	5	8	36	50
Regular section <31	80	8	3	7	2	25	22
Regular section >30	51	19	11	11	7	24	36
Course total 2005	**66**	**15**	**6**	**8**	**5**	**85**	**33**
Course total 2000 (% of enrollment)	66	13	10	7	4	87	32
Total Mnstrm Calculus I & II 2005	**64**	**16**	**7**	**8**	**5**	**286**	**32**
Total Mnstrm Calculus I & II 2000 (% of enrollment)	62	16	11	7	4	277	32

	Percentage of sections taught by			
Two-Year Colleges	Full-time %	Part-time %	Enrollment in 1000s	Average section size
Mainstream Calculus I 2005	**88**	**12**	**49**	**22**
Mainstream Calculus I 2000	84	16	53	23
Mainstream Calculus II 2005	**87**	**13**	**19**	**18**
Mainstream Calculus II 2000	87	13	20	20
Total Mnstrm Calculus I & II 2005	**87**	**13**	**68**	**21**
Total Mnstrm Calculus I & II 2000	85	15	73	22

Non-mainstream Calculus I sections taught by full-time faculty (TTE and OFT) was seven percentage points lower than the percentage of enrollment taught by those same faculty in fall 2000. However, such comparisons between percentage of sections and percentage of enrollment may be problematic.

A similar pattern held in two-year colleges, where 88% of Mainstream Calculus I sections were taught by full-time faculty (up slightly from fall 2000) compared to 73% of Non-mainstream Calculus I sections (down slightly from fall 2000).

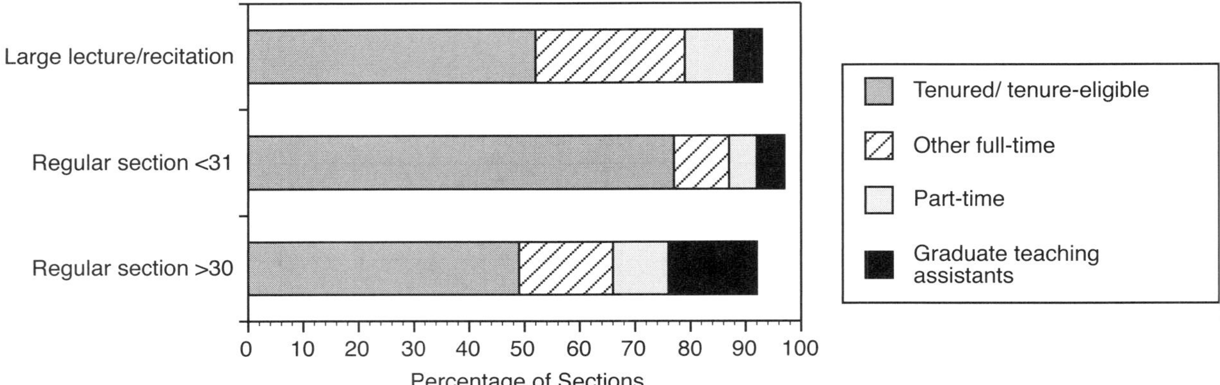

FIGURE S.7.1 Percentage of sections in Mainstream Calculus I taught by tenured/tenure-eligible, other full-time, part-time, and graduate teaching assistants in mathematics departments at four-year colleges and universities by size of sections in fall 2005. Deficits from 100% represent unknown instructors.

Table S.8 lists the percentage of unknown instructors in large lecture sections of Non-mainstream Calculus I as being 30%. An unknown percentage of 30% makes it impossible to draw any conclusions from the first row of Table S.8.

Between 1995 and 2005, a first-year course of growing importance in the mathematical sciences curriculum was Elementary Statistics (where the word "elementary" means "no Calculus prerequisite"). Table S.9 describes the situation in mathematics depart-

TABLE S.8 Percentage of sections in Non-Mainstream Calculus I and II taught by tenured/tenure-eligible faculty, postdoctoral and other full-time faculty, part-time faculty, graduate teaching assistants, and unknown in mathematics departments at four-year colleges and universities by size of sections, and percentage of sections taught by full-time and part-time faculty in mathematics programs at public two-year colleges in fall 2005. Also total enrollments (in 1000s) and average section sizes. Distance-learning sections are not included. (For four-year colleges and universities, data in parentheses show percentage of enrollments in 1995, 2000.)

	Percentage of sections taught by						
Four-Year Colleges & Universities	Tenured/ tenure-eligible %	Other full-time %	Part-time %	Graduate teaching assistants %	Unknown %	Enrollment in 1000s	Average section size
Non-Mainstream Calculus I							
Large lecture/recitation	19	33	9	9	30	28	64
Regular section <31	40	18	20	14	8	30	23
Regular section >30	36	24	26	13	2	50	44
Course total 2005 % of sections	35	23	21	13	9	108	37
Course total (1995,2000) % of enrollment	(57,44)	(10,21)	(18,19)	(15,12)	(--,4)	(97, 105)	(39,40)
Non-Mainstream Calculus II							
Course total 2005 % of sections	33	26	23	17	1	10	46
Course total (1995,2000) % of enrollment	(44,53)	(11,10)	(18,22)	(26,15)	(--,1)	(14,10)	(35,40)
Total Non-Mnstrm Calculus I & II 2005 % of Sections	35	23	21	13	8	118	38
Total Non-Mnstrm Calculus I & II (1995,2000) % of enrollment	(55,44)	(10,20)	(18,19)	(16,12)	(--,5)	(111, 115)	(38, 40)

Two-Year Colleges	Percentage of sections taught by			
	Full-time	Part-time		
Non-Mainstream Calculus I 2005 % of sections	73	27	20	23
Non-Mainstream Calculus I (1995,2000) % of sections	(77,74)	(23,26)	(26,16)	(26,22)
Non-Mainstream Calculus II 2005 % of sections	66	34	1	21
Non-Mainstream Calculus II (1995,2000) % of sections	(63,92)	(37,8)	(1,1)	(19,20)
Total Non-Mnstrm Calculus I & II 2005 % of sections	72	28	21	23
Total Non-Mnstrm Calculus I & II (1995,2000) % of sections	(76,76)	(24,24)	(27,17)	(26,22)

TABLE S.9 Percentage of sections in Elementary Statistics (no Calculus prerequisite) and Probability and Statistics (no Calculus prerequisite) taught by various types of instructors in mathematics departments at four-year colleges and universities by size of sections, and percentage of sections in Elementary Statistics (with or without Probability) taught by full-time and part-time faculty in mathematics programs at public two-year colleges in fall 2005. Also total enrollments (in 1000s) and average section sizes. Distance-learning enrollments are not included. (For four-year colleges and universities, data from 1995, 2000 show percentage of enrollments.)

	Percentage of sections taught by						
Mathematics Departments	Tenured/ tenure-eligible %	Other full-time %	Part-time %	Graduate teaching assistants %	Unknown %	Enrollment in 1000s	Average section size
Elementary Statistics (no calculus prerequisite)							
Large lecture/recitation	30	27	34	2	7	12	32
Regular section <31	56	12	28	2	2	54	24
Regular section >30	49	18	22	6	5	56	40
Course total 2005 % of sections	**51**	**16**	**27**	**3**	**4**	**122**	**31**
Course total (1995,2000) % of enrollment	(65,45)	(7,13)	(19,24)	(8,7)	(--,11)	(97, 114)	(33,42)
Probability & Statistics (no calculus prerequisite)							
Course total 2005 % of sections	**29**	**24**	**44**	**1**	**2**	**18**	**30**
Course total (1995,2000) % of enrollment	(61,50)	(6,28)	(15,23)	(19,0)	(--,0)	(18,13)	(31,25)
Total All Elem.Probability & Statistics courses 2005 % of sections	**48**	**17**	**29**	**3**	**3**	**140**	**31**
Two course total (1995,2000) % of enrollment	(64,46)	(7,14)	(18,24)	(10,6)	(na,10)	(115, 127)	(33,25)

Two-Year Colleges	Percentage of sections taught by Full-time	Part-time	Enrollment in 1000s	Average section size
Elementary Statistics (with or without probability)	**65**	**35**	**101**	**26**
Course total (1995,2000)	(69,66)	(31,34)	(69,71)	(28,25)

Note: 0 means less than one half of 1%.

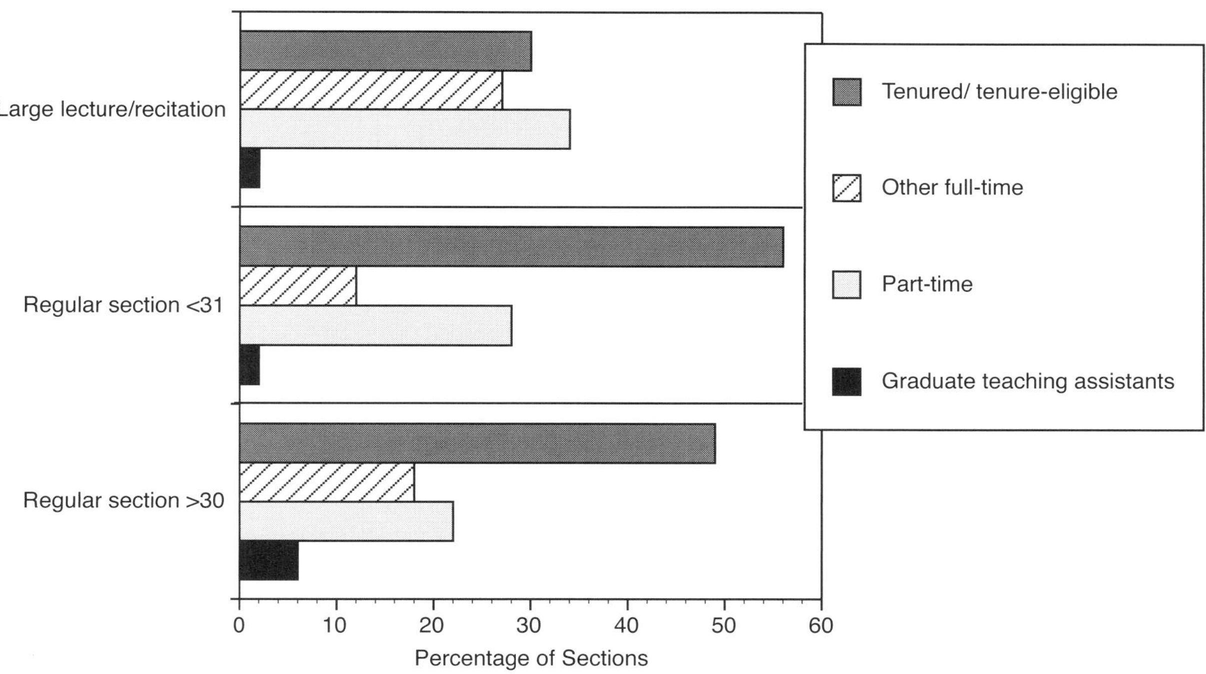

FIGURE S.9.1 Percentage of sections in Elementary Statistics (no Calculus prerequisite) taught by tenured/tenure-eligible, other full-time, part-time, and graduate teaching assistants in mathematics departments at four-year colleges and universities by size of sections in fall 2005.

ments of two- and four-year colleges and universities, while Table S.10 describes the situation in separate statistics departments. These two tables suggest that mathematics departments (which taught the vast majority of the nation's Elementary Statistics courses in fall 2005) devoted a much higher percentage of full-time faculty resources to the course than did statistics departments. In addition, the percentage of Elementary Statistics sections taught by TTE faculty (and by the combination of TTE and OFT faculty) in mathematics departments lies about midway between the corresponding percentages for Mainstream and Non-mainstream Calculus I sections. Also note that the average section size in Elementary Statistics courses taught in statistics departments increased between fall 2000 and fall 2005.

TABLE S.10 Percentage of sections in Elementary Statistics (no Calculus prerequisite) and Probability and Statistics (no Calculus prerequisite) taught by tenured/tenure-eligible, other full-time, part-time faculty, graduate teaching assistants, and unknown in statistics departments at four-year colleges and universities by size of sections in fall 2005. Also total enrollments (in 1000s) and average section sizes. Distance enrollments are not included. (Data from 1995,2000 show percentage of enrollments.)

	Percentage of sections taught by						
Statistics Departments	Tenured/ tenure-eligible %	Other full-time %	Part-time %	Graduate teaching assistants %	Unknown %	Enrollment in 1000s	Average section size
Elementary Statistics (no calculus prerequisite)							
Large lecture/recitation	19	27	16	17	21	28	82
Regular section <31	33	18	7	23	20	1	12
Regular section >30	33	14	18	30	5	13	50
Course total 2005 % of sections	**26**	**21**	**16**	**22**	**15**	**42**	**63**
Course total (1995,2000) % of enrollment	(47,36)	(15,17)	(10,22)	(29,19)	(--,6)	(35,40)	(51,65)
Probability & Statistics (no calculus prerequisite)							
Course total 2005 % of sections	**34**	**38**	**0**	**16**	**13**	**2**	**68**
Course total (1995,2000) % of enrollment	(32,18)	(4,12)	(2,13)	(61,32)	(--,25)	(8,4)	(48,55)
Total Elem. Probability & Statistics courses 2005 % of sections	26	22	15	22	15	44	64
Two course total (1995,2000) % of enrollment	(44,34)	(13,17)	(9,21)	(35,21)	(--,7)	(43,44)	(50,58)

Note: 0 means less than one half of 1%.

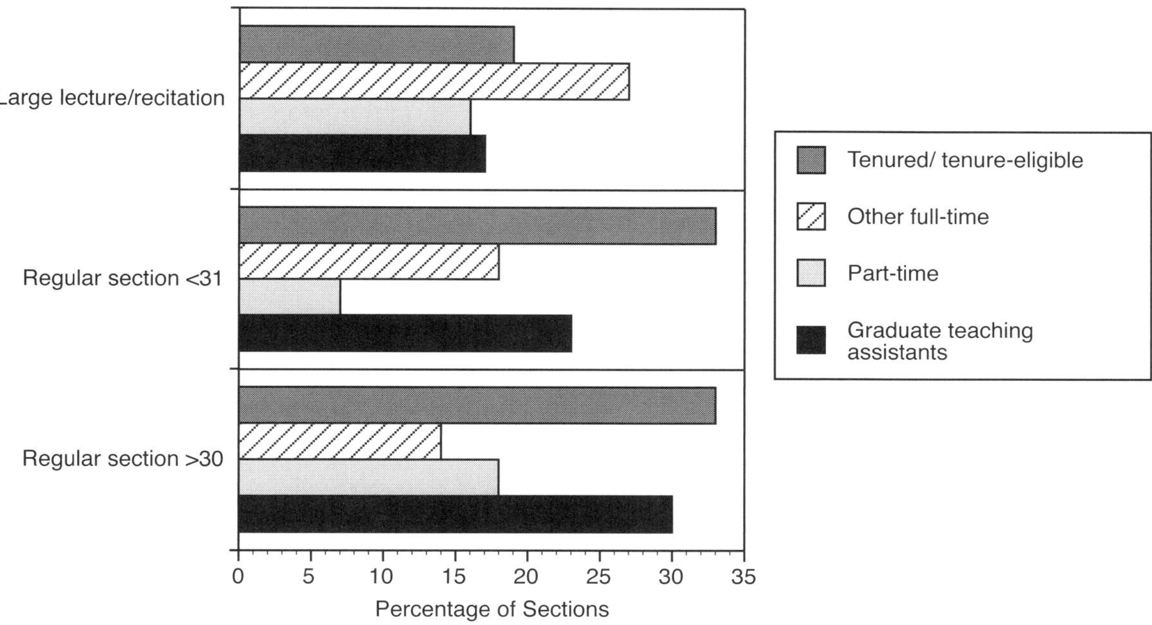

FIGURE S.10.1 Percentage of sections in Elementary Statistics (no Calculus prerequisite) taught by tenured/tenure-eligible faculty, other full-time faculty, part-time faculty, and graduate teaching assistants in statistics departments at four-year colleges and universities by size of sections in fall 2005.

How are first-year courses taught? (Tables S.11, S.12, and S.13)

The calculus-reform movement of the early 1990s stressed changes in how mathematics courses should be taught, as well as changes in their content. Starting in 1995, CBMS surveys tracked the spread of two broad families of pedagogical methods used to help students learn in their first-year courses. One family of techniques was technology-based, including the use of graphing calculators, computers, and computer assignments. The second family was sometimes described as "humanistic methods" and included the use of group projects and writing assignments. Tables S.11, S.12, and S.13 summarize the findings of CBMS2005 concerning use of these pedagogical methods in the nation's first-year courses in fall 2005. See the tables in Chapter 5 for more details, including presentation of this data based on the highest degree offered by the mathematics or statistics department that taught the course.

Tables S.11 and S.12 show that in four-year mathematics departments nationally, graphing calculators and computer assignments are widely (but far from universally) used in Mainstream Calculus courses, while the use of writing assignments almost never exceeded the fifteen percent level and the use of group projects was even lower. Calculator use in Non-mainstream Calculus I was somewhat higher than in Mainstream Calculus I, while the use of the other pedagogical methods in Non-mainstream Calculus I was in the single digits.

In both types of Calculus I courses, the percentage of two-year college sections that used any one of the four pedagogical techniques mentioned above exceeded the corresponding percentage for four-year mathematics departments.

CBMS2005 asked departments about the use of a new teaching tool in their first-year classes, namely the use of online homework and testing software that was offered by many textbook publishers (and others) in fall 2005. The two-year questionnaire described these online systems as using "commercial or locally produced online-response homework and testing systems", and the questionnaires sent to four-year mathematics and statistics departments described them as "online homework generating and grading packages." The results were somewhat surprising, given the apparent level of resources invested in such systems by textbook publishers. In almost every type of course, utilization percentages for such online resource systems were in the single digits. Of course, those percentages represent departmental responses, and perhaps students' voluntary use of the systems is higher.

Table S.13 investigates the use of the same five pedagogical tools in Elementary Statistics courses and reveals some marked differences between different types of departments. The percentage of sections of Elementary Statistics that used graphing calculators ranged from 73% in two-year colleges, to 36% in four-year mathematics departments, to only about 5% in statistics departments. The use of computer assignments in Elementary Statistics courses varied over a much smaller range, from 45% in two-year colleges to 58% in statistics departments, and Table S.13 suggests that almost 40% of Elementary Statistics sections taught in statistics departments use neither

TABLE S.11 Percentage of sections in Mainstream Calculus I and II taught using various reform methods in mathematics departments of four-year colleges and universities by size of sections, and percentage of sections taught using various reform methods in public two-year college mathematics programs in fall 2005 (For four-year colleges and universities, figures in parentheses show percentages of enrollments from 1995 and 2000.) Also total enrollments (in 1000s) and average section sizes. Distance-learning sections are not included.

	Percentage of sections taught using						
Four-Year Colleges & Universities	Graphing calculators %	Writing assignments %	Computer assignments %	On-line resource systems %	Group projects %	Enrollment in 1000s	Average section size
Mainstream Calculus I (Section %)							
Large lecture/recitation	48	13	24	6	12	80	46
Regular section <31	58	16	20	2	7	63	22
Regular section >30	43	10	20	6	13	58	35
Course total (section %)	**51**	**13**	**21**	**4**	**10**	**201**	**32**
(1995,2000) enrollment %	(37,51)	(22,27)	(18,31)	na	(23,19)	(192, 190)	(33,32)
Mainstream Calculus II (Section %)							
Large lecture/recitation	38	9	20	4	7	36	50
Regular section <31	47	13	24	2	5	25	21
Regular section >30	42	5	18	5	5	24	36
Course total (section %)	**43**	**9**	**21**	**3**	**6**	**85**	**33**
(1995,2000) enrollment %	(29, 48)	(24,18)	(17,27)	na	(20, 15)	(83,87)	(30,32)
Total Mnstrm Calculus I & II (Section %)	**49**	**12**	**21**	**4**	**9**	**285**	**32**
(1995, 2000) enrollment %	(35, 50)	(23, 24)	(18, 30)	na	(22,18)	(275, 277)	(32, 32)
Two-Year Colleges							
Mainstream Calculus I (Section %)	**79**	**19**	**20**	**5**	**19**	**49**	**22**
(1995, 2000) section %	(65, 78)	(20, 31)	(23, 35)	na	(22, 27)	(58,53)	(25,23)
Mainstream Calculus II (Section %)	**81**	**18**	**30**	**7**	**25**	**19**	**18**
(1995,2000) section %	(63, 74)	(13, 25)	(16, 37)	na	(18, 25)	(23,20)	(23,20)
Total Mainstream Calculus I & II (Section %)	**80**	**18**	**23**	**5**	**21**	**68**	**21**
(1995, 2000) section %	(65, 76)	(18, 28)	(24, 35)	na	(22, 27)	(81,73)	(24,22)

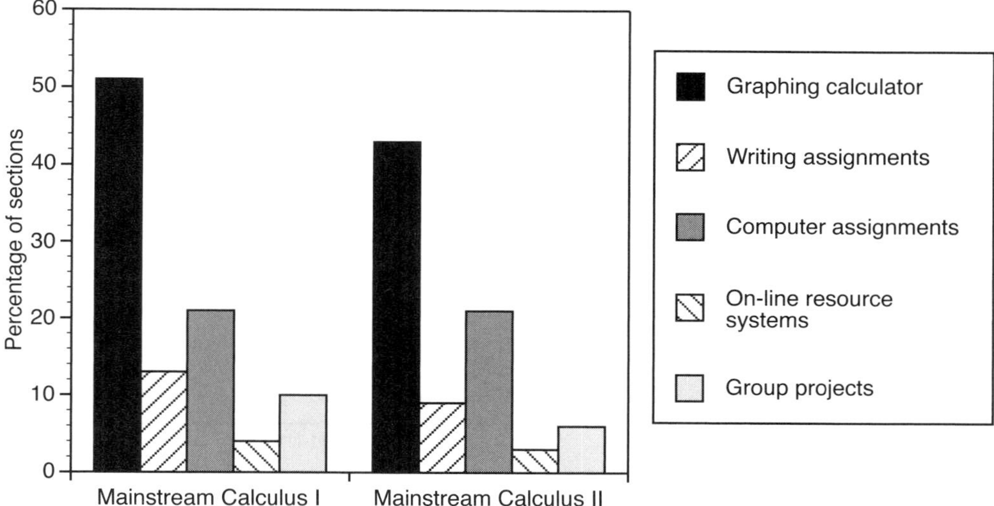

FIGURE S.11.1 Percentage of sections of Mainstream Calculus I and Mainstream Calculus II taught using various reform methods in mathematics departments at four-year colleges and universities in fall 2005.

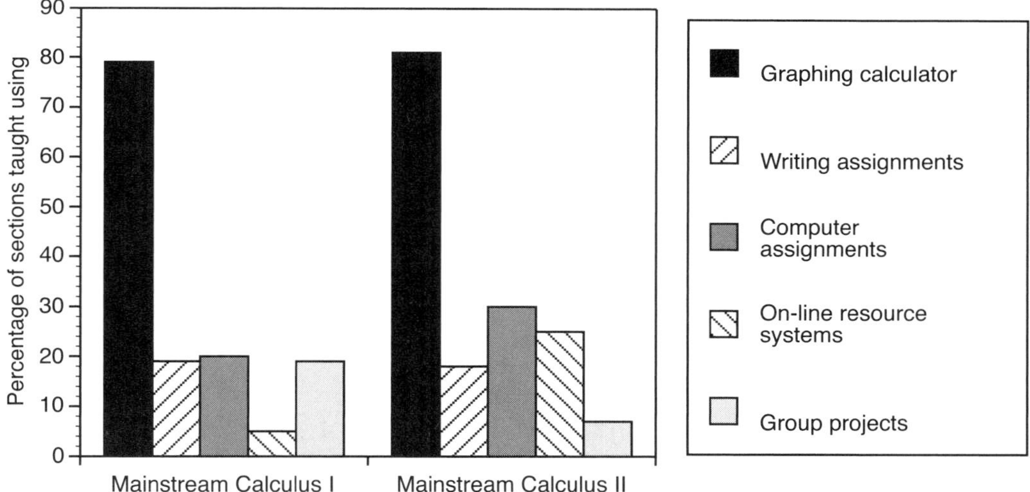

FIGURE S.11.2 Percentage of sections in Mainstream Calculus I and Mainstream Calculus II taught using various reform methods in mathematics programs at public two-year colleges in fall 2005.

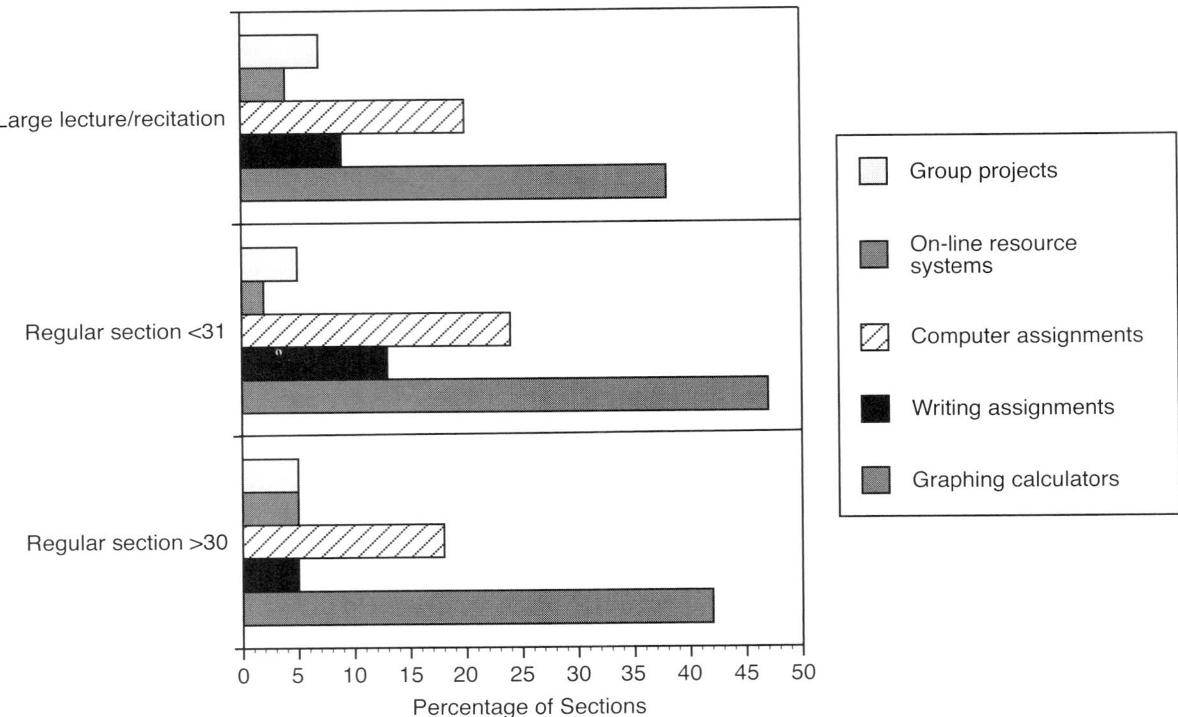

FIGURE S.11.3 Percentage of sections in Mainstream Calculus II taught using various reform methods in mathematics departments at four-year colleges and universities by size of sections in fall 2005.

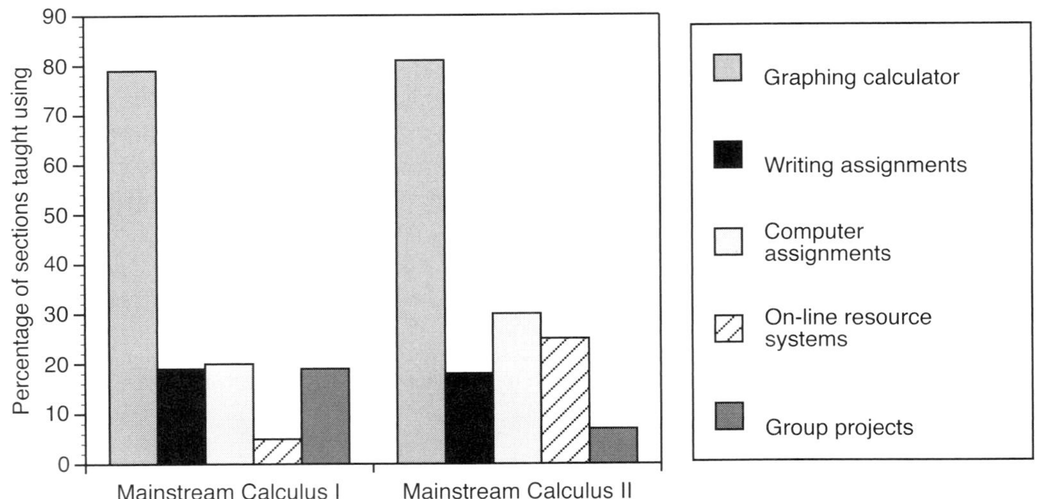

FIGURE S.11.4 Percentage of sections in Mainstream Calculus I and Mainstream Calculus II taught using various reform methods in mathematics programs at public two-year colleges in fall 2005.

TABLE S.12 Percentage of sections in Non-Mainstream Calculus I taught using various reform methods in mathematics departments at four-year colleges and universities by size of sections, and percentage of sections taught using various reform methods in mathematics programs at public two-year colleges, in fall 2005. Also total enrollments (in 1000s) and average section sizes. Distance-learning sections are not included. (For four-year colleges and universities, data from 1995 and 2000 show percentage of enrollments.)

	Percentage of sections taught using						
Four-Year Colleges & Universities	Graphing calculators %	Writing assignments %	Computer assignments %	On-line resource systems %	Group projects %	Enrollment in 1000s	Average section size
Non-Mnstream Calculus I							
Large lecture/recitation	60	7	8	7	4	28	64
Regular section <31	63	1	5	4	1	30	23
Regular section >30	37	7	4	5	6	50	44
Course total 2005 % of sections	**53**	**4**	**5**	**5**	**3**	**108**	**37**
(1995,2000) % of enrollment	(26,45)	(7,14)	(6,13)	na	(7,9)	(97, 105)	(39, 40)
Two-Year Colleges							
Non-Mnstream Calculus I 2005 % of sections	**77**	**14**	**9**	**3**	**14**	**20**	**23**
(1995,2000) % of sections	(44,72)	(17,20)	(8,15)	na	(20,20)	(26, 16)	(26,22)

Note: 0 means less than one-half of 1%.

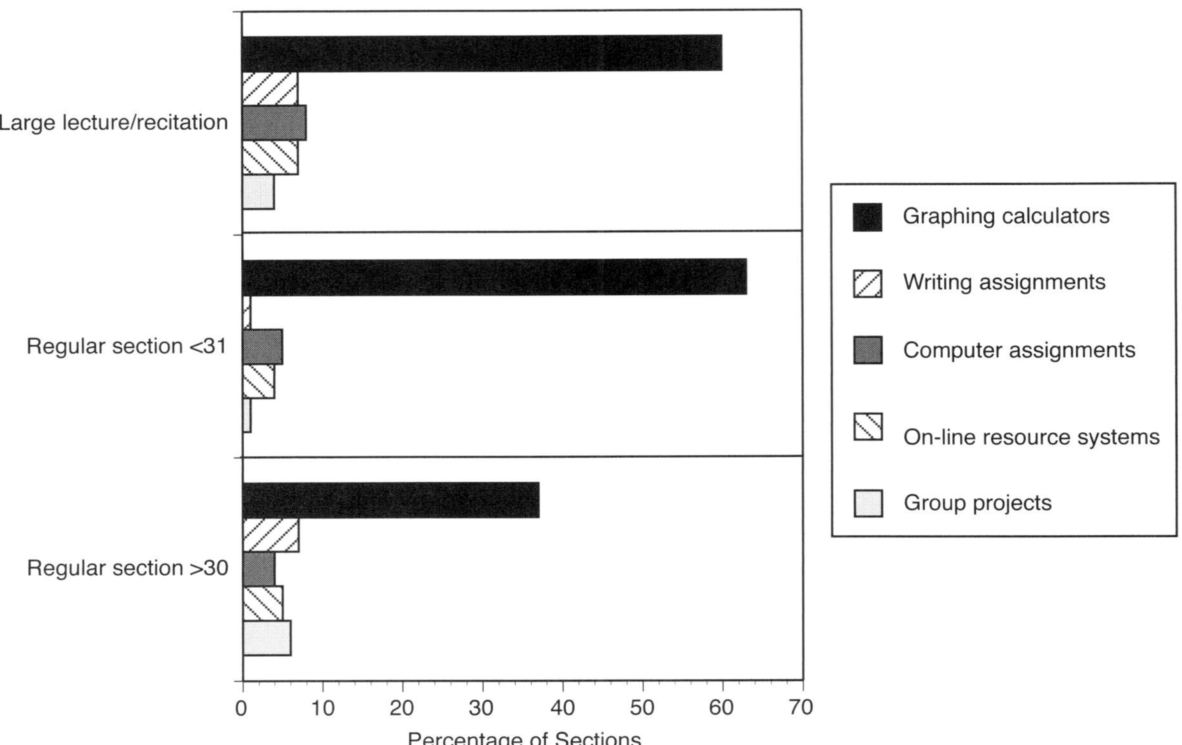

FIGURE S.12.1 Percentage of sections in Non-Mainstream Calculus I taught using various reform methods in mathematics departments at four-year colleges and universities by size of sections in fall 2005.

TABLE S.13 Percentage of sections in Elementary Statistics (no Calculus prerequisite) taught using various reform methods in mathematics and statistics departments in four-year colleges and universities, and percentage of sections in mathematics programs at public two-year colleges taught using various reform methods in fall 2005. Also total enrollment (in 1000s) and average section sizes. (Data from 1995,2000 show percentage of enrollments.)

	Percentage of sections taught using						
Elementary Statistics	Graphing calculators %	Writing assignments %	Computer assignments %	On-line resource systems %	Group projects %	Enrollment in 1000s	Average section size
Mathematics Departments							
Large lecture/recitation	42	48	83	0	38	12	32
Regular section <31	30	30	56	4	19	54	24
Regular section >30	44	21	46	2	5	56	40
Course total 2005 % of sections	36	28	55	3	16	122	31
Course total (1995,2000) % of enrollment	(na,47)	(na, 39)	(51,48)	na	(na,22)	(95, 114)	(33,42)
Statistics Departments							
Large lecture/recitation	9	42	59	26	30	28	82
Regular section <31	0	19	85	30	16	1	12
Regular section >30	1	57	52	1	22	13	50
Course total 2005 % of sections	5	46	58	16	26	42	63
Course total (1995,2000) % of enrollment	(na,13)	(na,23)	(59,63)	na	(na,43)	(35,40)	(51,65)
Two-year colleges							
Course total 2005 % of sections	73	44	45	10	24	101	26
Course total (1995,2000) % of sections	(na,59)	(na,50)	(46,46)	na	(na,35)	(69,71)	(28,25)

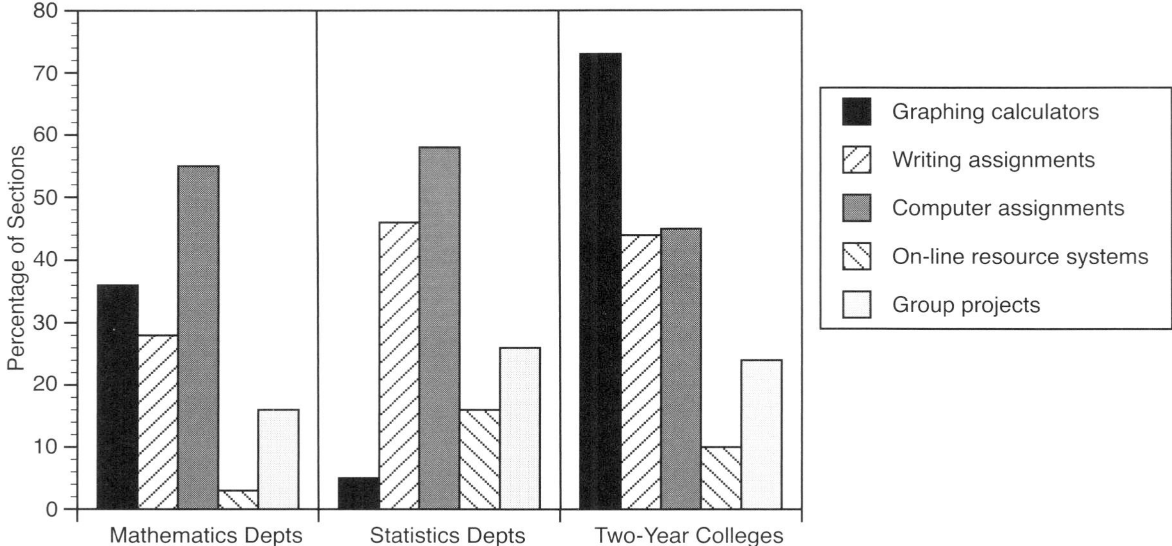

FIGURE S.13.1 Percentage of sections in Elementary Statistics (no Calculus prerequisite) taught using various reform methods in four-year colleges and universities and in two-year colleges, in fall 2005.

graphing calculators nor computer technology. Writing assignments were much more widely used in Elementary Statistics courses than in any Calculus course. Group projects, while not used in more than about one in four Elementary Statistics courses, were more widely used in that course than in Calculus. Statistics departments showed more interest in online resource systems than did either four-year mathematics departments or two-year college mathematics programs, with one in six statistics departments using such online resource systems in their Elementary Statistics courses.

Demographics of the Mathematical Sciences Faculty

The remaining tables in this chapter present a snapshot of faculty demographics in mathematics and statistics departments of four-year colleges and universities and in the mathematics programs of two-year colleges during fall 2005. Further details about four-year mathematics and statistics department faculty appear in Chapter 4, while additional information about two-year mathematics program faculty is given in Chapter 7.

Sources of demographic data

Data concerning two-year college mathematics faculty were collected, as in previous CBMS surveys, as part of the two-year-college questionnaire (see Sections D, E, F, and G of the 2005 questionnaire). In contrast, data concerning four-year college and university faculty came from a totally separate survey, conducted by the Joint Data Committee (JDC) of five professional societies (the American Mathematical Society, the American Statistical Association, the Institute of Mathematical Statistics, the Mathematical Association of America, and the Society for Industrial and Applied Mathematics).

Since 1957, the Joint Data Committee (JDC) has carried out annual departmental surveys of four-year mathematics and statistics departments for its own purposes. In fall 2000, department chairs objected strongly to answering almost the same faculty demographics questions on two separate surveys, one for JDC and the other for CBMS2000. Consequently, CBMS2005 and JDC made an agreement to use the JDC survey in fall 2005 as the basis for demographic estimates needed for the CBMS2005 report.

Using the JDC survey to obtain faculty data for CBMS2005 simplified the lives of department chairs but had two important drawbacks in terms of the faculty demographics sections of this report. The first concerned response rates. As can be seen from Appendix II, Part II, the JDC survey had strong response rates from doctoral departments, but response rates from bachelors departments were not as strong, and standard errors for the JDC estimates for bachelors-level departments were sometimes uncomfortably large. The second major drawback of using JDC data for faculty demographics sections of CBMS2005 was that JDC surveys do not include masters-level departments of statistics. Therefore, *the faculty demographic data concerning statistics departments in this chapter and in Chapter 4 describe only doctoral statistics departments, while earlier CBMS reports presented demographic data on both masters and doctoral statistics departments.* However, the data in Chapters 2, 3, and 5 on enrollments and curricular issues do include both masters and doctoral-level statistics departments.

In an attempt to make sure that historical data on faculty demographics in this report are internally consistent, *historical data on faculty demographics in CBMS2005 are taken from JDC data from previous years, rather than from earlier CBMS reports.* Therefore, historical faculty data in CBMS2005 may appear somewhat different from faculty data published in earlier CBMS reports.

Readers who compare CBMS2005 faculty demographic data on doctoral statistics departments with Joint Data Committee publications will see a difference between CBMS2005 data for doctoral statistics departments and what JDC publications call "Group IV." JDC's Group IV consists of doctoral statistics, biostatistics, and biometrics departments, some of which do not offer any undergraduate programs or courses. To make the faculty demographic data in this report fit into a study of the nation's undergraduate programs, only a subset of Group IV was used. This subset consisted of only those doctoral statistics departments with undergraduate programs, and excluded biometrics and biostatistics departments.

TABLE S.14 Number of full-time and part-time faculty in mathematics departments at four-year colleges and universities, in doctoral statistics departments at universities, and in mathematics programs at two-year colleges in fall 1995, 2000, and 2005. (Two-year college data for 2005 include only public two-year colleges.)

	1995	2000	2005
Four-Year Colleges & Universities			
Mathematics Departments			
Full-time faculty	19572	19779	21885
Part-time faculty	5399	7301	6536
Statistics Departments			
Full-time faculty	840	808	946
Part-time faculty	125	102	112
Two-Year College Mathematics Programs			
Full-time faculty	7742	7921	9403
Part-time faculty [1]	14266	14887	18227

[1] Paid by two-year colleges. In fall 2000, there were an additional 776 part-time faculty in two-year colleges who were paid by a third party (e.g., by a school district, in a dual-enrollment course) and in 2005 the number paid by a third party was 1915.

Note on data sources: Data on four-year mathematics and statistics departments in Table S.14 are taken from annual reports of the Joint Data Committee of AMS/ASA/IMS/MAA/SIAM, published in fall issues of the *Notices of the American Mathematical Society.* Combined data for statistics and biostatistics departments with Ph.D. programs are reported as Group IV data in those reports, and the figures reported in Table S.14 for statistics departments were obtained by removing all departments that do not have undergraduate programs from the Group IV totals.

The number of mathematical sciences faculty members (Table S.14)

Table S.14 shows that between fall 1995 and fall 2005 there were substantial increases in the number of full-time and part-time faculty in four-year mathematics departments. Over the decade there was a 12% increase in the number of full-time faculty in four-year mathematics departments, with almost all of that growth in the last half of the decade. The number of part-time faculty in four-year mathematics departments, which had grown by more than a third between 1995 and 2000, actually declined between fall 2000 and fall 2005 as four-year colleges increased their full-time staff, but part-time numbers still rose by nearly 21% over the decade 1995–2005. For comparison, recall that during the same period, total four-year college and university enrollments grew by 21% (see Table S.1) and enrollments in mathematics and statistics departments increased by about 8% (see Table S.2).

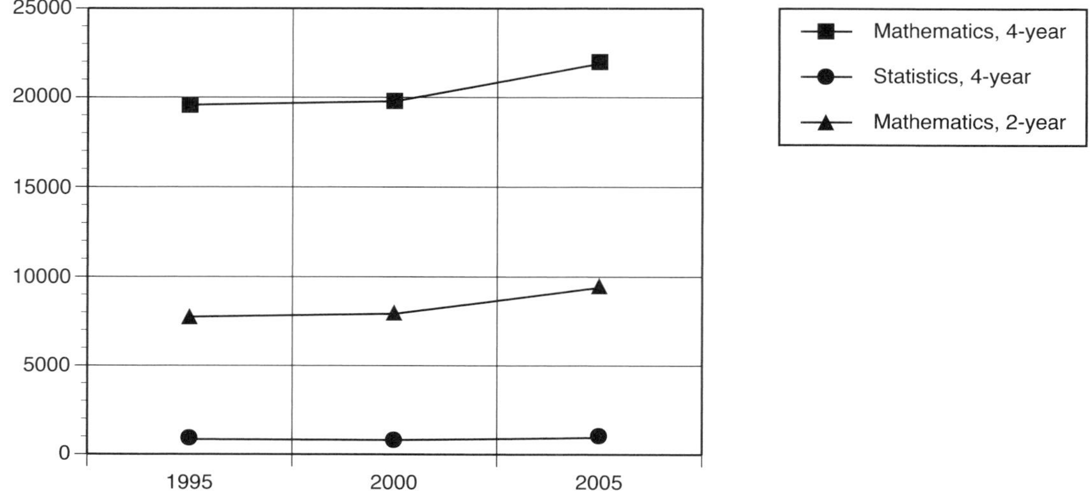

FIGURE S.14.1. Number of full-time faculty in mathematics departments of four-year colleges and universities, in doctoral statistics departments, and in mathematics programs at two-year colleges in fall 1995, 2000, and 2005.

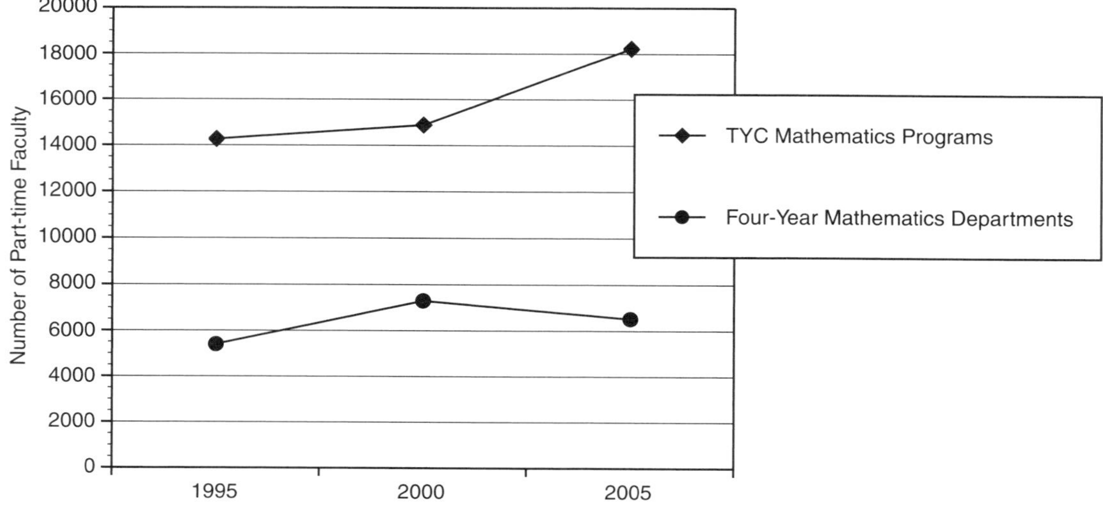

FIGURE S.14.2 Number of part-time faculty in mathematics departments at four-year colleges and universities and in mathematics programs at two-year colleges (TYCs) in fall 1995, 2000, and 2005.

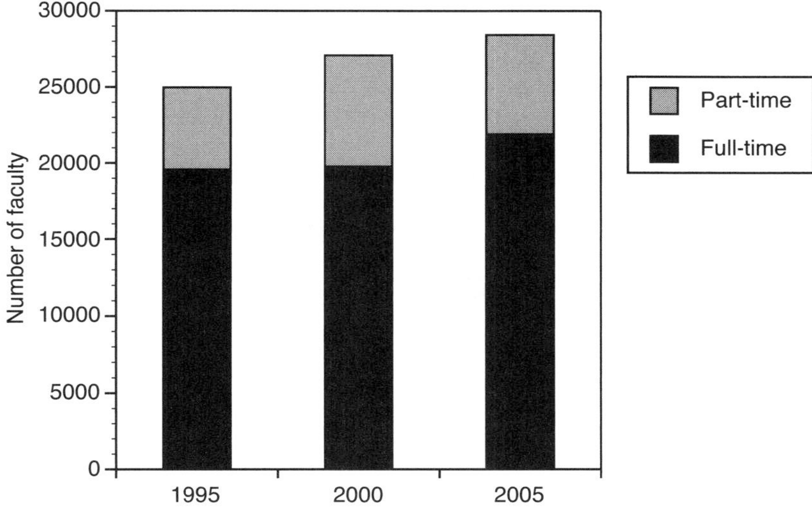

FIGURE S.14.3 Number of full-time and part-time faculty in mathematics departments of four-year colleges and universities in fall 1995, 2000, and 2005.

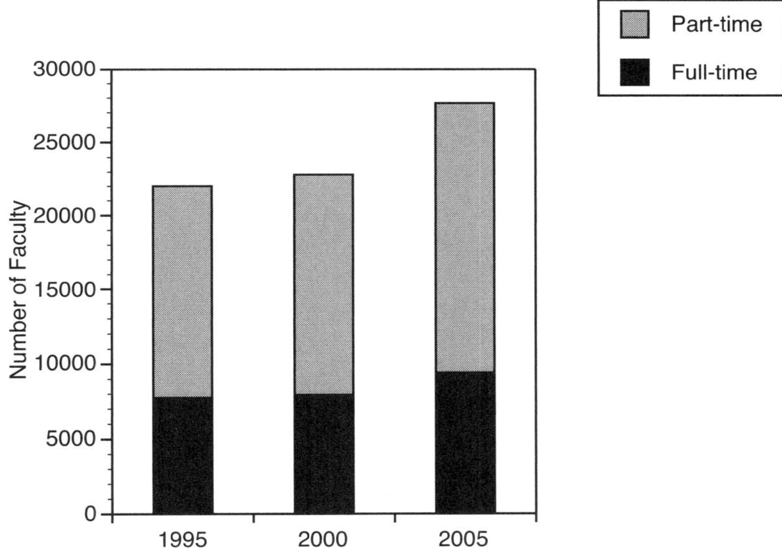

FIGURE S.14.4 Number of full-time and part-time faculty in mathematics programs at two-year colleges in fall 1995, 2000, and 2005.

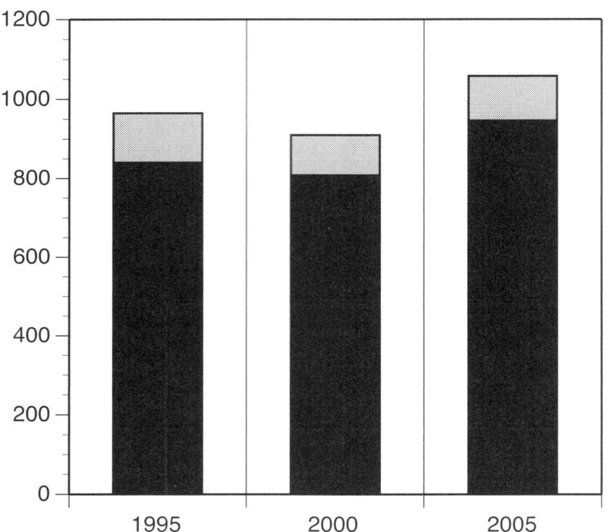

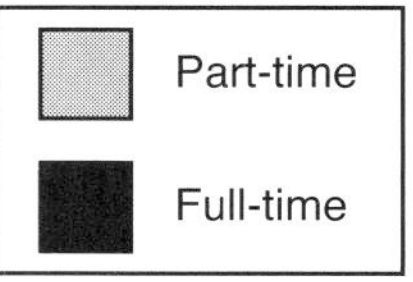

FIGURE S.14.5 Number of full-time and part-time faculty in doctoral statistics departments in fall 1995, 2000, and 2005.

The number of full-time faculty in doctoral statistics departments, which dropped between 1995 and 2000, rebounded substantially between 2000 and 2005, recording a roughly 13% growth during the 1995–2005 decade. The number of part-time faculty in doctoral statistics departments declined by about 10% during that same ten-year period. To compare faculty growth with enrollment growth in doctoral statistics departments, one needs to use Table E.2 of Chapter 3 rather than Table S.2. Table E.2 shows that undergraduate enrollments in doctoral statistics departments stood at 62,000 in fall 1995, and at 62,000 in fall 2005. The ten-year undergraduate enrollment growth in statistics departments that appears in Table S.2 was all in masters-level departments.

Two-year college mathematics programs saw a roughly 21% increase in full-time faculty between 1995 and 2005, an increase that matches the 21% growth in total TYC enrollment and also the 21% mathematics and statistics enrollment growth in TYCs that was mentioned earlier in this chapter.

The roughly 10% decline between fall 2000 and fall 2005 in the number of part-time faculty in four-year mathematics departments stands in contrast to the Table S.6 finding that the percentage of sections taught by part-time faculty in four-year mathematics departments held steady between fall 2000 and fall 2005, suggesting that the typical part-time faculty member in fall 2005 was teaching a larger number of courses than in fall 2000. CBMS2005 does not have data on the average teaching assignment of part-time faculty, but Table 22 of [NCES2] shows that the average part-time faculty member in natural science departments of four-year institutions spent about 6.7 hours per week in the classroom in fall 2003.

Part-time faculty comprised about 23% of all faculty in four-year mathematics departments in fall 2005. Compared with other disciplines, the 23% figure for part-time faculty is not particularly large. Federal data published by NCES in fall 2006 [NCES2] showed that, across all disciplines in four-year institutions, the percentage of part-time faculty among all faculty was about 43% in 2003, a figure that has held steady since at least 1992. Within the natural sciences, the category into which the NCES report places mathematics and statistics, the percentage of part-time faculty among all faculty was 23.5% in 2003.

Appointment type and degree status of the faculty (Tables S.15 and S.16)

The approximately 11% growth (see Table S.14) in the total number of full-time faculty in four-year mathematics departments between fall 2000 and fall 2005 consisted of a roughly 6% growth in tenured and tenure-eligible (TTE) faculty, coupled with a 31% growth in the number of full-time mathematics faculty who are outside of the TTE stream. Starting in 2003, the Joint Data Committee (JDC) of the mathematical sciences professional societies began collecting data on the number of postdoctoral (PD) faculty, a subsection of the OFT category, and this CBMS2005 report will present parallel data on the entire OFT category and on the subcategory of PD faculty.

Starting in 2003, the term "postdoctoral appointment" had a standard definition in JDC surveys. A postdoctoral (PD) appointment is a full-time, temporary position that is primarily intended to provide an opportunity to extend graduate training or to further research. Consequently, a department's sabbatical replacements, its senior visiting faculty, and its non-TTE instructors are not counted as PD appointees. CBMS2005 used the JDC definition.

Anecdotal evidence suggests that there was substantial growth in the number of postdoctoral appointments in mathematical sciences departments between 1995 and 2005, in large part due to the NSF VIGRE program. Table S.15 shows that in fall 2005, about one in six members of the combined OFT category in four-year mathematics departments were postdoctoral appointees.

TABLE S.15 Number of full-time faculty who are tenured and tenure-eligible (TTE), postdocs, and other full-time (OFT) in mathematics and doctoral statistics departments of four-year colleges and universities, and in mathematics programs at two-year colleges, in fall 2000 and fall 2005. (Postdocs are included in the Other full-time category.)

Four-Year Colleges and Universities	Fall 2000				Fall 2005			
	Total	TTE	Other full-time	Postdoc	Total	TTE	Other full-time	Posdoc
Mathematics Departments								
Full-time faculty	**19779**	**16245**	**3533**	**na**	**21885**	**17256**	**4629**	**819**
Having doctoral degree	16640	14978	1662	na	18071	15906	2165	813
Having other degree	3139	1267	1872	na	3814	1350	2464	6
Doctoral Statistics Departments								
Full-time faculty	**808**	**709**	**99**	**na**	**946**	**783**	**163**	**51**
Having doctoral degree	794	707	87	na	915	781	133	51
Having other degree	14	2	12	na	31	2	30	0
Total Math & Stat Depts	**20587**	**16954**	**3632**	**na**	**22831**	**18039**	**4792**	**870**
Two-Year College Mathematics	Total full-time faculty	Full-time permanent	Full-time temporary		Total full-time faculty	Full-time permanent	Full-time temporary	
Full-time faculty	**7921**	**6960**	**961**		**9403**	**8793**	**610**	
Grand Total	**28508**	**23914**	**4593**	**na**	**32234**	**26832**	**5402**	**870**

Note: Round-off may make marginal totals seem inaccurate.

Full-time faculty numbers in doctoral statistics departments fell between fall 1995 and fall 2000, and then rose by about 17% between fall 2000 and fall 2005. The number of OFT faculty in doctoral statistics departments rose by almost 65% between 2000 and 2005, while the number of TTE faculty grew by about 10%. Postdoctoral positions are more common in doctoral statistics than in mathematics departments; of the OFT faculty in doctoral statistics departments in fall 2005, almost one in three held postdoctoral appointments.

Two-year colleges usually do not have tenured and tenure-eligible faculty, and yet they make a distinction between faculty who are "permanent full-time" and "temporary full-time." The number of permanent full-time faculty in two-year college mathematics programs grew by about 26% between fall 2000 and fall 2005. That increase more than wiped out the 8% decline between fall 1995 and fall 2000 and resulted in a net increase in permanent full-time faculty of about 16% during the 1995–2005 decade (cf. Tables SF.6 in CBMS1995 and CBMS2000). The number of temporary full-time faculty in two-year college mathematics programs declined by about a third from the levels of fall 2000, but still almost quadrupled between 1995 and 2005.

In four-year mathematics departments, the percentage of TTE faculty holding doctorates rose from 90% in fall 1995 to 92% in fall 2000 and remained at the 92% level in fall 2005. The percentage of TTE faculty holding doctoral degrees varies considerably by the highest degree offered by the department, and the data on percentage of doctoral degrees by type of department appears in Chapter 4 of this report.

Table S.15 shows that in doctoral statistics departments, the percentage of Ph.D.-holding faculty among all TTE faculty was above 99% in fall 2000 and fall 2005. Table SF.6 of CBMS1995 presents data showing that about 91% of TTE faculty in statistics departments held doctoral degrees in 1995, but it is important to remember that CBMS1995 data included masters-level as well as doctoral statistics departments.

The percentage of doctoral faculty in the OFT category is understandably far lower than in the TTE category. Table SF.5 of CBMS1995 shows that in four-year mathematics departments the percentage was 43% in fall 1995, and the JDC data presented in Table S.15 of this report shows that the percentage remained steady at 47% in fall 2000 and fall 2005. Table S.15 of this report shows that among the OFT faculty in doctoral statistics departments, the percentage of Ph.D.-holding faculty actually declined between fall 2000 and fall 2005, in spite of the fact that in fall 2005, almost one out of three members of the OFT group were postdoctoral appointees. Perhaps this decline represented the addition of many masters-level full-time instructors in doctoral statistics departments.

Table S.16 shows the percentage of mathematics program permanent faculty in two-year colleges who are at various degree levels. There was not much variation between the percentages reported in 1990 and in 2005. The percentage of two-year college mathematics faculty holding doctorates held steady at the 16 to 17 percent level, and masters-degree faculty have slowly replaced bachelors-degree faculty in mathematics programs. Table S.16 contains an anomaly that will reappear many times in this report. CBMS studies before 2005 included both public and some private two-year colleges while CBMS2005 does not include any private two-year colleges. NCES data on enrollments in public and private two-year colleges can sometimes be used to estimate public two-year college numbers, as in the discussion of Table S.1 above, but the resulting estimates are rough, at best.

TABLE S.16 Percentage of full-time permanent faculty in mathematics programs at two-year colleges by highest degree in Fall 1990, 1995, 2000, and 2005. (Data for 2005 include only public two-year colleges.)

Highest degree of TYC permanent mathematics faculty	Percentage of full-time permanent faculty			
	1990 %	1995 %	2000 %	2005 %
Doctorate	17	17	16	16
Masters	79	82	81	82
Bachelors	4	1	3	2
Number of full-time permanent faculty	7222	7578	6960	8793

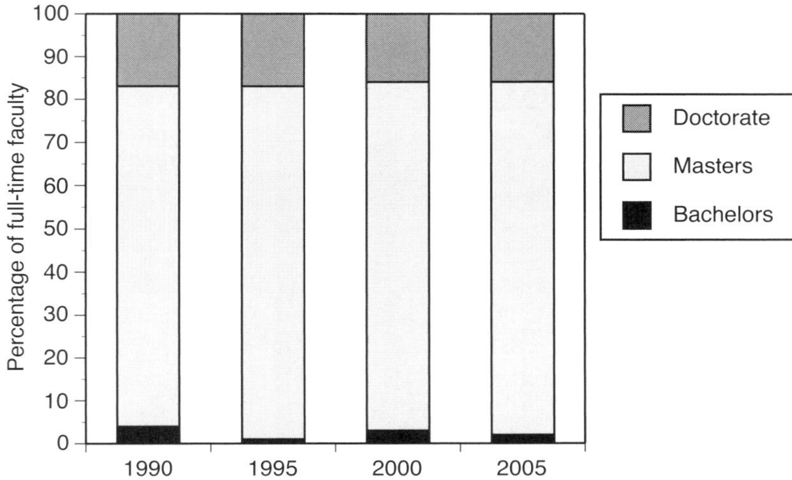

FIGURE S.16.1 Percentage of full-time permanent faculty in mathematics programs at two-year colleges by highest degree in fall 1990, 1995, 2000, and 2005. Data for 2005 include only public two-year colleges.

Gender, Age, and Ethnicity Among the Mathematical Sciences Faculty (Tables S.17 to S.23)

JDC surveys show that the percentage of women in mathematical sciences departments has been rising for many years, and Table S.17 shows that the percentage of women in the nation's mathematics and statistics faculty rose again between fall 2000 and fall 2005.

In four-year mathematics departments, 15% of the tenured faculty were women in fall 2000, a figure that rose to 18% in fall 2005. The percentage of women among tenure-eligible mathematics department faculty was 29% in both fall 2000 and fall 2005, and in the OFT category, the percentage of women rose by three points, to 44%. Because women held only 23% of the PD positions in mathematics departments in fall 2005, that three percentage point increase must have been concentrated in the non-postdoctoral OFT category. In estimating future trends, the fact that women received 30% of mathematics and statistics doctorates between 2000 and 2005 suggests that the percentage of women among mathematics department faculty will continue to rise.

The figures in Table S.17 do not tell the whole story about the percentage of women among mathematics department faculty in the U.S. Tables in Chapter 4 present this data on the basis of the highest degree offered by the department, and show considerable variation in the percentage of women faculty between, for example, doctoral mathematics departments and mathematics departments that offer only bachelors degrees. For example, Table F.1 of Chapter 4 shows that between fall 2000 and fall 2005, the percentage of women among tenured faculty in doctoral mathematics departments rose from about 7% to about 9%, percentages that are only half as large as the corresponding percentages for all mathematics departments in Table S.17.

Doctoral statistics departments also saw an increase in the percentage of women faculty between fall 2000 and fall 2005. In fall 2000, 9% of tenured faculty in doctoral statistics departments were women, while in fall 2005 the percentage was 13%. The percentage of women in tenure-eligible positions also rose, from 34% to 37%, and 31% of postdoctoral faculty in doctoral statistics departments were women.

In recent years, women have held a greater proportion of positions in mathematics programs at two-year colleges than in mathematics departments of four-year colleges and universities. In fall 2000, women held 49% of mathematics program positions in two-year colleges, and by fall 2005 that percentage had risen to 50%.

Tables S.18 and S.19 present data on the age of tenured and tenure-eligible mathematical sciences faculty members, by gender. The average age data for fall 2000 is taken from the CBMS2000 report, and data for fall 2005 about four-year mathematics and statistics departments come from surveys by the JDC. Information about age distribution among two-year college mathematics faculty was collected as part of the CBMS2005 survey.

In four-year mathematics departments, the average age of tenured men and women rose between fall 2000 and fall 2005, presumably because senior faculty are delaying retirement. The average age of tenure-eligible-but-not-tenured men and women also increased, possibly reflecting the fact that many new Ph.D.s spent time in postdoctoral positions or other visiting positions before entering their first tenure-

TABLE S.17 Gender among full-time faculty in mathematics and doctoral statistics departments of four-year colleges and universities by type of appointment, and among permanent full-time faculty in mathematics programs at two-year colleges in fall 2000 and fall 2005. Also gender among doctoral and masters degree recipients. (Postdocs are included in the Other full-time category.)

Four-Year Colleges and Universities	Fall 2000					Fall 2005				
Mathematics Departments	Total	Tenured	Tenure-eligible	Other full-time	Postdoc	Total	Tenured	Tenure-eligible	Other full-time	Postdoc
Full-time faculty	19779	12959	3287	3533	na	21885	12874	4382	4629	819
Number of women	4346 (22%)	1941 (15%)	954 (29%)	1450 (41%)	na	5641 (26%)	2332 (18%)	1250 (29%)	2059 (44%)	191 (23%)
Doctoral Statistics Departments										
Full-time faculty	808	572	137	99	na	946	604	179	163	51
Number of women	140 (17%)	51 (9%)	47 (34%)	42 (42%)	na	211 (22%)	79 (13%)	66 (37%)	66 (40%)	16 (31%)

	July 1, 1980-June 30, 2005	July 1, 2000-June 30, 2005
Number of PhDs from US Math & Stat Depts[1]	25019	5365
Number of women among new PhDs[1]	5702 (23%)	1607 (30%)

Two-Year College Mathematics Programs	Total full-time	Full-time age<40	Total full-time	Full-time age<40
Full-time faculty	6960	1392	8793	2326
Number of women	3423 (49%)	626 (45%)	4373 (50%)	1148 (49%)

Masters degrees in mathematics and statistics granted in the U.S. in 2003-04[2]	4191
Number of women among new masters recipients[2]	1889 (45%)

[1] Second Annual Reports of the AMS-ASA-IMA-MAA-SIAM Joint Data Committee, Tables 3-E through 3-G, AMS Notices, 1980-2005.

[2] 2005 Digest of Educational Statistics, NCES, Table 262, available at http://nces.ed.gov/programs/digest/d05/tables/dt05_252.asp

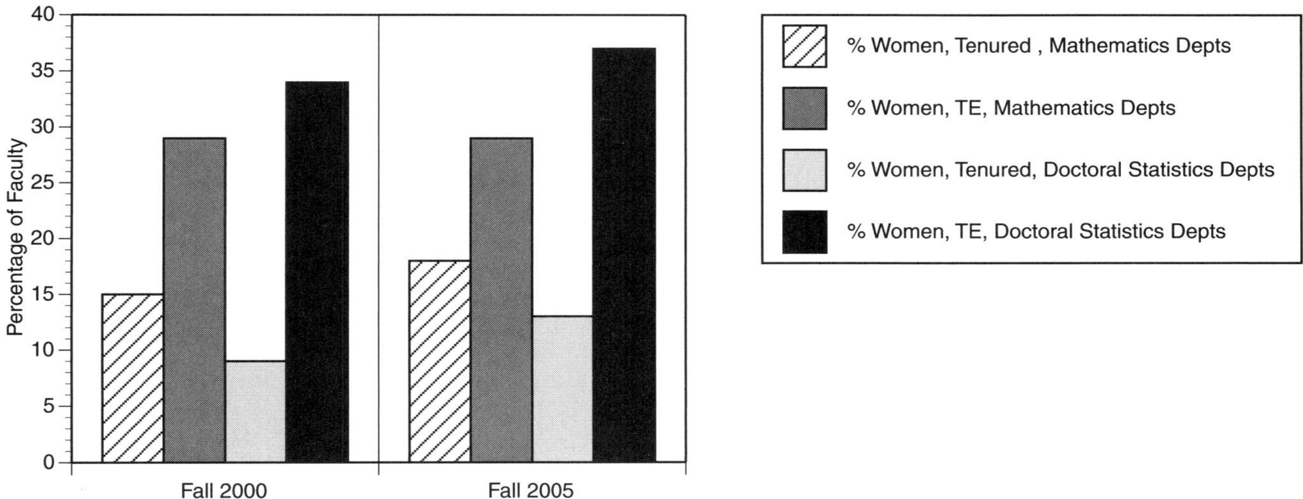

FIGURE S.17.1 Percentage of women in tenured and tenure-eligible(TE) categories in mathematics departments of four-year colleges and universities and doctoral statistics departments, in fall 2000 and 2005.

TABLE S.18 Percentage of all tenured and tenure-eligible faculty in mathematics departments of four-year colleges and universities in various age groups, and average age, by gender in fall 2005. Percentage full-time permanent faculty in mathematics programs at public two-year colleges, by age, and average ages in fall 2005. Also, historical data from fall 2000.

Four-Year College & University Mathematics Departments	Percentage of tenured/tenure-eligible faculty										Average age 2000	Average age 2005
	<30	30-34	35-39	40-44	45-49	50-54	55-59	60-64	65-69	>69		
Tenured men	0%	1%	4%	8%	9%	10%	11%	11%	5%	2%	52.4	53.7
Tenured women	0	0	1	3	2	3	2	1	0	0	49.6	50.2
Tenure-eligible men	1	6	5	3	1	1	1	0	0	0	36.6	38.9
Tenure-eligible women	1	2	2	1	1	0	0	0	0	0	37.8	38.6
Total tenured & tenure-eligible faculty	**2**	**9**	**13**	**14**	**13**	**14**	**14**	**13**	**6**	**2**		
	Percentage of permanent full-time faculty											
Two-Year College Mathematics Programs	<30	30-34	35-39	40-44	45-49	50-54	55-59	>59				
Full-time permanent faculty	**5**	**8**	**12**	**13**	**15**	**18**	**17**	**11**			47.6	47.8

Note: 0 means less than half of 1%. Round-off may cause some marginal totals to appear inaccurate.

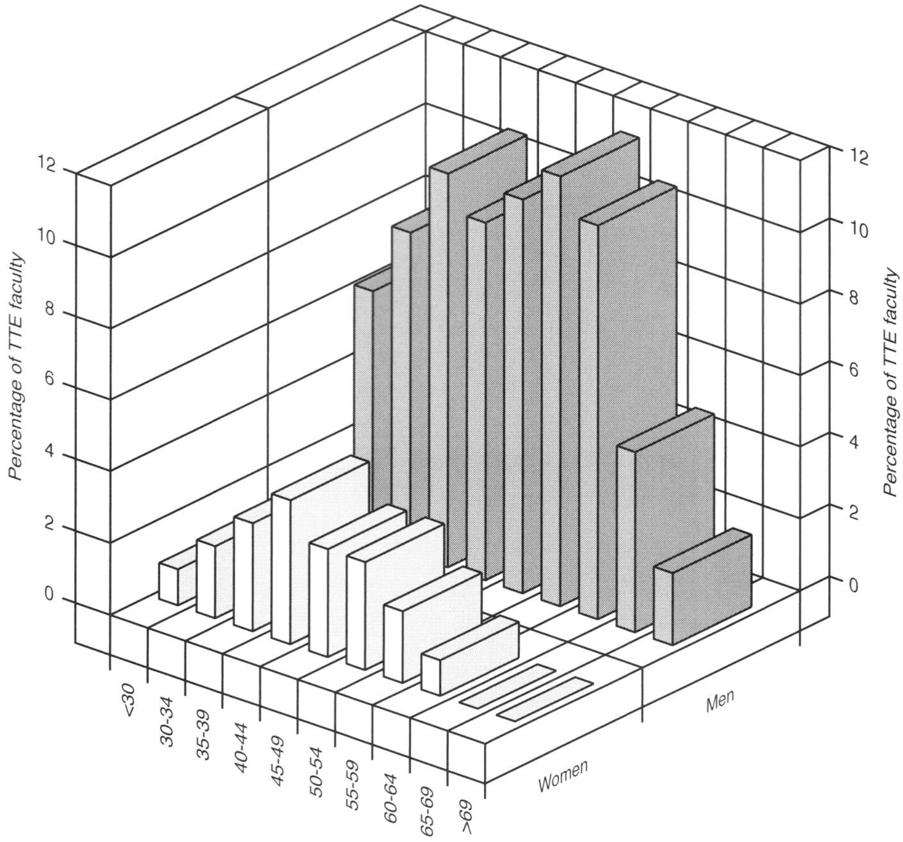

FIGURE S.18.1 Percentage of all tenured and tenure-eligible (TTE) faculty in mathematics departments at four-year colleges and universities belonging to various age groups, by gender, in fall 2005.

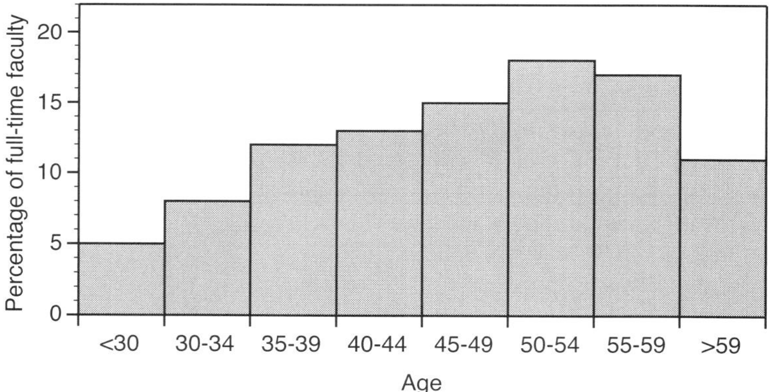

FIGURE S.18.2 Percentage of permanent full-time faculty in various age groups in mathematics programs at public two-year colleges in fall 2005.

TABLE S.19 Percentage of tenured and tenure-eligible faculty belonging to various age groups in doctoral statistics departments at universities by gender, and average ages in fall 2005. Also average ages for doctoral and masters statistics departments (combined) in fall 2000.

Doctoral Statistics Departments	Percentage of tenured/tenure-eligible faculty										Average age 2000[1]	Average age 2005
	<30	30-34	35-39	40-44	45-49	50-54	55-59	60-64	65-69	>69		
Tenured men	0%	1%	6%	8%	10%	11%	11%	9%	6%	2%	52.6	52.7
Tenured women	0	1	2	3	2	1	1	1	0	0	48.3	45.6
Tenure-eligible men	2	8	5	1	0	0	0	0	0	0	34.4	33.7
Tenure-eligible women	2	4	2	0	0	0	0	0	0	0	38.0	33.2
Total tenured & tenure-eligible faculty	**5**	**15**	**15**	**12**	**12**	**12**	**12**	**9**	**6**	**2**		

Note: 0 means less than half of 1%. Roundoff may cause some marginal totals to appear inaccurate.

[1]Average ages for fall 2000 from CBMS2000 Table F.5.

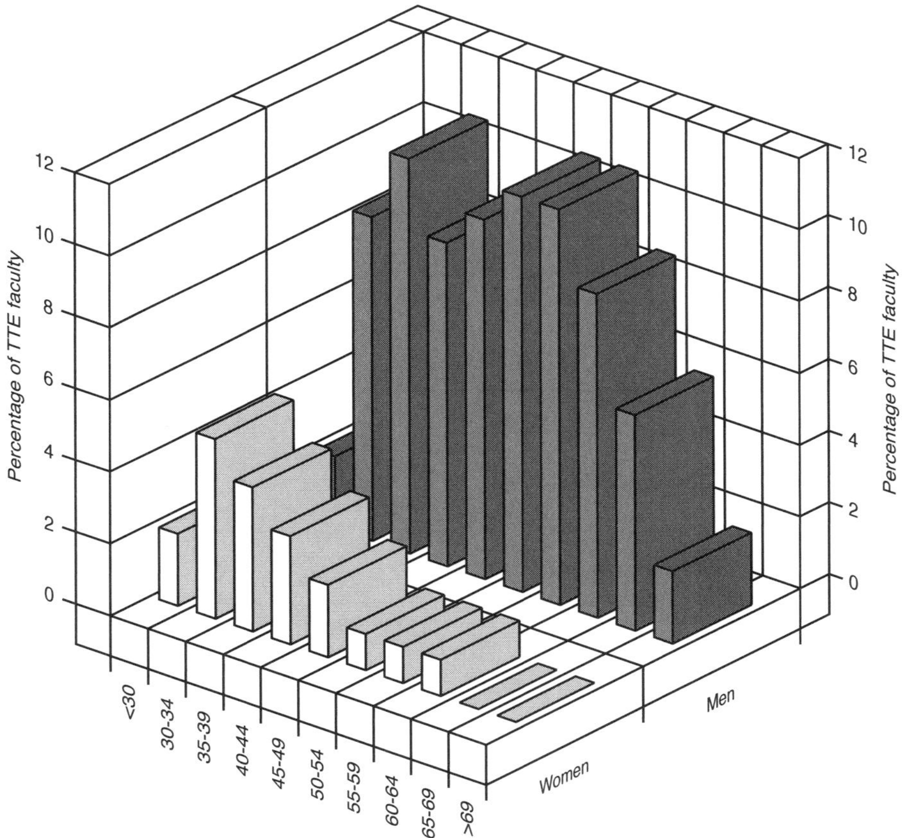

FIGURE S.19.1 Percentage of tenured and tenure-eligible faculty in various age groups, by gender, in doctoral statistics departments in fall 2005.

eligible positions. Table S.19 shows similar increases in average ages in doctoral statistics departments, with the exception of tenure-eligible-but-not-tenured women faculty, whose average age actually declined slightly between fall 2000 and fall 2005. The average ages of faculty in two-year college mathematics programs also increased between fall 2000 and fall 2005, but only marginally.

For some reason, the average ages of each of the four faculty groups studied in Tables S.18 and S.19 are lower in doctoral statistics departments than in mathematics departments. Table F.4 in Chapter 4 shows that this average age difference persists even if doctoral statistics departments are compared with doctoral mathematics departments rather than with all mathematics departments.

For a study of the age distribution of mathematics program faculty in two-year colleges, see Tables TYF.16 and TYF.17 in Chapter 7 of this report.

Data on the ages of faculty is becoming difficult to obtain from departmental surveys, and some departments reported that they were prohibited by university policy from obtaining such data. There may be federal sources for this age-distribution data.

Table S.20 presents the distribution of all full-time mathematical sciences faculty among various ethnic groups. The CBMS2005 questionnaires used the ethnic categories and descriptions that appear in contemporary federal surveys. Because the percentage of mathematical sciences faculty in several of the federal categories rounded to zero, Tables S.20 and S.21 combine some of the smaller categories into a column titled "unknown/other".

Comparisons of Table S.20 with fall 2000 data in CBMS2000 Table SF.11 show that the percentage of four-year mathematics department faculty listed as "White, not Hispanic" declined from 84% in fall 2000 to 80% in fall 2005. The percentage of Asians among

TABLE S.20 Percentage of gender and of racial/ethnic groups among all tenured, tenure-eligible, postdoctoral, and other full-time faculty in mathematics departments of four-year colleges and universities in fall 2005.

Mathematics Departments	Asian	Black, not Hispanic	Mexican American/ Puerto Rican/ other Hispanic	White, not Hispanic	Not known/ other
Tenured men	5%	1%	1%	39%	1%
Tenured women	1	0	0	9	0
Tenure-eligible men	2	0	0	11	0
Tenure-eligible women	1	0	0	4	0
Postdoctoral men	1	0	0	2	0
Postdoctoral women	0	0	0	1	0
Full-time men not included above	1	0	0	7	1
Full-time women not included above	1	0	0	7	0
Total full-time men	9	2	2	59	2
Total full-time women	3	1	1	21	1

Note: 0 means less than half of 1% and this may cause apparent column sum inconsistencies.

Note: The "Not known/other" category includes the federal categories Native American/Alaskan Native and Native Hawaiian/Other Pacific Islander.

the four-year mathematics faculty grew from 10% in fall 2000 to 12% in fall 2005. The percentage of faculty classified as "Black, not Hispanic" and "Mexican American, Puerto Rican, or Other Hispanic" did not change much between 2000 and 2005.

Table S.21 shows the distribution of doctoral statistics faculty among various ethnic groups. Consequently, the table should be compared with Table F.7 of Chapter 4 in the CBMS2000 report, rather than with any Chapter 1 table from CBMS2000. The percentage of doctoral statistics department faculty listed as "White, not Hispanic" declined from 75% in fall 2000 to 71% in fall 2005 while the percentage listed as "Asian" rose from 21% in fall 2000 to 25% in fall 2005.

The distribution of mathematics program faculty in public two-year colleges among various ethnic groups is studied in Tables TYF.10 through TYF.15 of Chapter 7 of this report.

TABLE S.21 Percentage of gender and of racial/ethnic groups among all tenured, tenure-eligible, postdoctoral, and other full-time faculty in doctoral statistics departments at universities in fall 2005.

Doctoral Statistics Departments	Asian	Black, not Hispanic	Mexican American/ Puerto Rican/ other Hispanic	White, not Hispanic	Not known/ other
Tenured men	10%	0%	1%	41%	1%
Tenured women	2	0	0	6	0
Tenure-eligible men	6	0	0	7	0
Tenure-eligible women	3	0	0	4	0
Postdoctoral men	1	0	0	2	1
Postdoctoral women	1	0	0	1	0
Full-time men, not included above	1	0	0	5	0
Full-time women, not included above	0	0	0	4	0
Total full-time men	18	1	1	55	2
Total full-time women	7	1	0	16	1

Note: 0 means less than half of 1%; roundoff causes apparent column sum inconsistencies.

Note: The column "Not known/other" includes the federal categories Native American/Alaskan Native and Native Hawaiian/Other Pacific Islander.

Table S.22 summarizes data on faculty members who left mathematical sciences departments due to death or retirement between September 1, 2004 and August 31, 2005. Historical comparisons can be based on Tables SF.15 in the CBMS1995 and CBMS2000 reports. Four-year mathematics departments lost 2.7%, 3.0%, and 2.9% of their TTE faculty to deaths and retirements in the 1994–1995, 1999–2000, and 2004–2005 academic years respectively, while mathematics programs at two-year colleges lost 3.6%, 2.3%, and 3.3% of permanent full-time faculty during those same academic years. Statistics departments lost 3.6%, 1.8%, and 1.8% of their TTE faculty in those three academic years, but when comparing those three percentages, readers must keep in mind that the tables in CBMS1995 and CBMS2000 present data on all statistics departments, while CBMS2005 presents data on doctoral statistics departments only.

TABLE S.22 Number of deaths and retirements of tenured/tenure-eligible faculty from mathematics departments and from doctoral statistics departments by type of school, and of full-time permanent faculty from mathematics programs at two-year colleges between September 1, 2004 and August 31, 2005. Historical data is included when available. (Two-year college data for 2005 includes only public two-year college data. Historical data on statistics departments includes both masters and doctoral statistics departments.)

Four-Year College & University	1989-1990	1994-1995	1999-2000	2004-2005	Number of tenured/ tenure-eligible faculty 2005
Mathematics Departments					
Univ(PhD)	135	172	174	139	5652
Univ(MA)	68	132	165	140	3563
Coll(BA)	119	137	123	219	8041
Total deaths and retirements in all Mathematics Departments	322	441	462	499	17256
Doctoral Statistics Departments:Total deaths and retirements	17	33	16	14	783
Two-Year College Mathematics Programs					Number of full-time permanent faculty 2005
Total deaths and retirements in all TYC Mathematics Programs	na	274	163	292	8793

Table S.23 summarizes CBMS2005 findings about teaching assignments in four-year mathematical sciences departments of various types. The CBMS2000 table with comparable data for four-year colleges and university mathematics departments is Table SF.16. For data on teaching assignments in the mathematics programs of two-year colleges, see Table TYF.2 in Chapter 7 of this report, and for historical comparisons of two-year college teaching assignments, see Table TYR.18 of CBMS2000.

Among doctoral mathematics departments, about two-thirds had typical fall-term teaching assignments of at most six contact hours while 91% had typical teaching assignments of at most eight contact hours. Slightly more than half of all masters-level mathematics departments had typical fall-term teaching assignments of at most eleven contact hours, while almost all masters-level departments assigned at most twelve contact hours. Among bachelors-level

TABLE S.23 Percentage of four-year college and university mathematics and statistics departments having various weekly teaching assignments in classroom contact hours for tenured and tenure-eligible faculty in spring 2005 and fall 2005, by type of department. Also average assignment by type of department.

	< 6 hrs %	6 hrs %	7–8 hrs %	9–11 hrs %	12 hrs %	>12 hrs %	Average assignment
Mathematics Departments							
Univ (PhD) Fall	24	42	25	5	2	2	6.3
Univ (PhD) Spring	26	40	26	4	2	2	6.2
Univ (MA) Fall	0	4	5	44	48	0	10.3
Univ (MA) Spring	0	7	2	40	51	0	10.3
College (BA) Fall	0	0	3	30	53	14	11.3
College (BA) Spring	0	0	3	28	53	16	11.5
Statistics Departments							
Univ (PhD) Fall	48	45	4	0	4	0	5.3
Univ (PhD) Spring	50	40	4	2	4	0	5.3

departments, the majority reported teaching assignments of twelve contact hours per term.

Anecdotal evidence suggested that teaching assignments in four-year college and university mathematics departments declined between 2000 and 2005. Comparing Table S.23 with CBMS2000 Table SF.16 shows that, on the national scale, any teaching assignment changes between 2000 and 2005 were marginal.

CBMS also investigated spring-term teaching assignments by asking departments to report their average teaching assignments for spring 2005 as well as for fall 2005. The actual differences detected were minor. For example, consider doctoral mathematics departments. Twenty-four percent of doctoral mathematics departments reported average fall-term teaching assignments of less than six contact hours, while 26% of those departments reported average spring-term teaching assignments of less than six contact hours. Sixty-six percent of doctoral mathematics departments reported fall-term teaching assignments less than or equal to six contact hours, and the corresponding spring-term percentage was also 66%. Among bachelors-level departments, there appears to be a marginal increase in spring-term teaching assignments when compared to fall. These conclusions are reflected in the "Average assignment" column of Table S.23.

Among doctoral statistics departments, just less than half reported typical fall-term teaching assignments of at most six contact hours, while essentially all reported typical fall teaching assignments of at most eight contact hours. For comparison, in CBMS2000 only 34% of doctoral statistics departments reported average fall-term teaching assignments less than or equal to six contact hours, a percentage that rose to 48% in CBMS2005. In both CBMS2000 and CBMS2005, almost all doctoral statistics departments reported typical teaching assignments of at most eight contact hours. As was the case in mathematics departments, there was no major difference between fall- and spring-term teaching assignments in doctoral statistics departments.

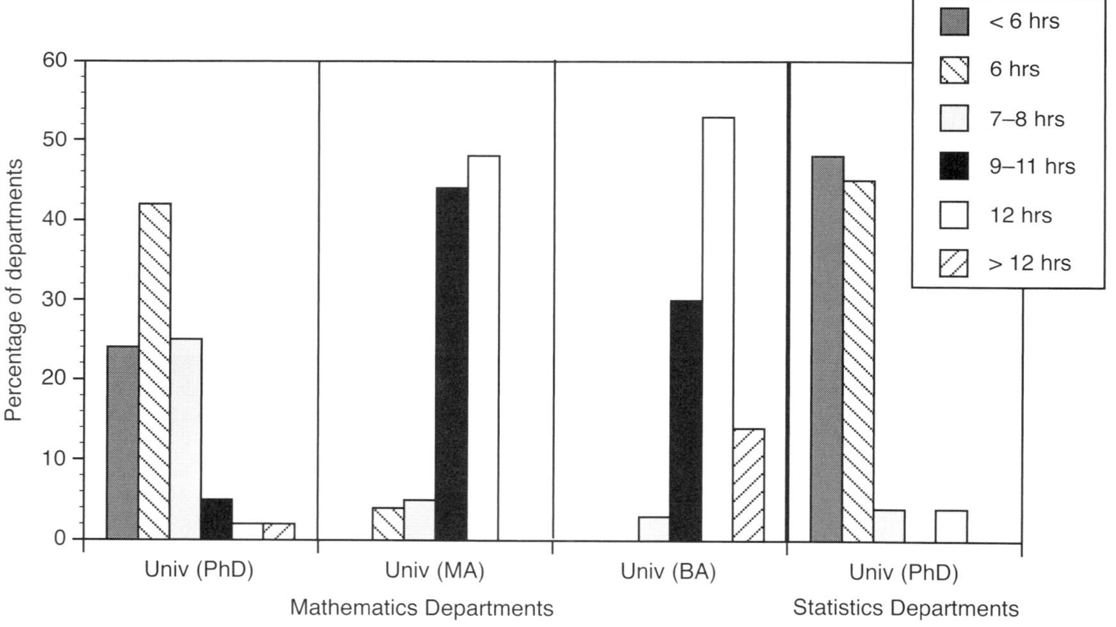

FIGURE S.23.1 Percentage of mathematics departments and doctoral statistics departments in four-year colleges and universities having various weekly teaching assignments (in classroom contact hours) for tenured and tenure-eligible faculty, by type of department, in fall 2005.

Chapter 2

CBMS2005 Special Projects

Each CBMS survey accepts proposals for special projects from various professional society committees. Special projects chosen for one CBMS survey might, or might not, be continued in the next CBMS survey. This chapter presents data from the special projects of CBMS2005:

- The mathematical education of pre-college teachers (Tables SP.1 to SP.10)
- Academic resources available to undergraduates (Tables SP.11 to SP.15)
- Dual enrollments in mathematics (Tables SP.16 and SP.17)
- Mathematics and general education requirements (Table SP.18)
- Requirements in the national major in mathematics and statistics (Tables SP.19 to SP.24)
- Assessment in mathematics and statistics departments (Table SP.25).

Terminology: Recall that in CBMS2005, the term "mathematics department" includes departments of mathematics, applied mathematics, mathematical sciences, and departments of mathematics and statistics. Experience shows that mathematics departments may offer a broad spectrum of courses in mathematics education, actuarial science, and operations research as well as in mathematics, applied mathematics, and statistics. Computer science courses are sometimes also offered by mathematics departments. The term "statistics department" refers to departments of statistics or biostatistics that offer undergraduate statistics courses. Courses and majors from separate departments of computer science, actuarial science, operations research, etc., are not included in CBMS2005. Departments are classified by highest degree offered. For example, the term "masters-level department" refers to a department that offers a masters degree but not a doctoral degree.

Tables SP.1 to SP.10: The Mathematical Education of Pre-college Teachers

In 2001, the American Mathematical Society (AMS) and the Mathematical Association of America (MAA) jointly published a CBMS study entitled *The Mathematical Education of Teachers* [MET] that made recommendations concerning the amount and kind of undergraduate mathematics and statistics that pre-service teachers should study. MET also called for closer collaboration between mathematicians and mathematics educators in the design of the undergraduate mathematics and statistics courses that pre-service teachers take. CBMS2000 provided baseline data about the extent to which the MET recommendations were already in place in fall 2000 and CBMS2005 provided five-year-later data to track further implementation of the MET report.

Table SP.1 shows that, in fall 2005, about 87% of mathematics departments and 44% of statistics departments reported belonging to a college or university that offered a teacher certification program for some or all of grades K–8. This compares to percentages of 84% for mathematics departments and 58% for statistics departments in fall 2000. The meaning of the fourteen point drop among statistics departments is not clear.

TABLE SP.1 Percentage of mathematics departments and statistics departments whose institutions offer a certification program for some or all of grades K–8, by type of department, in fall 2005. (Data from fall 2000 in parentheses).

	Percentage whose institutions have a K-8 teacher certification program
Mathematics Departments	
Univ (PhD)	78 (72)
Univ (MA)	92 (87)
Coll (BA)	88 (85)
Total Math Depts	**87 (84)**
Statistics Departments	
Univ (PhD)	40 (58)
Univ (MA)	59 (63)
Total Stat Depts	**44 (58)**

At the time of CBMS2000, teacher certification programs were almost entirely limited to four-year colleges and universities. By fall 2005 that had changed. Table SP.2 shows the percentages of public two-year colleges with programs allowing three types of students to complete their entire mathematics certification requirements at the two-year college. The three types of students mentioned in the table are undergraduates without a bachelors degree (called "pre-service teachers"), in-service teachers who already have certification in some other subject, and people who leave a first career to enter a second career in pre-college teaching (called "career-switchers"). The percentages in Table SP.2 are not large, but given the large number of two-year colleges in the U.S., it is clear that two-year colleges could make a major contribution to educating the next generation of teachers. Table SP.2 shows that two-year college credentialing programs tended to focus on producing K–8 teachers.

TABLE SP.2 Percentage of mathematics programs at public two-year colleges (TYCs) having organized programs that allow various types of pre- and in-service teachers to complete their entire mathematics course or licensure requirements, in fall 2005.

	Percentage of TYCs with an organized program in which students can complete their entire mathematics course or licensure requirements
Pre-service elementary teachers	30
Pre-service middle-school teachers	19
Pre-service secondary teachers	3
In-service elementary teachers	16
In-service middle school teachers	15
In-service secondary teachers	2
Career-switchers aiming for elementary teaching	19
Career-switchers aiming for middle school teaching	14
Career-switchers aiming for secondary teaching	6

To what extent did mathematics and statistics departments in four-year colleges and universities cooperate with their schools of education in teacher certification programs in fall 2005? One mark of such cooperation is for the department to have a seat on the committee that governs the certification program. Table SP.3 shows that about 80% of all mathematics departments were represented on that governing committee in fall 2005 (with considerable variation by type of department). Fewer statistics departments (about 28%) had members on the governing committees. Table SP.3 shows that the fall 2005 percentages were substantially larger than the corresponding percentages in CBMS2000, which reported 69% for mathematics departments and 0% for statistics departments (see CBMS2000 Table PSE.2).

Another mark of a department's involvement in K-8 teacher education is the existence of special mathematics (or statistics) courses or course sequences designed for K-8 pre-service teachers. Table SP.3 shows that the percentage of mathematics departments having such sequences rose from 77% in fall 2000 to 86% in fall 2005. The percentage of statistics departments with a special course for pre-service K-8 teachers was smaller in fall 2005 than the percentage for mathematics departments, but was higher than in fall 2000.

TABLE SP.3 Percentage of mathematics and statistics departments in universities and four-year colleges offering K-8 certification programs that are involved in K–8 teacher certification in various ways, by type of department, in fall 2005. (Data from fall 2000 in parentheses).

	Percentage of departments in schools offering K–8 certification programs that		
	Have a department member on the certification program's control committee	Offer a special course or course sequence for K-8 teachers	Designate special sections of regular courses for K-8 teachers
Mathematics Departments			
Univ (PhD)	58 (63)	81 (79)	31 (11)
Univ (MA)	86 (74)	96 (92)	45 (13)
Coll (BA)	82 (68)	85 (73)	21 (4)
Total Math Depts	80 (69)	86 (77)	25 (7)
Statistics Departments			
Univ (PhD)	29 (0)	11 (4)	0 (0)
Univ (MA)	25 (0)	33 (0)	0 (0)
Total Stat Depts	28 (0)	16 (4)	0 (0)

Note: 0 means less than one-half of 1%.

Table SP.4 shows a clear trend away from special mathematics courses for pre-service teachers in two-year college curricula, with the percentage of two-year colleges offering such courses in fall 2005 being less than one-fourth of the corresponding percentage reported for fall 2000 by CBMS2000. This decrease stands in marked contrast to the situation in four-year colleges and universities.

TABLE SP.4 Percentage of public two-year colleges (TYCs) that are involved with K-8 teacher preparation in various ways, in fall 2005.

	Percentage of TYCs
Assign a mathematics faculty member to coordinate K–8 teacher education in mathematics	38
Offered a special mathematics course for preservice K–8 teachers in 2004–2005 or 2005–2006	11
Offer mathematics pedagogy courses in the mathematics department	9
Offer mathematics pedagogy courses outside of the mathematics department	10

How many mathematics courses were required for a student seeking K–8 certification in fall 2005? That is a complicated question because of the wide variety of certification programs in the U.S. In fall 2005, some colleges and universities offered a single-track program for K–8 certification, while others divided K–8 certification into two sub-tracks (one for early grades and one for later grades), and still others further subdivided their later-grades track into discipline-specific later-grade certification programs. (In a discipline-specific later-grades program, a student might become certified to teach in some cluster of disciplines, say mathematics and science, in the later grades.) CBMS2005 addressed that diversity by dividing universities with K–8 certification programs into those that had a single set of mathematics requirements for K–8 certification, and those that had different mathematics requirements for early and later grade certification.

But even the meaning of "early grades" and "later grades" is complicated, because in fall 2005, different states, colleges, and universities divided K–8 certification in different ways. Some, for example, had an undivided K–8 certification, others put grades 4, 5, 6, 7, and 8 together in a single certification category, and still others put only grades 6, 7, and 8 together. In an attempt to make a single questionnaire fit all of the certification patterns, the CBMS2005 questionnaire defined the term "early grades certification" to mean the certification that included grades K–3, and defined the term "later grades certification" to be the certification that included grades 5 and 6.

Table SP.5 shows that the majority (56%) of departments with K–8 certification programs do not distinguish between early and later grades in terms of mathematics requirements, and also shows how many mathematics courses are required for various certifications. Comparisons with CBMS2000 data are possible, at least for programs that have different requirements for early and later grades. In each type of mathematics department, the number of mathematics courses required for K–8 teacher certification rose between fall 2000 and fall 2005. Chapter 2 of *The Mathematical Education of Teachers* recommended that K–3 teachers take at least nine semester hours of mathematics, which translates into three one-semester courses, and that prospective teachers of the middle grades should take at least 21 semester hours, which translates into seven semester courses. For CBMS2005, all reported data on course requirements were translated into semester courses, and Table SP.5 shows that while MET's course recommendations had not been completely implemented by fall 2005, the nation was closer to them than in the base-year study in fall 2000.

TABLE SP.5 Among all four-year colleges and universities with K-8 certification programs, the percentage that have different requirements for early grades (K–3) certification and for later grades (including 5 and 6) certification in terms of semester courses, including the number of semester courses required, and the percentage that have the same requirements for their combined K-8 certification program, including the number of courses required, in fall 2005. Also the average number of semester mathematics department courses required for various teacher certifications in those colleges and universities offering K–8 certification programs, by certification level and type of department, in fall 2005. (Data for fall 2000 in parentheses).

	Having different mathematics requirements for early & later grades certification		Having the same mathematics requirements for early & later grades certification
Percentage of mathematics departments with K-8 certification programs	44%		56%
	Percentage of departments with K-8 certification programs that require various numbers of mathematics courses		Percentage of departments with K-8 certification programs that require various numbers of mathematics courses
Number of mathematics courses required for certification	for early grades	for later grades	for all K-8 grades
0 required	11 (8)	16 (7)	4 (na)
1 required	17 (17)	7 (12)	26 (na)
2 required	31 (45)	5 (42)	37 (na)
3 required	17 (14)	2 (12)	22 (na)
4 required	17 (11)	11 (10)	11 (na)
5 or more required	8 (6)	58 (18)	0 (na)
Type of mathematics department	Avg number of courses required	Avg number of courses required	Avg number of courses required in combined K-8 certification program
Univ(PhD)	3.3 (2.2)	5.5 (2.5)	2.4 (na)
Univ(MA)	3.3 (3.3)	6.9 (4.1)	2.5 (na)
Coll(BA)	2.5 (2.3)	5.3 (2.8)	2 (na)
All mathematics departments	2.7 (2.4)	5.6 (3)	2.1 (na)

In fall 2005, which mathematics courses did pre-service K–8 teachers take? Table SP.6 records departmental responses to the question "In your judgment, which three of the following courses in your department are most likely to be taken by pre-service K–8 teachers?" The responses recorded in SP.6 can be compared with Table PSE.5 of CBMS2000. It would have been desirable to pose a more precise question, such as "Of all students receiving certification for part or all of grades K–8 between July 1, 2004 and June 30, 2005, what percentage actually took each of the following courses?" The CBMS2005 project directors decided that the data retrieval work required for a department to answer the more precise question would cut into CBMS2005 survey response rates in a major way, so the less precise question was used. This may limit the utility of Table SP.6. With that caveat in place, Table SP.6 suggests some conclusions. It suggests that in fall 2005 there were clear differences between the mathematical expectations for early and later-grade certification programs, that the mathematics requirements for K–3 certification seemed to center on a multi-term course (e.g., a two-semester sequence) for elementary education majors and a course in College Algebra, and that the mathematics requirements for later-grades certification seemed to focus on Calculus, Geometry, and Elementary Statistics. (See Table SP.8, below, for a discussion of when pre-service K–8 teachers begin their mathematics and statistics studies.)

TABLE SP.6 Among mathematics departments at four-year colleges and universities having different requirements for early and later grades certification, the percentage identifying a given course as one of the three mathematics courses most likely to be taken by pre-service teachers preparing for K–3 teaching or for later grades teaching (including 5 and 6) by type of department, in fall 2005.

	Most likely for K–3 certification			Most likely for later grades certification		
Among Mathematics Departments With Different Early and Later Grades Requirements	Univ (PhD) Math	Univ (MA) Math	Coll (BA) Math	Univ (PhD) Math	Univ (MA) Math	Coll (BA) Math
Multi-term course for elementary education majors	59	70	64	28	47	38
Single term course for elementary education majors	21	37	33	16	10	12
College algebra	41	40	56	21	40	23
Precalculus	15	6	46	13	13	15
Intro to mathematical modeling	5	0	0	8	0	0
Mathematics for liberal arts	28	30	25	8	7	2
Finite mathematics	23	7	15	10	7	8
Mathematics history	5	0	0	31	23	18
Calculus	21	6	12	64	50	77
Geometry	10	24	0	43	47	53
Elementary Statistics	31	26	27	41	44	55

Yet another mark of departmental involvement in K–8 teacher education is the appointment of a department member to coordinate the program. Table SP.4 shows that about 38% of two-year colleges appointed such a coordinator in fall 2005, up from 22% in fall 2000 reported in CBMS2000 Table PSE.3. CBMS2005 posed a different question to four-year mathematics departments in fall 2005. Four-year mathematics departments that offered multiple sections of their elementary mathematics education course were asked whether they appointed a department member to coordinate the multi-section course. Table SP.7 shows that the percentage varied from 90% among doctoral departments that offered multiple sections of their elementary education course to 69% among bachelors-level mathematics departments. Of the course coordinators, the majority were tenured or tenure-eligible, and in all types of departments, at least 90% of the coordinators were either tenured, tenure-eligible, or a full-time department member with a Ph.D.

TABLE SP.7 Among mathematics departments with multiple sections of their elementary mathematics education course, the percentage that administer their multiple sections in various ways, by type of department. Also, among departments with a course coordinator, the percentage with coordinators of various kinds, by type of department, in fall 2005.

	Mathematics Departments		
Departments with multiple sections of their Elementary Mathematics Education course	Univ (PhD)	Univ (MA)	College (BA)
Number with multiple sections	81	143	335
Percentage using same text for all sections	97%	91%	100%
Percentage with course coordinator	90%	82%	69%
Status of Course Coordinator			
a) Tenured/Tenure eligible	65%	81%	68%
b) Postdoc	0	0	0
c) Full-time visitor	2	9	0
d) Full-time, with Ph.D., not (a),(b),(c)	28	9	32
e) Full-time, without PhD, not (a),(b),(c)	2	0	0
f) Part-time	3	0	0
g) Graduate teaching assistant	0	0	0

TABLE SP.8 Percentage of mathematics departments estimating when K-8 pre-service teachers take their first mathematics education course, by type of department, in fall 2005.

	Mathematics Departments		
When Students Take K-8 Mathematics Education Course	Univ (PhD)	Univ (MA)	College (BA)
Freshman year	23%	43%	23%
Sophomore year	45	36	64
Junior year	27	17	13
Senior year	5	4	0

The final two tables in this part of Chapter 2 give data about other ways that departments participated in teacher education programs. Table SP.9 shows the number of departments of various types that offered *secondary* mathematics certification programs, and shows where students in those programs learned about the history of mathematics in fall 2005. Table SP.10 shows the extent to which mathematics and statistics departments were involved in graduate teacher education programs, either inside or outside of the department.

TABLE SP. 9 Number and percentage of mathematics departments in universities and four year colleges with secondary mathematics certification programs whose pre-service secondary teachers learn mathematics history in various ways, by type of department, in fall 2005.

	Mathematics Departments		
Mathematics Departments with Secondary Certification Programs	Univ (PhD)	Univ (MA)	College (BA)
Number	151	170	833
Percentage with a required mathematics history course for secondary certification	58%	69%	41%
Percentage with mathematics history only in other required courses for secondary certification	22	25	43
Percentage with no mathematics history requirement for secondary certification	19	7	16

TABLE SP.10 Degree of participation by mathematics and statistics departments in graduate mathematics education programs of various kinds, by type of department, in fall 2005.

	Mathematics Departments			Statistics Departments	
Participation in a Graduate Mathematics Education Program	Univ (PhD)	Univ (MA)	College (BA)	Univ (PhD)	Univ (MA)
Percentage with no graduate mathematics education courses	43	21	89	58	56
Percentage with mathematics education courses that are part of a degree program in their own department	29	35	2	23	29
Percentage with mathematics education courses that are part of a degree program in another department	28	44	9	19	15

Tables SP.11 to SP.15: Academic Resources Available to Undergraduates

In fall 2005, as in fall 2000, almost all two-year colleges reported using placement testing for incoming students. In CBMS2000, 67% of two-year colleges reported that their placement test led to mandatory placement. The CBMS2005 survey changed the question somewhat, and found that in fall 2005, 88% of public two-year colleges had mandatory placement based on the placement test or based on the placement test and other information. Table SP.11 also shows the source of placement tests used by public two-year colleges with placement testing programs. The use of locally written placement tests declined, falling from 99% of two-year colleges in fall 2000 to 11% in fall 2005. Because many two-year colleges indicated that they used placement tests from several sources, the percentages in Table SP.11 do not add to 100%.

TABLE SP.11 Percentage of public two-year colleges that have placement testing programs and use them in various ways, and the source of the placement tests, in fall 2005. (Data from fall 2000 in parentheses.)

	Percentage of two-year colleges %
That offer placement tests	97 (98)
That usually require placement tests of first-time enrollees	97 (98)
That require students to discuss placement scores with advisors	90 (79)
That use placement tests as part of mandatory placement	88 (na)
That periodically assess the effectiveness of their placement tests	81 (85)
Source of Placement Test	
Written by department	11 (99)
Provided by ETS	22 (30)
Provided by ACT	51 (34)
Provided by professional society	12 (3)
Provided by other external source	25 (26)

Table SP.12 shows that most mathematics departments in two-year colleges, and most mathematics and statistics departments in four-year colleges and universities, offered labs or tutoring centers for their students in fall 2005. The only major change since fall 2000 was the increase in the percentage of statistics departments that offered labs or tutoring centers (up from six out of ten to eight out of ten). Table SP.13 shows the types of assistance available in mathematics and statistics labs and tutoring centers. Among mathematics departments of four-year colleges and universities, the emphasis on computer use in the labs declined from the levels observed in fall 2000, while it increased in both statistics departments and two-year colleges. The use of para-professional and part-time faculty as tutors declined between 2000 and 2005, while tutoring by full-time faculty increased.

TABLE SP.12 Percentage of mathematics and statistics departments in four-year colleges and universities, and mathematics programs in public two-year colleges, that operate a lab or tutoring center in their discipline in fall 2005. (Fall 2000 data in parentheses)

Percentage with Lab or Tutoring Center	Mathematics Departments	Statistics Departments	Two-Year College Mathematics Programs
Univ (PhD)	96 (90)	79 (61)	--
Univ (MA)	91 (95)	85 (50)	--
Coll (BA)	88 (89)	--	--
All departments	89 (89)	80 (59)	95 (98)

TABLE SP.13 Among mathematics and statistics departments in four-year colleges and universities and mathematics programs in public two-year colleges that operate labs or tutoring centers, the percentage that offer various services, by type of department, in fall 2005. (Fall 2000 data in parentheses.)

Percentage Offering Various Services in Labs & Tutoring Centers	Computer-aided instruction %	Computer software %	Media such as video tapes %	Tutoring by students %	Tutoring by para-professional staff %	Tutoring by part-time faculty %	Tutoring by full-time faculty %	Internet resources %
Mathematics Departments								
Univ (PhD)	33	48	20	98	29	22	27	38
Univ (MA)	33	55	40	96	43	23	28	37
Coll (BA)	25	33	27	99	20	9	19	21
Total Mathematics Departments	27 (38)	38 (62)	27 (24)	98 (99)	24 (35)	13 (18)	21 (16)	25 (33)
Statistics Departments								
Univ (PhD)	44	68	13	96	13	9	17	27
Univ (MA)	51	83	17	100	17	0	17	69
Total Statistics Departments	46 (36)	71 (63)	14 (17)	97 (93)	14 (37)	7 (11)	17 (3)	37 (23)
Two-Year College Mathematics Programs	75 (68)	72 (69)	68 (74)	94 (96)	67 (68)	48 (48)	51 (42)	77 (53)

Note: 0 means less than one-half of 1%.

Tables SP.14 and SP.15 show the extent to which departments of various kinds made a spectrum of academic enrichment opportunities available to their undergraduates in fall 2005. These tables expand upon Table AR.12 in CBMS2000. With few exceptions, the percentage of departments offering a given academic opportunity increased between 2000 and 2005. Perhaps the most notable exception in Table SP.14 is the decline from 47% to 34% in the number of four-year mathematics departments that offer opportunities for their undergraduates to become involved with K–12 schools. The difference between mathematics and statistics departments in terms of the availability of the senior thesis option in fall 2005 (76% in mathematics departments, compared to 31% among statistics departments) may also be noteworthy.

TABLE SP.14 Percentage of mathematics programs at public two-year colleges, and of mathematics and statistics departments in four-year colleges and universities, that offer various kinds of special opportunities for undergraduates, by type of department, in fall 2005. (Fall 2000 data in parentheses.)

Percentage with Special Opportunities for Undergraduates	Honors sections of courses for majors %	Math or Stat club %	Special programs for women %	Special programs for minorities %	Math or Stat contests %	Special Math or Stat colloquia for undergrads %	Outreach in K–12 schools %
Mathematics Departments							
Univ (PhD)	70	88	15	10	92	70	51
Univ (MA)	44	92	21	23	68	71	63
Coll (BA)	18	66	4	6	62	37	26
Total Mathematics Depts	28 (20)	72 (61)	8 (9)	8 (7)	67 (63)	46 (54)	34 (47)
Statistics Departments							
Univ (PhD)	27	27	0	7	22	47	11
Univ (MA)	41	29	0	0	29	44	15
Total Statistics Depts	30 (46)	27 (25)	0 (2)	6 (2)	23 (28)	46 (41)	12 (7)
Two-Year College Mathematics Programs	24 (20)	22 (14)	7 (4)	15 (4)	37 (28)	6 (9)	25 (20)

Note: 0 means less than one-half of 1%.

TABLE SP.15 Percentage of mathematics programs in public two-year colleges, and of mathematics and statistics departments in four-year colleges and universities, that offer various additional special opportunities for undergraduates, by type of department, in fall 2005. (Fall 2000 data, where available, in parentheses.)

Percentage with Additional Opportunities for Undergraduates	Undergrad. research opportunity %	Indep. studies opportunity %	Assigned advisors in dept. %	Senior thesis opportunity %	Math career day %	Graduate school advising %	Internship opportunity %	Senior seminar opportunity %
Mathematics Departments								
Univ (PhD)	90	95	85	62	24	49	47	39
Univ (MA)	74	91	97	53	15	61	55	46
Coll (BA)	54	79	88	48	10	45	35	38
Total mathematics depts	62 (59)	83 (80)	89 (82)	50	12	47	39	39
Statistics Departments								
Univ (PhD)	60	62	73	27	15	56	47	15
Univ (MA)	59	100	85	44	15	59	71	29
Total statistics depts	60 (58)	70 (67)	76 (71)	31	15	57	52	18
Two-Year College Mathematics Programs	9 (4)	38 (25)	40 (33)	na	na	na	na	na

Tables SP.16 and SP.17: Dual Enrollments—College Credit for High School Courses

Dual-enrollment courses are courses taught in high school by high school instructors for which high school students receive both high school and college credit. This arrangement is *not* the same as obtaining college credit based on AP or IB examination scores. Dual enrollment is encouraged by many state governments as a way to utilize state-wide educational resources more efficiently.

In fall 2000, most dual-enrollment courses involved an agreement between a high school, where the course was taught, and a local two-year college that awarded college credit for the course. In many states, public four-year colleges and universities were required to count such dual-enrollment credits toward their graduation requirements. Based on CBMS2000 findings, the Mathematical Association of America Board of Governors called for careful tracking of dual-enrollment growth and related quality-control issues, and CBMS2005 agreed to study dual-enrollment issues in fall 2005 in both two- and four-year colleges and universities.

Table SP.16 shows that dual-enrollment courses were widespread among two-year colleges in fall 2005, with about 50% of all public two-year colleges awarding college credit for some dual-enrollment courses. In fall 2005 there were about 58,000 enrollments in Precalculus at two-year colleges, and about 14,000 dual-enrollments in high school versions of that same course, meaning that just over 19% of all credit in Precalculus awarded by two-year colleges was earned in dual-enrollment courses. Also, there were about 51,000 enrollments in Calculus I courses taught in two-year colleges, and about 11,000 enrollments in the dual-enrollment version of that same course. Consequently, about 18% of all Calculus I credit awarded by two-year colleges was through dual enrollments.

Comparing enrollment percentages for fall 2005 with data from CBMS2000 is somewhat problematic because the CBMS2000 survey asked two-year colleges to report the number of dual-enrollment sections rather than the number of dual enrollments. Nevertheless, it may be worth noting that CBMS2000 found that in fall 2000, about 18% of two-year college sections in Precalculus and about 15% of two-year college Calculus I sections were dual-enrollment sections.

In fall 2000, anecdotal evidence suggested that few of the nation's four-year colleges and universities were involved in granting dual-enrollment credit for high school mathematics and statistics courses, so that four-year departments were not asked to report on their dual-enrollment activity. Table SP.16 of CBMS2005 shows that in fall 2005, about one in seven mathematics departments, and one in twelve statistics departments, at four-year colleges and universities had entered into dual-enrollment agreements with high schools. However, in fall 2005 the number of dual-enrollment registrations in four-year colleges and universities was small compared to the number of traditional enrollments. For example, the number of dual enrollments in College Algebra and in Calculus I were only about 4% of the number of regular enrollments in those courses. In statistics departments, the number of dual enrollments in Elementary Statistics was about 3% of traditional enrollments in that same course.

A major concern in dual-enrollment courses is the degree of quality control exercised by the two-year or four-year department through which college-level credit for the courses is awarded. Table SP.16 examines several types of quality control that college-level departments might have had over their dual-enrollment courses in fall 2005, and presents comparison data for dual-enrollment programs of two-year colleges from fall 2000. (Comparable data from fall 2000 do not exist for dual-enrollment programs at four-year colleges and universities.) CBMS2000 showed that in fall 2000, 79% of two-year colleges reported that they always controlled the choice of the textbook used in their dual-enrollment courses. By the fall of 2005, that percentage dropped slightly, to 74%, and the corresponding percentage of "never control the textbook" responses grew from 10% in fall 2000 to 14% in fall 2005. Both final exam design and the choice of instructor in dual-enrollment courses seemed to drift away from two-year colleges' control between 2000 and 2005, with the largest change occurring in the degree of control over the final examination. Only in the area of syllabus design or approval did the degree of control by two-year colleges in dual-enrollment courses seem to increase between fall 2000 and fall 2005. Four-year college and university mathematics departments that were involved in dual-enrollment programs in fall 2005 exercised a degree of course control roughly similar to that of two-year college mathematics programs, except in terms of the choice of textbook, an area in which four-year departments had considerably less control than two-year departments.

Monitoring teaching quality is another opportunity for quality-control in dual-enrollment courses. About two-thirds of two-year colleges monitored the teaching of dual-enrollment instructors, while among four-year mathematics departments the number was closer to one in six. The findings reported in Table SP.16 will not be reassuring to those who expect two- and four-year colleges and universities to control the content and depth of courses for which they are granting college credit.

TABLE SP.16 Percentage of departments offering dual-enrollment courses taught in high school by high school teachers, enrollments in various dual-enrollment courses in spring 2005 and fall 2005, compared to total of all other enrollments in fall 2005, and (among departments with dual enrollment programs) percentage of various departmental controls over dual-enrollment courses, by type of department. (Fall 2000 data in parentheses.)

	Four-year Mathematics			Two-year Mathematics			Four-year Statistics		
Percentage of Departments with Dual-Enrollment Courses	14%			50%			8%		
Number of Dual Enrollments	Dual enrollments spring 2005	Dual enrollments fall 2005	Other enrollments fall 2005	Dual enrollments spring 2005	Dual enrollments fall 2005	Other enrollments fall 2005	Dual enrollments spring 2005	Dual enrollments fall 2005	Other enrollments fall 2005
College algebra	2673	8046	201000	9913	11362	206000	na	na	na
Precalculus	2944	597	93000	14650	13801	58000	na	na	na
Calculus I	5540	8490	201000	8218	11188	51000	na	na	na
Statistics	340	981	124000	3648	2440	111000	1563	1295	43000
Other	3470	723	na	5452	3045	na	0	0	na
Dept. Control of Dual Enroll. Courses Taught by H S Teachers	Never	Sometimes	Always	Never	Sometimes	Always	Never	Sometimes	Always
Textbook choice	41%	15%	44%	14% (10)	12% (12)	74% (79)	36%	30%	34%
Syllabus design/approval	2%	6%	92%	4% (8)	7% (11)	89% (82)	36%	0%	64%
Final exam design	40%	30%	30%	36% (15)	28% (28)	37% (57)	100%	0%	0%
Choice of instructor	32%	20%	48%	35% (19)	13% (20)	52% (61)	36%	0%	64%
Departmental teaching evaluations required in dual enrollment courses			16%			64% (67)			0%

Table SP.17 describes a relatively new phenomenon, in which colleges and universities send their own faculty members out into high schools to teach courses that grant both high school and college credit. About one in twenty-five mathematics departments in four-year colleges and universities had such programs in fall 2005, as did about one in eight public two-year colleges. The number of students involved in these programs was small compared to the number of dual-enrollment students taught by high school teachers.

TABLE SP. 17 Percentage of departments in four-year colleges and universities and in public two-year colleges that assign their own full-time or part-time faculty members to teach courses in a high school that award both high school and college credit, and number of students enrolled, in fall 2005.

	Four-year Mathematics Departments	Two-year Mathematics Departments	Statistics Departments
Assign their own members to teach dual-enrollment courses	4%	12%	0%
Number of students enrolled	2874	2008	0

Table SP.18: Mathematical Sciences and General Education Requirements

Table SP.18 examines the role of mathematics and statistics courses in the general education requirements of U.S. colleges and universities in fall 2005. Because of the wide variety of academic structures in U.S. universities, CBMS2005 began by asking each department whether its own academic unit had a quantitative requirement for bachelors degrees granted through that academic unit. The phrase "its own academic unit" was designed to address a situation, widespread in universities, in which a mathematics department belonged to a college (say the Arts and Sciences College), and all students of that college were required to take a quantitative course of some kind, even though students in some of the university's other colleges (say the College of Fine Arts) did not need to do so.

Table SP.18 shows that in almost nine out of ten cases, the academic unit to which the four-year mathematics and statistics departments belonged did have a quantitative requirement in fall 2005. In a majority of those cases, the mathematics department reported that the only way for a student to fulfill the quantitative requirement was by taking a course in the mathematics department. About one-quarter of the time, any mathematics course was adequate to fulfill the requirement, and in the other cases only certain mathematics courses fulfilled the requirement. Asked which departmental courses could satisfy general education requirements, departments most frequently mentioned Calculus, followed closely by Elementary Statistics, College Algebra, Precalculus, and a special general education course in the department. Among the several freshman mathematics course options proposed in the CBMS2005 questionnaire, all but one seemed to satisfy general education requirements in a majority of mathematics departments, the exception being "a mathematical models course." In statistics departments, the elementary statistics course was the primary general education course in the department.

TABLE SP.18: Percentage of four-year mathematics and statistics departments whose academic units have various general education requirements, and the department's role in general education, by type of department in fall 2005.

	Four-year Mathematics Departments			Statistics Departments	
General Education	Univ (PhD) %	Univ (MA) %	College (BA) %	Univ (PhD) %	Univ (MA) %
There is a quantitative requirement in the department's college	87	98	91	86	88
The quantitative requirement must be taken in the department	51	68	61	8	0
Any freshman course in the department fulfills the quantitative requirement	26	28	32	27	17
Only certain departmental courses fulfill quantitative requirement	74	72	69	73	83
Departmental courses satisfying the quantitative requirement					
College algebra or Precalculus	56	61	62	na	na
Calculus	97	87	86	na	na
Mathematical models	23	11	13	na	na
A probability/statistics course	55	60	66	94	60
Statistical literacy	na	na	na	27	20
A special general education course in the department	52	73	55	0	0
Some other course(s) in the department	50	71	57	33	20

Tables SP.19 to SP.25: Curricular Requirements of Mathematics and Statistics Majors in the U.S.

In the CBMS2000 report, Table SE.5 presented data on the percentage of mathematics and statistics departments that offered certain upper-division courses in the 2000–2001 academic year. Based on course availability, CBMS2000 concluded that in fall 2000, there were large differences between the kind of mathematical sciences major available to students in doctoral-level departments and in bachelors-level departments. In response to a request from the MAA Committee on the Undergraduate Program in Mathematics, CBMS2005 collected data about specific requirements of majors, about course-offering patterns for all upper-division mathematics and statistics courses during the two-year window consisting of the 2004–2005 and 2005–2006 academic years, and about the extent to which a student could use interdisciplinary components from another mathematical science (e.g., upper-division courses in statistics and computer science) to fulfill the requirements of a mathematics major.

Obtaining national data on the requirements of the mathematics major in fall 2005 was complicated because most mathematics departments offer several different tracks within the mathematics major, each with its own set of requirements. For example, there might be an applied mathematics track, another track for students intending to teach mathematics in high school, another track that focuses on probability and statistics, another designed for students planning for mathematics graduate school, etc., etc. (Some departments refer to these tracks as being separate majors, but in this report we will refer to them as separate tracks within the departmental major.)

In fall 2005, was there any course seen as so central to mathematics that it was required in all of a department's potentially many tracks? Table SP.19 shows that a computer science course comes closest of all to being a universal requirement for U.S. mathematics majors. Real Analysis I, Modern Algebra I, and a statistics course were essentially tied for second place, with about a third of departments reporting that these courses were required in each track of their majors. Capstone experiences (e.g., senior project, thesis, seminar, internship) were widespread requirements in masters- and bachelors-level departments, but not in doctoral departments.

Long ago, many mathematics majors required two semesters of analysis and two semesters of modern algebra. CBMS2005 asked departments whether all, some, or none of the tracks within their major required Modern Algebra I plus another upper-division algebra course, and posed an analogous question about Real Analysis I plus another upper-division analysis course. A large majority of departments reported that in fall 2005, none of the tracks within their majors required two semesters of modern algebra courses, and that none of the tracks within their majors required two semesters of upper-division analysis courses. More specifically, at least seven out of ten bachelors departments reported that none of their tracks required two semesters of analysis, and that none of their tracks required two semesters of algebra. Even among doctoral departments, the majority reported that no track within the department required two semesters of algebra.

TABLE SP.19: Percentage of four-year mathematics departments requiring certain courses in all, some, or none of their majors, by type of department, in fall 2005.

	Required in all majors			Required in some but not all majors			Not required in any major		
Mathematics Department Requirements	Univ (PhD) %	Univ (MA) %	College (BA) %	Univ (PhD) %	Univ (MA) %	College (BA) %	Univ (PhD) %	Univ (MA) %	College (BA) %
Modern Algebra I	24	48	56	59	42	36	18	10	8
Modern Algebra I plus another upper division algebra course	5	8	8	40	28	17	55	63	75
Real Analysis I	36	39	46	49	54	29	15	7	25
Real Analysis I plus some other upper division analysis course	10	4	8	49	36	20	41	60	71
At least one computer science course	55	76	64	27	16	14	18	8	22
At least one statistics course	32	56	32	40	32	32	28	11	35
At least one upper division applied mathematics course	16	23	21	52	41	25	32	36	54
A capstone experience (senior project, thesis, seminar, internship)	27	52	59	23	13	8	50	35	33
An exit exam (written or oral)	8	8	29	4	16	3	88	76	68

Table SP.20 shows that in fall 2005, at least three-quarters of all doctoral statistics departments required three semesters of calculus, including multi-variable calculus, plus Linear Algebra, for all tracks of their majors. At the other end of the spectrum, almost two-thirds of all statistics departments reported that they do not require any applied mathematics course (beyond calculus courses and Linear Algebra) in any track of their majors.

TABLE SP.20 Percentage of statistics departments requiring certain courses in all, some, or none of their majors, by type of department, in fall 2005.

	Required in all majors		Required in some but not all majors		Not required in any major	
Percentage of Statistics Departments that Require	Univ (PhD) %	Univ (MA) %	Univ (PhD) %	Univ (MA) %	Univ (PhD) %	Univ (MA) %
(a) Calculus I	92	86	4	0	4	14
(b) Calculus II	87	86	4	0	8	14
(c) Multivariable Calculus	78	51	9	17	13	31
(d) Linear algebra/Matrix theory	84	69	3	0	13	31
(e) at least one Computer Science course	72	86	16	0	12	14
(f) at least one applied mathematics course, not incl. (a), (b), (c), (d)	24	14	12	17	64	69
(g) a capstone experience (e.g., a senior thesis or project, seminar, or internship)	34	51	9	17	57	31
(h) an exit exam(oral or written)	0	0	0	17	100	83

In fall 2005, to what extent did the nation's mathematics majors include interdisciplinary linkages with computer science and statistics? As noted above, an introductory computer science course was perhaps the most universal course requirement for a mathematics major. But were any upper-division courses in computer science allowed to count toward a track within the mathematics department major? If CBMS2005 data are interpreted conservatively, some answers are possible. For example, Table SP.21 shows that 69% of all doctoral mathematics departments allow some upper-division computer science course from another department to count toward one of their mathematics major tracks. In addition, 17% of doctoral mathematics departments teach upper-division computer science courses themselves, and it is reasonable to suppose that some mathematics major tracks in such departments might include some of the department's own upper-level computer science courses. Therefore, between 69% and 86% of doctoral mathematics departments allow upper-division computer science courses to count toward the requirements of some of their mathematics major tracks, while at least 14% do not allow any upper-division computer science courses to fulfill requirements of their majors. Table SP.21 shows that between 42% and 64% of bachelors-level mathematics departments allow upper-level computer science courses to count toward their requirements for some tracks, leaving at least 36% that do not.

The percentages in Table SP.21 suggest that in fall 2005, a large majority of mathematics departments allowed upper-level statistics courses (either from their own department or from another department) to count toward the requirements of one of their majors.

Table SP.21 shows that among doctoral statistics departments, 55% allowed upper-level computer science courses from other departments to count towards a track within the statistics major, and four percent taught upper-level computer science courses of their own. Consequently, about 40% of doctoral statistics departments did not allow any upper-division computer science courses to count toward their departmental statistics major. Table SP.21 also shows that two out of three doctoral statistics departments allowed some upper-division mathematics courses to count toward the requirements of some statistics major track.

TABLE SP.21 Percentage of mathematics departments and statistics departments that allow upper division courses from other departments to count toward their undergraduate major requirements, by type of department, in fall 2005.

	Four-year Mathematics Departments			Statistics Departments	
Percentage of Departments that	Univ (PhD) %	Univ (MA) %	College (BA) %	Univ (PhD) %	Univ (MA) %
Teach upper level computer science	17	25	42	4	29
Allow upper level CS courses from other depts. to count toward their major	69	31	22	55	100
Teach upper level statistics	64	94	87	na	na
Allow upper level statistics courses from other depts. to count toward their major	55	12	15	na	na
Allow upper division mathematics courses to count toward their major	na	na	na	66	86

Table SP.22 examines the availability of many upper-division courses in mathematics departments during the two-year window consisting of the consecutive academic years 2004–2005 and 2005–2006 (which we abbreviate as 2004–2005–2006). Analogous data for a smaller course list during the single academic years 1995–1996 and 2000–2001 appears in Table SE.5 of the CBMS2000 report. All other things being equal, one would expect to see a larger percentage of departments offering a given course during a two-year window than during a one-year window, and in most cases that is what Table SP.22 shows.

It is somewhat surprising that only about 61% of all four-year college and university mathematics departments offered Modern Algebra during the two-year window 2004–2005–2006, compared to a 71% figure for mathematics departments offering the same course during the single academic year 2000–2001 and a 77% figure for Modern Algebra in the single academic year 1995–1996. Similarly surprising is the percentage of all mathematics departments that offered a course called Real Analysis/Advanced Calculus: 70% for the 1995–1996 academic year, 56% for the 2000–2001 academic year, and 66% for the two-academic-year window 2004–2005–2006. These percentages, combined with the course-requirement data in Table SP.19, suggest that Modern Algebra and Real Analysis no longer hold the central position in the undergraduate mathematics major that they once did.

It may be worth noting that the percentage of bachelors-level mathematics departments offering Number Theory and Combinatorics was larger in 2004–2005–2006 than in 2000–2001, but the importance of this observation is tempered by the fact that less than a third of bachelors-level departments offered these courses in 2004–2005–2006.

Table SP.22 reinforces the tentative conclusion from CBMS2000 that there was a real difference between the mathematics major available to students in doctoral departments and in bachelors departments. For example, during the academic year 2000–2001, 87% of doctoral mathematics departments offered a Modern Algebra course, compared to 63% of bachelors departments. During the two-year window 2004–2005–2006, 86% of doctoral mathematics departments offered a Modern Algebra course, compared to 52% of bachelors-level departments. The situation for Real Analysis is similar: in 2000–2001, about 90% of doctoral mathematics departments offered Real Analysis, compared to 45% of bachelors-level departments, and during the two-year window 2004–2005–2006, 95% of doctoral departments and 57% of bachelors departments offered the course. The course-availability gaps between doctoral and bachelors departments for Geometry and Number Theory were larger, and specialized courses such as Combinatorics and Logic/Foundations were four times as likely to be available in doctoral mathematics departments than in bachelors-level departments.

Table SP.23 examines the analogous question for upper-level statistics courses taught in mathematics or in statistics departments. Among mathematics departments, for example, the percentage offering Mathematical Statistics in the two-year window 2004–2005–2006 was 38%, compared to a figure of 52% for the same course during the single academic year 2000–2001. The percentage of statistics departments that offered Mathematical Statistics in 2000–2001 was 90% and dropped to 76% in the two-year window 2004–2005–2006. Indeed, of the thirteen upper-division statistics courses in Table SP.23, ten were offered less frequently in statistics departments during the two-year window 2004–2005–2006 than during the one-year window 2000–2001. The exceptions were probability courses, biostatistics courses, and statistics senior seminars.

Tables SP.22 and SP.23 provide availability data for a broad spectrum of upper-division mathematics and statistics courses and could serve as baseline data for a future study of the evolution of the national mathematics and statistics curriculum between 2004–2005–2006 and 2009–2010–2011.

TABLE SP.22 Percentage of mathematics departments offering various upper-division mathematics courses at least once in the two academic years 2004-2005 and 2005-2006, plus historical data on the one year period 2000-2001, by type of department.

		Academic Years 2004-2005 & 2005-2006			
	All Math Depts 2000-01 %	All Math Depts 2004-5 & 2005-6 %	PhD Math %	MA Math %	BA Math %
Upper-level Mathematics Courses					
Modern Algebra I	71	61	86	87	52
Modern Algebra II	na	21	40	40	15
Number Theory	33	37	61	61	29
Combinatorics	18	22	55	38	14
Actuarial Mathematics	na	11	24	23	6
Foundations/Logic	16	11	27	16	7
Discrete Structures	na	14	27	22	10
History of Mathematics	na	35	43	68	28
Geometry	56	55	81	89	44
Math for secondary teachers	42	37	41	50	35
Adv Calculus/ Real Analysis I	56	66	95	86	57
Adv Calculus/Real Analysis II	na	26	62	44	17
Adv Mathematics for Engineering/Physics	na	16	50	28	7
Advanced Linear Algebra	na	19	52	42	9

TABLE SP.22, continued

Upper-level Math, Continued	All Math Depts 2000-01 %	Academic Years 2004-2005 & 2005-2006			
		All Math Depts 2004-5 & 2005-6 %	PhD Math %	MA Math %	BA Math %
Vector Analysis	na	9	21	6	7
Advanced Differential Equations	na	13	45	28	5
Partial Differential Equations	na	19	57	29	11
Numerical Analysis I and II	na	47	83	76	36
Applied Math/Modeling	24	26	48	47	18
Complex Variables	na	37	80	53	26
Topology	22	32	61	33	26
Mathematics of Finance	na	8	24	8	5
Codes & Cryptology	na	8	17	8	7
Biomathematics	na	8	24	9	4
Intro to Operations Research	13	12	17	20	10
Intro to Linear Programming	na	6	19	21	1
Math senior seminar/Ind study	58	45	61	48	42

TABLE SP.23 Percentage of mathematics and statistics departments offering various undergraduate statistics courses at least once in academic year 2000-2001 and at least once in the two academic years 2004-2005 and 2005-2006, by type of department.

		AY 2004-05 & 2005-06					AY 2004-05 & 2005-06		
Upper Level Statistics Courses	All Math Depts 2000-01 %	All Math Depts %	PhD Math %	MA Math %	BA Math %	All Stat Depts 2000-01 %	All Stat Depts %	PhD Stat %	MA Stat %
Mathematical Statistics	52	38	52	63	31	90	76	73	88
Probability	40	51	72	69	43	75	86	90	73
Stochastic Processes	6	6	21	13	2	46	43	42	44
Applied Statistical Analysis	13	13	26	32	7	72	65	63	73
Experimental Design	10	6	14	23	2	74	54	49	73
Regression & Correlation	9	6	20	12	3	82	62	55	88
Biostatistics	5	4	11	13	2	20	25	28	15
Nonparametric Statistics	4	2	6	8	0	45	38	33	59
Categorical Data Analysis	1	1	5	3	1	39	21	19	29
Sample Survey Design	3	4	13	8	1	52	49	43	73
Stat Software & Computing	5	3	11	7	1	48	43	35	73
Data Management	1	0	0	0	0	13	5	6	0
Statistics Senior Sem/Ind Study	5	3	8	8	1	34	41	36	59

Note: 0 means less than one-half of one percent.

Table SP.24 summarizes responses from mathematics and statistics departments about the career plans of their bachelors graduates from the 2004–2005 academic year. Departments were asked to give their best estimates of the percentages of their graduates who chose this or that post-college path; the question did not ask departments to do follow-up studies of the previous year's graduates. Consequently, the first four rows should be taken with a grain of salt, and the table does not answer the question "What did mathematics majors (statistics majors) do after graduation?" But it may say something about the extent to which mathematics and statistics departments know their graduating seniors.

TABLE SP.24 Departmental estimates of the percentage of graduating mathematics or statistics majors from academic year 2004-2005 who had various post-graduation plans, by type of department in fall 2005.

	Mathematics Departments			Statistics Departments	
Departmental Estimates of Post-college Plans	Univ (PhD)	Univ (MA)	College (BA)	Univ (PhD)	Univ (MA)
Students who went into pre-college teaching	16%	44%	32%	1%	0%
Students who went to graduate or professional school	21	16	19	18	29
Students who took jobs in business, government, etc.	19	21	29	16	36
Students who had other plans known to the department	4	1	2	0	6
Students whose plans are not known to the department	39	18	17	65	28

Table SP.25: Assessment Activities in Mathematics and Statistics Departments.

During the ten-year period leading up to 2005, state governments, national accrediting agencies, and professional organizations such as the Mathematical Association of America all placed great emphasis on departmental assessment studies [MAAGuidelines], [M], [CUPM], [GKM]. For further information, see http://www.maa.org/saum/index.html.

Table SP.25 summarizes departmental responses about their assessment activities during the period 1999–2005. Surveying departmental graduates was the most widely used assessment technique among masters- and bachelors-level mathematics departments and was also used by six out of ten doctoral mathematics departments. Other recommended assessment techniques were less widely used. Less than half of all mathematics departments used outside reviewers as part of their assessment efforts, perhaps because of cost issues. Less than half of all departments consulted "client departments," i.e., departments whose courses use mathematics or statistics courses as prerequisites, to see whether the client departments were satisfied with what their students had learned in mathematics courses. Less than half of all departments did follow-up studies to determine how well the department's courses prepared the department's own students for later departmental courses. But whatever assessment techniques were or were not used, Table SP.25 reports that in three quarters of mathematics departments, assessment efforts led departments to change their undergraduate programs.

TABLE SP.25 Percentage of four-year mathematics and statistics departments undertaking various assessment activities during the last six years, by type of department, in fall 2005.

	Four-year Mathematics Departments			Statistics Departments	
Percentage Using Various Assessment Tools	Univ (PhD) %	Univ (MA) %	College (BA) %	Univ (PhD) %	Univ (MA) %
Consult outside reviewers	47	45	29	37	59
Survey program graduates	62	81	74	54	71
Consult other departments	51	41	35	29	56
Study data on students' progress in later courses	45	52	38	30	56
Evaluate placement system	72	72	51	5	15
Change undergraduate program due to assessment	76	72	76	69	29

Chapter 3

Mathematical Sciences Bachelors Degrees and Enrollments in Four-Year Colleges and Universities

Mathematics and statistics departments in the nation's four-year colleges and universities offer a wide spectrum of undergraduate mathematical sciences courses and majors, sometimes including mathematics education, actuarial science, operations research, and computer science as well as mathematics and statistics. This chapter's fourteen tables describe

- the number of bachelors degrees awarded through the nation's mathematics and statistics departments (Table E.1),
- enrollments in mathematical sciences courses (Tables E.2–E.4),
- the kinds of instructors who teach undergraduate courses in mathematics and statistics departments (Tables E.5–E.12), and
- average class sizes and average sizes of recitation sections used in lecture/recitation classes (Tables E.13–E.14).

Because there is considerable variation among departmental practices based on highest degree offered, we present the data by type of department as well as by level and type of course.

The tables in this chapter expand upon Tables S.2 and S.4 of Chapter 1, and Chapter 5 provides additional detail about first-year courses. Mathematics and statistics courses and enrollments in two-year colleges are discussed in Chapter 6.

Highlights

- The total number of mathematical sciences bachelors degrees granted through the nation's mathematics and statistics departments in the 2004–2005 academic year was about five percent below the number granted five years earlier. This was caused by sharp declines in bachelors degrees in mathematics education and computer science that were granted through mathematics and statistics departments, declines that more than offset increases in the numbers of mathematics and statistics majors. See Table E.1.
- Hidden within the five percent decrease in overall mathematical sciences bachelors degrees was a major shift in the source of mathematical sciences bachelors degrees. In the 2004–2005 academic year, the number of bachelors degrees granted through doctoral mathematics departments was 41% larger than the number granted during 1999–2000, while the number granted through masters- and bachelors-level departments declined by 27% and 19% respectively from the levels of 1999–2000. However, bachelors-only departments continued to grant the largest number of mathematical sciences bachelors degrees. See Table E.1.
- The percentage of mathematical sciences bachelors degrees granted to women declined from 43% in academic year 1999–2000 to 40% in 2004–2005. See Table E.1.
- Total 2005 fall enrollments in the nation's mathematics and statistics departments declined by about 3% from the levels of fall 2000 and yet remained 8% above the levels of fall 1995. That 3% decline resulted from substantial enrollment losses in masters-level departments that more than offset enrollment gains in doctoral departments. Enrollments in bachelors-level departments remained essentially unchanged from fall 2000. If only mathematics and statistics courses are considered, i.e., if computer science courses are excluded, then enrollments in fall 2005 were essentially the same as in fall 2000 and were about 11% above the levels of fall 1995. See Table E.2.
- Total enrollments in calculus-level courses (which include courses in linear algebra and differential equations as well as calculus courses of various kinds) rose by about 3% from the levels of fall 2000 and were about 9% above the levels of fall 1995. See Table E.2.
- Combined enrollments in advanced mathematics and advanced statistics courses rose by about 8% over the levels of fall 2000 and by about 21% over the levels of fall 1995. That 8% increase over fall 2000 included a remarkable 22% increase in advanced mathematics and advanced statistics enrollments in doctoral mathematics departments and a roughly 31% increase over corresponding doctoral department enrollment levels in fall 1995. See Table E.2.
- In fall 2005, distance education, also called distance learning, was used much more widely in

two-year colleges than in four-year colleges and universities. (CBMS studies, including CBMS2005, have defined distance education as any teaching method in which at least half of the students in a course receive the majority of their instruction in situations where the instructor is not physically present.) About two-tenths of one percent of enrollments in Calculus I courses in four-year colleges and universities in fall 2005 were taught using distance education techniques, compared to about 5% of Calculus I enrollments in two-year colleges. In elementary statistics courses, about two percent of enrollments in the mathematics and statistics departments of four-year colleges and universities were taught using distance learning, compared to over 8% of corresponding enrollments in two-year colleges. See Table E.4.

- The decline in the percentage of mathematical science courses taught by tenured and tenure-eligible faculty that was observed in CBMS2000 continued, coupled with an increase in the percentage of courses taught by "other full-time faculty," a category that includes postdocs, visiting faculty, and a large cohort of non-doctoral full-time faculty. See Tables E.5 through E.12.
- Except in advanced-level courses, average section sizes in mathematical science courses declined slightly from the levels recorded in CBMS2000 but remained above the size recommended by Mathematical Association of America guidelines [MAAGuidelines]. See Table E.13.
- CBMS2005 presents data on the size of recitation sections used in calculus and elementary statistics courses taught in the lecture/recitation format (see Table E.14), and distinguishes between doctoral and non-doctoral faculty in a study of who teaches freshman and sophomore courses. See Tables E.6 through E.12.

Terminology: The two preceding CBMS survey reports are called CBMS1995 and CBMS2000.

Recall that in CBMS2005, the term "mathematics department" includes departments of mathematics, applied mathematics, mathematical sciences, and departments of mathematics and statistics. The term "statistics department" refers to departments of statistics that offer undergraduate statistics courses. The term "mathematical sciences courses" covers all courses that are taught by the nation's mathematics and statistics departments and includes courses in mathematics education, actuarial sciences, and operations research taught in a mathematics or statistics department, as well as courses in mathematics, applied mathematics, and statistics. Computer science courses (and majors) are included in CBMS2005 totals when the courses (and majors) are taught in (granted through) a mathematics or statistics department. CBMS2005 data does not include any courses or majors that are taught in, or granted through, separate departments of computer science, actuarial science, operations research, etc. Departments are classified on the basis of highest degree offered. For example, the term "bachelors-level department" refers to one that does not offer masters or doctoral degrees.

Table E.1: Bachelors degrees granted between July 1, 2004 and June 30, 2005

CBMS2000 revealed a one percent decrease in the number of bachelors degrees awarded through the nation's mathematics and statistics departments between the 1994–1995 academic year and the 1999–2000 academic year. CBMS2005 found a continuation of that trend, with the total number of bachelors degrees granted through the nation's mathematics and statistics departments dropping from 22,614 in the 1999–2000 academic year to 21,440 in the 2004–2005 academic year, a decline of about 5%.

If one looks only at the nation's mathematics departments (which granted about 97% of the 21,440 U.S. bachelors degrees in mathematics and statistics), one sees a variety of bachelors degree programs in a broad range of mathematical sciences—mathematics, applied mathematics, statistics, actuarial science, mathematics education, and (particularly among departments in four-year colleges) also computer science. The total number of bachelors degrees granted through the nation's mathematics departments declined slightly (about one-half of 1%) between the 1995 and 2000 CBMS surveys and fell by another 6% between 2000 and 2005, with the result that the total number of bachelors degrees granted through mathematics departments in the 2004–2005 academic year was about 94% of the number granted in the 1994–1995 academic year.

The number of statistics majors receiving their bachelors degrees through statistics departments in the 2004–2005 academic year rose by about 56% from the levels reported in CBMS2000 for 1999–2000 and was about 9% above the 1994–1995 level. Although this growth rate is impressive, it does not have a major impact on the total number of mathematical sciences bachelors degrees produced in the U.S. because bachelors degrees awarded through statistics departments make up less than 3% of the nation's total number of mathematics and statistics majors.

Table E.1 presents data on several subcategories of the broad mathematical sciences major within mathematics departments. Mathematics education, statistics, and computer science are listed separately, with all other majors granted through mathematics departments lumped into the mathematics category. The number of majors in that remainder category rose by about 7% over CBMS2000 levels and was about 2% higher in 2004–2005 than in 1994–1995. That 7% increase was counterbalanced by decreases in each of the other surveyed bachelors-degree categories (statistics, mathematics education, and computer science) in mathematics departments. For example, the number of mathematics education majors in mathematics departments decreased from 4,991 reported in CBMS2000 to 3,370 in CBMS2005, a decline of about 32%, and the number of computer science majors graduating from mathematics departments fell from 3,315 in the 1999–2000 academic year to 2,604 in the 2004–2005 year, a decline of about 21%. See Figure E.1.2.

Table E.1 in CBMS1995, CBMS2000, and CBMS2005 can be used to study the gender distribution of mathematical sciences bachelors degrees. In the 1994–1995 academic year, about 42% of the mathematical sciences bachelors degrees granted through mathematics and statistics departments were awarded to women, about 43% in 1999–2000, and about 40% in the 2004–2005 academic year. There is some variation based on type of department. For example, the percentage of bachelors degrees awarded to women by doctoral mathematics departments declined from 43% in 1994–1995 to 40% in 1999–2000, and to 37% in 2004–2005. The corresponding percentages in masters-only and bachelors-only mathematics departments bounced around between 1994–1995 and 2004–2005 and do not reveal a steady trend. The percentage of mathematics education degrees awarded to women through mathematics departments rose from 49% in 1994–1995 to about 60% in 2004–2005 (with most of the increase occurring between 1994–1995 and 1999–2000). Among computer science bachelors degrees granted through mathematics departments in 2004–2005, only 18% went to women, down from 24% in 1999–2000. In the nation's statistics departments, about 38% of bachelors degrees were awarded to women in 1994–1995, about 43% in 1999–2000, and about 42% in 2004–2005. In mathematics departments, women accounted for about 48% of all bachelors degrees awarded in 2004–2005, down from 59% in 1999–2000. See also Figure E.1.2.

Table E.1 reveals a potentially important shift in the kinds of mathematics departments through which mathematical sciences majors earned their bachelors degrees. Figure E.1.3 shows a jump in the percentage of all bachelors degrees from mathematics departments that were awarded through doctoral mathematics departments, with a corresponding drop in the percentage of bachelors degrees awarded by non-doctoral departments between 1999–2000 and 2004–2005. The declines for masters-level mathematics departments are particularly large; the number of majors produced by those departments dropped 27% from levels reported in CBMS2000. Some of that decline may have been a consequence of changes between 2000 and 2005 in the American Mathematical Society (AMS) departmental classification that was the basis for CBMS studies in 2000 and 2005. However, CBMS2005 is not the first CBMS survey to report a major decline in the number of bachelors degrees granted through masters-level mathematics departments; CBMS2000 reported a 17% decline in bachelors degrees granted through masters-level departments between the academic years 1994–1995 and 1999–2000.

As separate departments of computer science are created, mathematics departments lose computer science enrollments and majors. Consequently, it makes sense to track the number of bachelors degrees awarded through mathematics departments, excluding computer science degrees, in order to study bachelors degree productivity of mathematics departments. CBMS1995 showed that in the 1994–1995 academic year, 19,593 non-computer-science bachelors degrees were awarded through the nation's mathematics departments. CBMS2000 and CBMS2005 show that total dropped by about 4% between the 1994–1995 and 1999–2000 academic years, and by another 4% between the 1999–2000 and 2004–2005 academic years, reaching 18,222 in academic year 2004–2005 for a total decline of about 7% from ten years earlier.

Data from CBMS1995, CBMS2000, and CBMS2005 show that bachelors-level mathematics departments consistently produced at least 40% of the non-computer-science bachelors degrees granted through mathematics departments, with doctoral departments' percentage rising from 31% in 1995 to 40% in 2005. The percentage of non-computer-science bachelors degrees granted through masters-level mathematics departments dropped from 30% in 1995, to 20% in 2000, to 19% in 2005. A graph of these percentages closely resembles the graph in Figure E.1.3.

TABLE E.1 Bachelors degrees in mathematics, mathematics education, statistics, and computer science in mathematics departments and in statistics departments awarded between July 1, 2004 and June 30, 2005, by gender of degree recipient and type of department.

	Mathematics Departments				Statistics Departments			
Bachelors degrees in Math and Stat Depts	Univ (PhD)	Univ (MA)	Coll (BA)	**Total Math Depts**	Univ (PhD)	Univ (MA)	**Total Stat Depts**	**Total Math & Stat Depts**
Mathematics majors (including Act Sci, Oper Res, and joint degrees)								
Men	4112	1350	3358	**8820**				**8820**
Women	2282	1027	2482	**5791**				**5791**
(Percentage of women)	(36%)	(43%)	(43%)	**(40%)**				**(40%)**
Total Math degrees	**6393**	**2377**	**5839**	**14610**				**14610**
Mathematics Education majors								
Men	296	401	645	**1341**				**1341**
Women	470	628	930	**2028**				**2028**
	(61%)	(61%)	(59%)	**(60%)**				**(60%)**
Total Math Ed degrees	**766**	**1029**	**1575**	**3369**				**3369**
Statistics majors								
Men	64	44	17	**125**	237	120	**357**	**482**
Women	69	41	6	**116**	184	73	**257**	**373**
	(52%)	(48%)	(26%)	**(48%)**	(44%)	(38%)	**(42%)**	**(44%)**
Total Stat degrees	**133**	**85**	**23**	**241**	**421**	**193**	**614**	**855**
Computer Science majors								
Men	413	314	1412	**2139**				**2139**
Women	58	72	335	**465**				**465**
	(12%)	(19%)	(19%)	**(18%)**				**(18%)**
Total CS degrees	**471**	**386**	**1747**	**2603**				**2603**
Total degrees - Men	**4884**	**2109**	**5431**	**12424**	**237**	**120**	**357**	**12780**
Total degrees - Women	**2879**	**1768**	**3752**	**8399**	**184**	**73**	**257**	**8656**
	(37%)	**(46%)**	**(41%)**	**(40%)**	**(44%)**	**(38%)**	**(42%)**	**(40%)**
Total all degrees	**7763**	**3877**	**9183**	**20823**	**421**	**193**	**614**	**21437**

Note: Round-off may make row and column sums seem inaccurate.

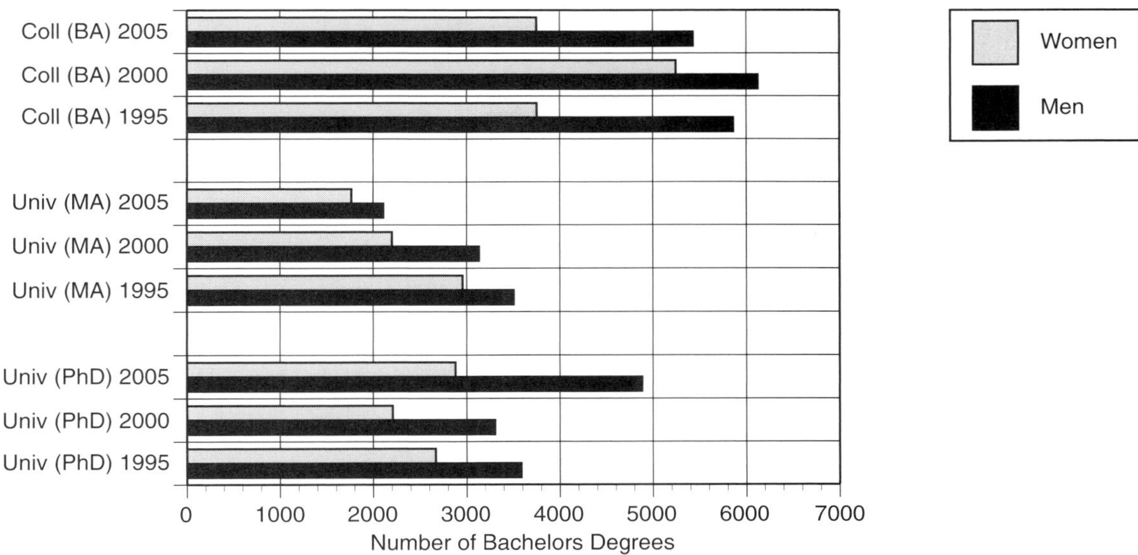

FIGURE E.1.1 Bachelors degrees in mathematics departments awarded between July 1 and June 30 in the academic years 1994-1995, 1999-2000, and 2004-2005, by gender and type of department.

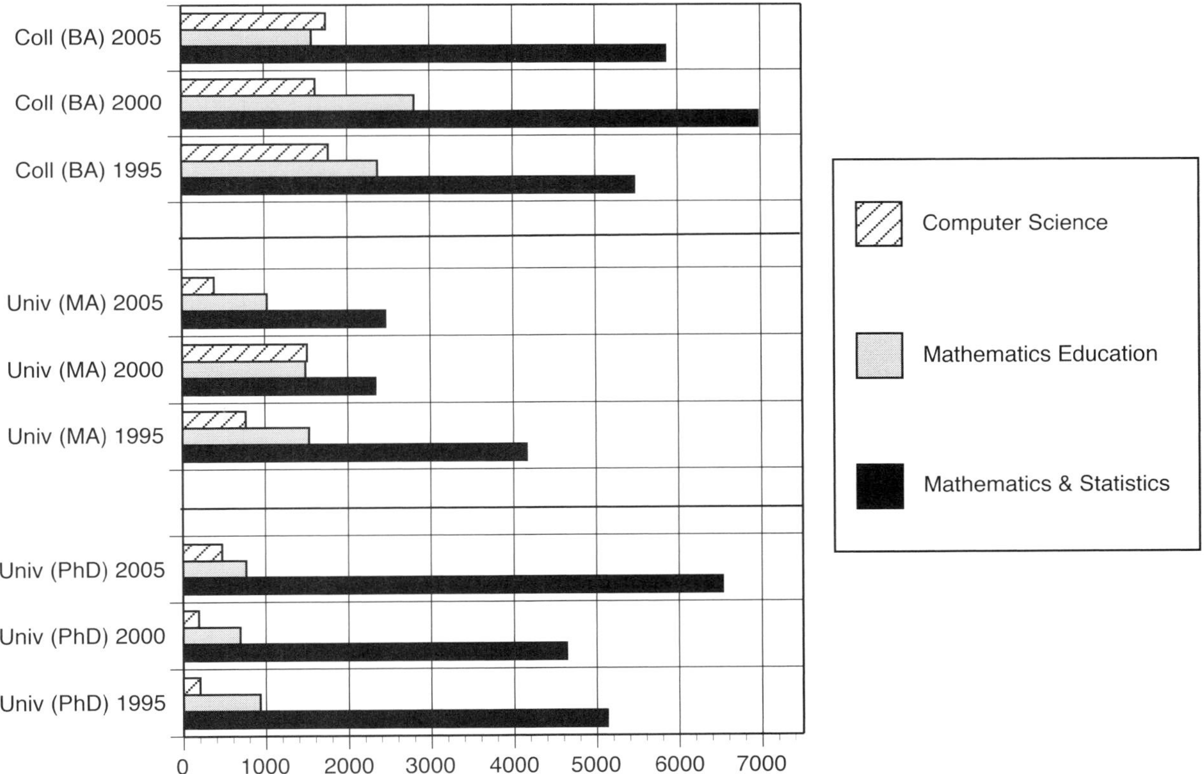

FIGURE E.1.2 Number of bachelors degrees granted in academic years 1994-1995, 1999-2000, and 2004-2005 by type of major and type of department.

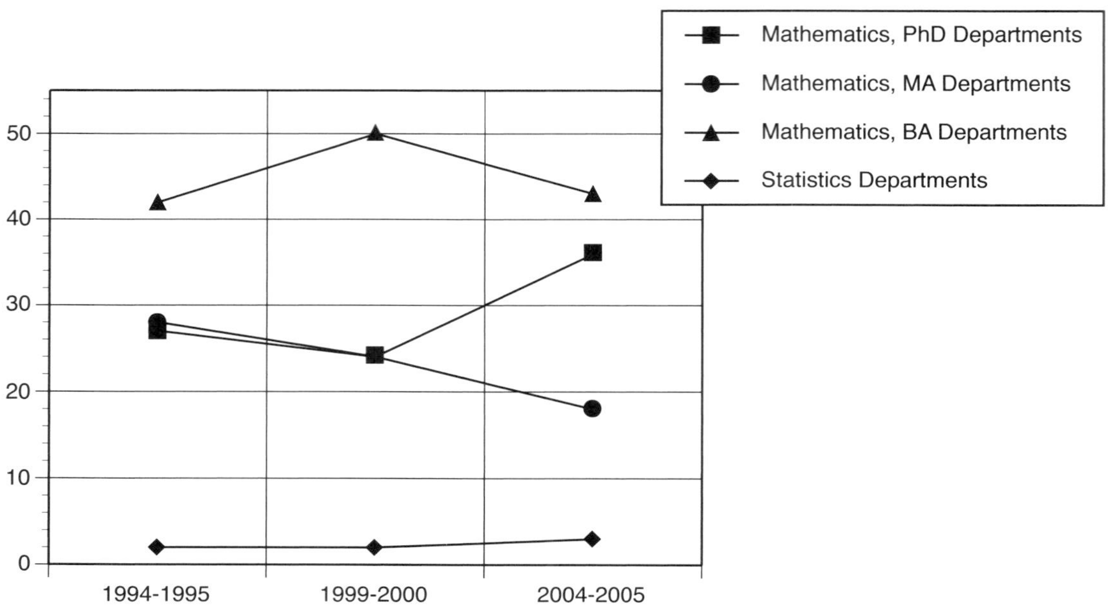

FIGURE E.1.3 Percentage of mathematical sciences bachelors degrees (including computer science) awarded through mathematics and statistics departments of various kinds in academic years 1994-1995, 1999-2000, and 2004-2005.

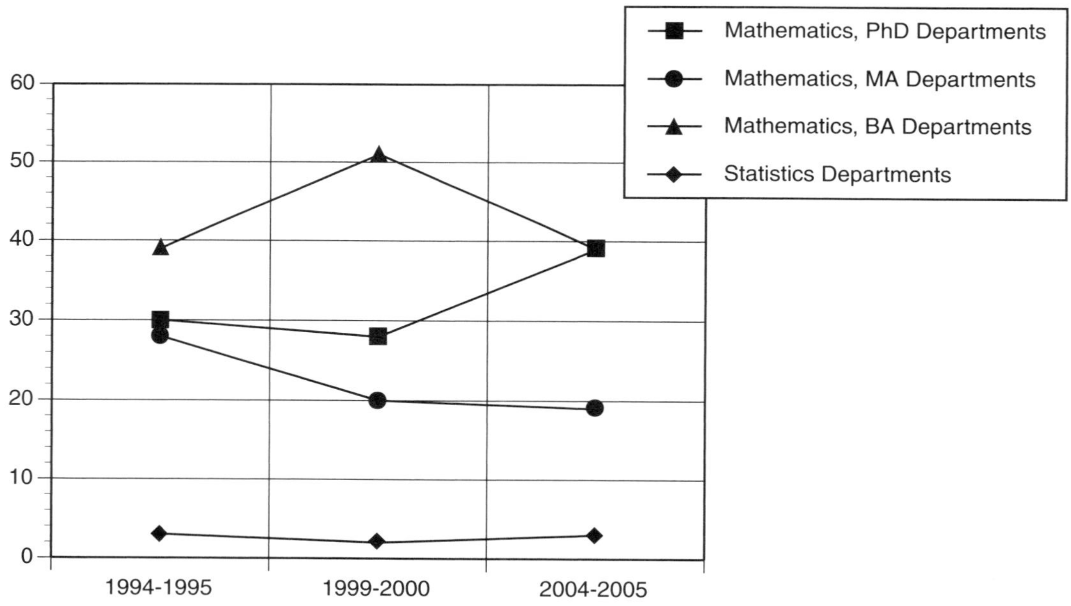

FIGURE E.1.4 Percentage of mathematics and statistics bachelors degrees (excluding computer science) awarded through mathematics and statistics departments of various kinds in academic years 1994-1995, 1999-2000, and 2004-2005.

Tables E.2 and E.3: Undergraduate enrollments and number of sections offered in mathematics and statistics departments

CBMS2005 Table E.2 divides mathematical sciences department enrollments into three broad categories: mathematics courses, statistics courses, and computer science courses. Total enrollments in all fall-term courses in mathematics and statistics departments at four-year colleges and universities declined by about 3% from levels recorded in CBMS2000. This was due to a pronounced decline in the number of computer science enrollments in mathematics departments, from 123,000 in fall 2000 to 57,000 in fall 2005. Statistics enrollments in mathematics and statistics departments increased by about 6%, and mathematics enrollments held essentially steady at fall 2000 levels. The decline in computer science enrollments more than offset slight enrollment increases in the combination of all mathematics and statistics courses. Even though total enrollments dropped from fall 2000 levels, they were about 8% above the levels of fall 1995.

Table E.2 reveals that the change in total enrollments varied considerably among departments of different kinds. Figure E.2.3 shows that enrollment growth in doctoral mathematics departments outstripped enrollment growth in bachelors-level mathematics departments, while in masters-level departments, there was a decline. Between fall 2000 and fall 2005, for example, enrollment in doctoral mathematics departments grew by about 7% (from 720,000 to 769,000), while total enrollments in masters-level departments dropped by over 20% (from 534,000 to 417,000), and total enrollment in bachelors-level departments increased marginally (from 654,000 to 659,000) . The reported 22% enrollment decline in masters-level departments may be misleading. As noted above, some of the decrease was due to changes made in the American Mathematical Society departmental classification system between 2000 and 2005.

Combined fall-term statistics enrollments in mathematics and statistics departments grew by about 6% between 2000 and 2005, compared to an 18% increase between 1995 and 2000. The majority (about 70%) of all statistics course enrollments were in mathematics departments, and the majority of statistics enrollments in mathematics departments were in bachelors-level departments. (See Figure E.2.2.) Statistics course enrollments in mathematics departments grew by 20% between fall 1995 and fall 2000, and by 6% between fall 2000 and fall 2005.

Total enrollments in calculus-level courses are sometimes used as a predictor for growth in the number of science, technology, engineering, and mathematics (STEM) professionals. Previous CBMS studies included linear algebra and differential equations courses as calculus-level courses, and CBMS2005 continued that practice. (Separate enrollment totals for individual calculus courses are given in Appendix I of this report.) The nation's combined calculus-level enrollments grew by about 6% between fall 1995 and fall 2000, and grew by another 3% between fall 2000 and fall 2005. That growth was concentrated primarily in doctoral-level mathematics departments. In fall 2005, calculus-level enrollments in doctoral departments were up 14% from the level of fall 2000, and up almost 30% from the level of fall 1995. By contrast, calculus-level enrollments in masters departments dropped by almost a third between CBMS2000 and CBMS2005, and in fall 2005 were about 29% below the levels of fall 1995. Once again we note that some of this decrease may have been an artifact of changes in the AMS departmental classification system. Bachelors-level departments saw their calculus-level enrollments rebound to 1995 levels, after a marked decrease between fall 1995 and fall 2000.

The combination of all advanced mathematics and upper-level statistics enrollments in mathematics and statistics departments is another predictor for the number of future STEM professionals, and is also a predictor for the number of mathematics and statistics majors. Combined upper-level enrollments rose to 169,000 in fall 2005, an almost 8% increase over figures reported in CBMS2000 and an almost 21% increase over corresponding figures in CBMS1995. The largest gains were in doctoral mathematics departments, where the combination of advanced mathematics and upper-level statistics enrollments rose by about 22% from the levels of fall 2000 and by about 31% when compared with fall 1995. Masters-level mathematics departments saw an 8% decline in the number of upper-division mathematics and statistics enrollments between 2000 and 2005, and a roughly 9% decline from the levels of fall 1995. In bachelors-level mathematics departments, advanced mathematics and upper-level statistics enrollments were essentially unchanged from fall 2000 levels, and were up by about 12% compared to fall 1995. In statistics departments, upper-level enrollments grew by about 15% between fall 2000 and fall 2005, with almost all of the growth occurring in doctoral statistics departments. Compared to fall 1995, upper-level enrollment in statistics departments in fall 2005 rose by almost 44%.

Table E.3 reflects departmental teaching effort in fall 2005 in a different way, by showing the number of sections offered rather than the total enrollment. The total number of sections offered by the nation's mathematics and statistics departments dropped by about 2% (as did total enrollments). The number of sections offered by doctoral mathematics departments rose by about 9% between fall 2000 and fall 2005, while the number of sections offered by masters-level mathematics departments dropped by about 23%. The number of sections offered by bachelors-level mathematics departments rose by more than 3% between fall 2000 and fall 2005, as did the number of sections offered by statistics departments. The number of sections of calculus-level courses grew by about 14% between fall 2000 and fall 2005 in the nation's doctoral and bachelors-level mathematics departments, and there was a 29% drop in the number of calculus-level sections offered by masters-level mathematics departments (compared to

TABLE E.2 Enrollment (in thousands) in undergraduate mathematics, statistics, and computer science courses (including distance-learning enrollments) in mathematics and statistics departments by level of course and type of department, in fall 2005. (Numbers in parentheses are (1995,2000) enrollments.)

	Fall 2005 (1995,2000) enrollments (1000s)						
	Mathematics Departments				Statistics Departments		
	Univ (PhD)	Univ (MA)	Coll (BA)	**Total Math Depts**	Univ (PhD)	Univ (MA)	**Total Stat Depts**
Mathematics courses							
Precollege	55 (60,59)	60 (84,59)	87 (78,101)	**201 (222,219)**			
Introductory (incl. Precalc)	269 (222,258)	190 (193,227)	248 (198,238)	**706 (613,723)**			
Calculus	345 (264,302)	88 (124,131)	154 (150,137)	**587 (538,570)**			
Advanced Mathematics	52 (41,43)	24 (25,24)	36 (30,35)	**112 (96,102)**			
Total Math courses	**720 (587,662)**	**362 (426,441)**	**525 (456,511)**	**1607 (1469,1614)**			
Statistics courses							
Elementary Statistics	30 (23,38)	32 (35,35)	86 (57,63)	**148 (115,136)**	42 (46,46)	13 (3,8)	**54 (49,54)**
Upper Statistics	15 (10,12)	9 (7,12)	10 (11,11)	**34 (28,35)**	20 (16,17)	3 (0,3)	**24 (16,20)**
Total Stat courses	**44 (33,50)**	**42 (42,47)**	**96 (68,74)**	**182 (143,171)**	**62 (62,63)**	**16 (3,11)**	**78 (65,74)**
CS courses							
Lower CS	3 (4,5)	11 (18,33)	30 (52,52)	44 (74,90)	0 (0,0)	1 (1,1)	**2 (1,1)**
Middle CS	1 (0,1)	1 (3,7)	6 (10,9)	**8 (13,17)**	0 (0,0)	0 (0,0)	**0 (0,0)**
Upper CS	1 (2,2)	1 (4,6)	3 (6,8)	**5 (12,16)**	0 (0,0)	0 (0,0)	**0 (0,0)**
Total CS courses	**5 (6,8)**	**13 (25,46)**	**39 (68,69)**	**57 (99,123)**	**0 (0,0)**	**2 (1,1)**	**2 (1,1)**
Total all courses	**769 (626,720)**	**417 (493,534)**	**659 (592,654)**	**1845 (1711,1908)**	**62 (62,63)**	**18 (4,12)**	**80 (66,75)**

Note: Due to round-off, row and column sums may appear inaccurate.

a 23% enrollment decline in calculus-level courses in such departments). The number of advanced mathematics and statistics sections in doctoral mathematics departments grew by about 18% (compared with a 22% enrollment increase). The number of advanced sections in masters-level departments dropped by about 9% (compared to an 8% enrollment decrease), and the number of advanced sections offered by bachelors-level mathematics departments grew by about 3% even though enrollment was unchanged from fall 2000.

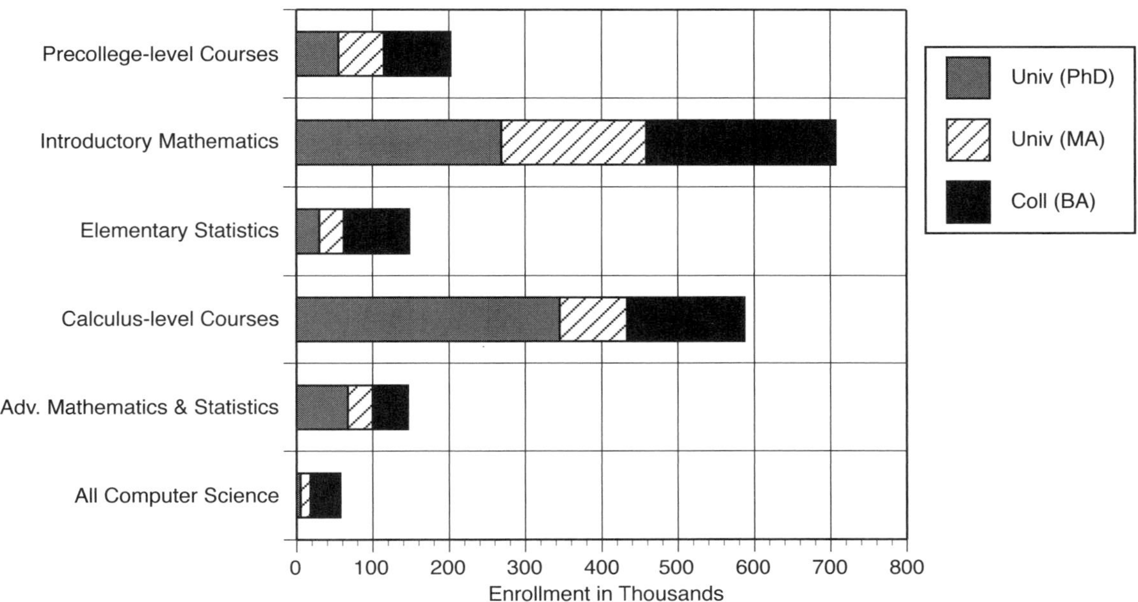

FIGURE E.2.1 Enrollment (thousands) in undergraduate mathematics, statistics, and computer science courses in four-year college and university mathematics departments by type of course and type of department in fall 2005.

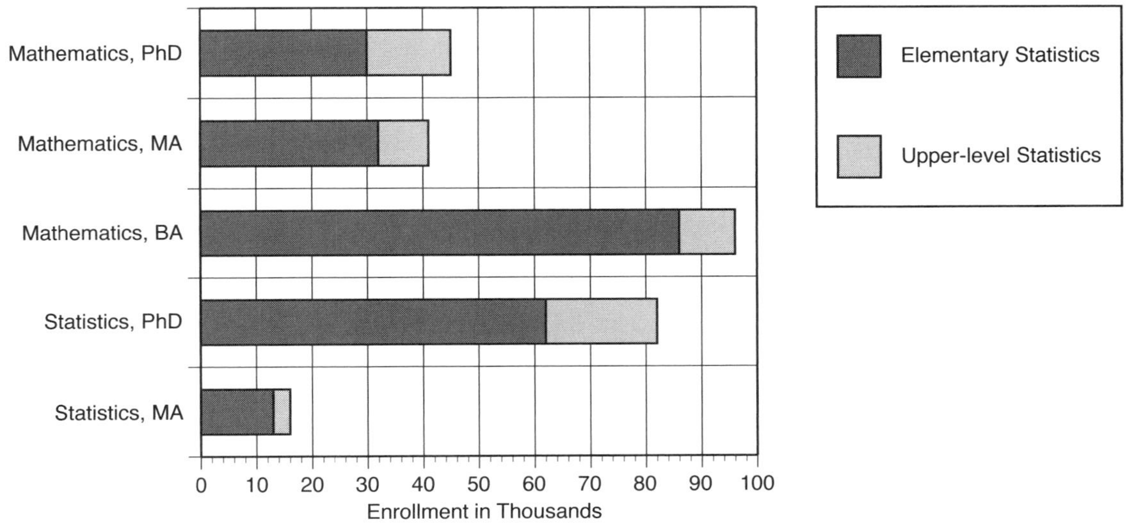

FIGURE E.2.2 Enrollment (thousands) in undergraduate statistics courses by level of course and type of department in fall 2005.

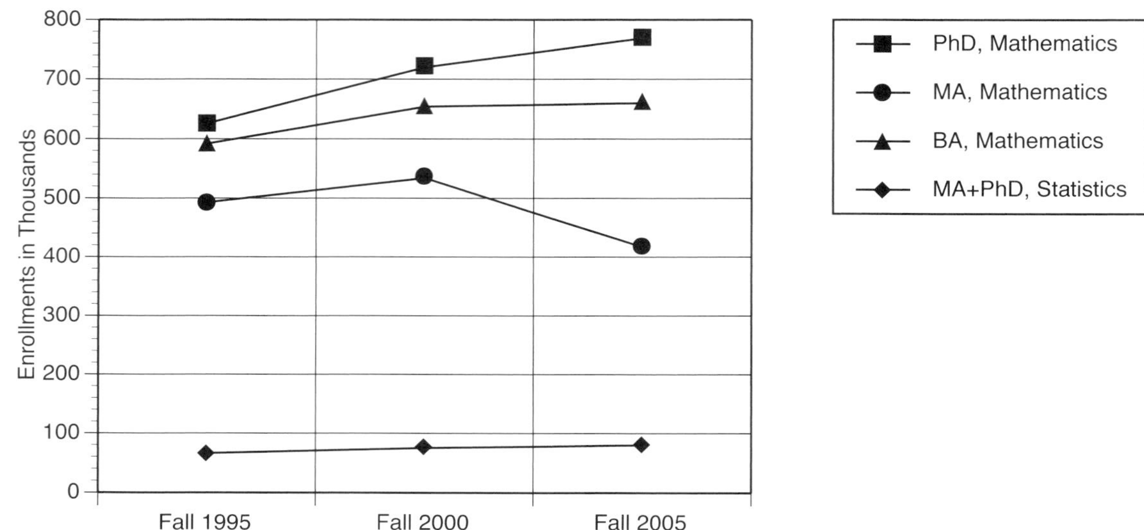

FIGURE E.2.3 Undergraduate enrollment (in thousands) in doctoral, masters, and bachelors mathematics departments, and in a combination of all masters and doctoral-level statistics departments, in fall 1995, fall 2000, and fall 2005.

TABLE E.3 Number of sections (not including distance-learning) of undergraduate mathematics, statistics, and computer science courses in mathematics and statistics departments, by level of course and type of department, in fall 2005 with fall 2000 figures in parentheses. (CBMS2000 data from Table E.10.)

	Number of sections: Fall 2005 (Fall 2000)						
	Mathematics Departments				Statistics Departments		
Mathematics courses	Univ (Phd)	Univ (MA)	Coll (BA)	**Total Math Depts**	Univ (PhD)	Univ (MA)	**Total Stat Depts**
Precollege level	1363 (1493)	1902 (1772)	3862 (4388)	**7126 (7653)**			
Introductory (incl. Precalc)	5518 (5032)	5543 (6506)	9895 (8987)	**20955 (20525)**			
Calculus	7696 (6768)	3237 (4551)	7388 (6438)	**18321 (17757)**			
Advanced Mathematics	2625 (2392)	1622 (1936)	3507 (3415)	**7754 (7743)**			
Total Math courses	17202 (15685)	12303 (14765)	24652 (23228)	**54157 (53678)**			
Statistics courses							
Elementary Statistics	629 (827)	924 (1064)	3191 (2372)	**4744 (4263)**	696 (786)	186 (123)	**882 (909)**
Upper Statistics	869 (580)	714 (638)	771 (728)	**2354 (1946)**	499 (476)	156 (122)	**654 (598)**
Total Stat courses	1498 (1407)	1638 (1702)	3962 (3100)	**7098 (6209)**	1195 (1262)	342 (245)	**1537 (1507)**
CS courses							
Lower CS	114 (92)	512 (1553)	1629 (2557)	**2254 (4202)**	11 (4)	22 (12)	**33 (16)**
Middle CS	61 (24)	121 (465)	739 (590)	**921 (1079)**	2 (0)	14 (2)	**16 (2)**
Upper CS	61 (98)	83 (527)	444 (868)	**587 (1493)**	0 (0)	0 (8)	**0 (8)**
Total CS courses	236 (214)	715 (2545)	2811 (4015)	**3762 (6774)**	13 (4)	36 (22)	**49 (26)**
Total all courses	**18935 (17306)**	**14656 (19012)**	**31425 (30343)**	**65017 (66661)**	**1208 (1266)**	**378 (267)**	**1586 (1533)**

Note: Round-off may make row and column sums seem inaccurate.

Table E.4: Distance education in four-year colleges and universities

The terms "distance education" and "distance learning" have been broadly defined in recent CBMS studies to mean any learning format in which the majority of students receive at least half of their instruction in situations where the instructor is not physically present. This includes, for example, correspondence courses (electronic or paper), courses that use broadcast lectures, and courses taught via the internet. Some universities have experimented with teaching their calculus courses in large computer labs, where students interact with sophisticated tutorial programs in lieu of interacting with an instructor.

CBMS2000 asked about the number of sections of a given course taught using distance-learning methods, and follow-up calls in fall 2000 revealed that to be the wrong question. In some cases, all distance-learning students were enrolled in a single section of a course, with the result that average section size estimates may have been inflated in the CBMS2000 report. With that in mind, CBMS2005 asked departments to report separately the number of students enrolled in distance-learning sections of a given course and the number of students enrolled in non-distance-learning sections. Table E.4 summarizes the results for the types of courses most frequently taught using distance education in fall 2005 and shows that, in fall 2005, distance education was not widely used in four-year colleges and universities. Among four-year mathematics departments, only in elementary statistics courses did distance enrollments exceed 2% of total enrollments, and in Calculus I courses the percentage was insignificant. The middle column of Table E.4 allows comparisons with the situation in two-year colleges, where distance education is more common. For example, at two-year colleges, distance-education enrollments were about five percent of total enrollment in certain precalculus and Calculus I courses, and accounted for more than 8% of total enrollments in elementary statistics courses. For more details on the use of distance education in two-year colleges, see Chapter 6.

TABLE E.4 Enrollments in distance-learning courses (meaning at least half of the students receive the majority of their instruction in situations where the instructor is not physically present) and in other sections for various freshman and sophomore courses, by type of department, in fall 2005.

	Four-year Mathematics Departments		Two-year Mathematics Departments		Statistics Departments	
	Distance-learning Enrollments	Other Enrollments	Distance-learning Enrollments	Other Enrollments	Distance-learning Enrollments	Other Enrollments
Precollege Level	2489	198760	37036	927697	--	--
College Algebra. Triginometry, & Precalculus	5856	352591	15721	298081	--	--
Calculus I	593	308518	3620	68919	--	--
Calculus II	577	94858	270	20003	--	--
Differential Equations & Linear Algebra	238	82034	83	7423	--	--
Elementary Statistics	3075	140077	9894	107304	990	44303

Note: For some distance-learning enrollments in this table, the Standard Error (SE) was very large. See the SE Appendix.

Tables E.5 to E.12: Who taught undergraduate mathematics and statistics in fall 2005?

Chapter 3 of the CBMS2000 report contained several sets of tables, all produced from the same data set. CBMS2000 Tables E.4 to E.9 presented results as percentages of *enrollments*, e.g., the percentage of introductory-level enrollments taught by tenured or tenure-eligible faculty. Tables E.12 through E.18 of that report presented the same information in terms of the *number of sections*. Because the data transformation needed to produce percentage-of-enrollment tables from responses to CBMS2000 questionnaires made certain problematic assumptions, standard error (SE) values for Tables E.4 to E.9 were not calculated. This concern led the CBMS2005 project directors to present 2005 data in terms of numbers and percentages of sections of various kinds. As long as one is careful to compare the percentage-of-sections tables in CBMS2005 with percentage-of-sections tables from CBMS2000, historical trends can be studied, and the heading of Tables E.5 to E.12 in CBMS2005 contains a reference to the proper comparison table from CBMS2000. For example, Table E.5 of CBMS2005 should be compared with Table E.12 of CBMS2000.

The faculty categories used in CBMS2005 Tables E.5 to E.12 are tenured and tenure-eligible (TTE) faculty, other full-time faculty (OFT), which is the set of all full-time faculty who are not in the TTE category, part-time (PT) faculty, and graduate teaching assistants (GTAs). In cases where departmental responses did not account for all sections of a given type of course, there is also an "unknown" column. For example, postdoctoral faculty and scholarly visitors who teach courses would be included in the OFT category.

Table E.12 of the CBMS2000 study reported marked changes between fall 1995 and fall 2000 in the percentage of sections taught by various types of faculty in mathematics and statistics departments. CBMS2000 reported that, when compared with fall 1995 data, the percentage of sections taught in fall 2000 by tenured and tenure-eligible (TTE) faculty had dropped, sometimes by a large amount, with a corresponding increase in the percentage of sections taught by other full-time (OFT) faculty, a category that includes scholarly visitors, postdocs, full-time instructors and lecturers, and an increase in the number of sections taught by part-time faculty. CBMS2000 also found a pronounced drop in the number of sections taught by graduate teaching assistants (GTAs) between fall 1995 and fall 2000. (See also [LM].) (In CBMS surveys, to say that a GTA teaches a section means that she or he is the instructor of record for that section. Teaching assistants who supervise recitation sections for a larger lecture course are not counted as teaching their own section of the course.)

Table E.5 in the current report shows that between fall 2000 and fall 2005, the decline in the percentage of sections taught by TTE faculty continued, except among sections of computer science courses. For mathematics courses as a whole, the percentage taught by TTE faculty dropped by six percentage points, from 52% in fall 2000 to 46% in fall 2005. At the same time, the percentage of mathematics sections taught by OFT faculty rose by six points, and the percentage of mathematics sections taught by GTAs rose by two percentage points, from 7% to 9%. The percentage of statistics courses taught by TTE faculty dropped by eleven and ten percentage points in mathematics and statistics departments respectively, with a corresponding rise in teaching by OFT faculty. Only in computer science sections was there a marked increase in the percentage of sections taught by TTE faculty.

In some cases the change in the percentage of sections taught by TTE faculty was surprisingly large. For example, between fall 2000 and fall 2005, the percentage of statistics sections taught by TTE faculty in doctoral mathematics departments dropped from 63% to 39%, and the analogous percentage in masters-level mathematics departments dropped from 72% to 49%. Figures E.4.1, E.4.2, and E.4.3 show the percentages of various types of courses taught by different kinds of instructors in fall 2005.

CBMS2005 Tables E.6 through E.12 examine the fine structure of the global data in Table E.5, presenting data on courses at various levels of the curriculum (pre-college-level, introductory-level, and calculus-level, elementary statistics, introductory-level computer science, middle-level computer science, and advanced-level mathematics and statistics courses). The tables show the numbers of sections taught by different types of instructors, and they include important new data: the category of OFT faculty is subdivided into those who had a doctoral degree and those who did not. In order to allow comparisons with previous CBMS studies, one column of the tables presents the number of sections taught by all OFT faculty, independent of degree earned, and a second column shows the number of sections taught by doctoral OFT faculty. This refinement was introduced to make a distinction between sections taught by postdocs and scholarly visitors on the one hand, and by non-doctoral full-time instructors on the other. For example, Table E.6 shows that of the 7,126 sections of pre-college-level courses offered in mathematics departments in fall 2005, about 9% were taught by TTE faculty, 4% by doctoral OFT faculty, 21% by non-doctoral OFT faculty, etc. (It is also of interest to note that the number of pre-college sections dropped between fall 2000 and fall 2005, from 7,653 to 7,126.) By contrast, Table E.8 shows that of the 18,321 sections of calculus-level courses taught in mathematics departments, about 61% were taught by TTE faculty, about 10% by doctoral OFT faculty, and about 7% by non-doctoral OFT faculty.

CBMS2000 reported that between fall 1995 and fall 2000, the percentage of mathematics department sections taught by graduate teaching assistants (GTAs) declined, often to a pronounced degree. CBMS2005 data suggests a reversal of that trend. For example, in fall 2000, about 9.5% of precollege sections were taught by GTAs, while in fall 2005 the percentage was 14.6%. In introductory-level courses (including College Algebra, Precalculus, Mathematics for Liberal Arts, etc.), the percentage of sections taught by GTAs was essentially unchanged from fall 2000 levels. In calculus-level sections, the percentage rose from 6.4% to 7.6%. Only in elementary statistics and lower-level computer science was there a decline in the percentage of sections taught by GTAs. In elementary statistics, the percentage dropped from about 9% of all elementary statistics sections taught in mathematics and statistics departments combined to about 6% (Table E.9).

Tables E.5 and E.6 contain what appears to be anomalous data; they report that some mathematics sections in bachelors-only departments are taught by GTAs. Follow-up telephone calls to various bachelors-level mathematics departments revealed that some departments "borrow" GTAs from graduate departments at their own universities, and some departments classified as bachelors-level when the CBMS2005 sample frame was set up subsequently created masters programs, often Master of Arts in Teaching programs, and were using their new GTAs to teach courses in fall 2005. This anomaly will reappear in Chapter 5, which looks at first-year courses in considerable detail.

Table E.12 in CBMS2005 is new. Earlier CBMS studies made the assumption that all upper-division sections were taught by tenured and tenure-eligible (TTE) faculty. To test that assumption, CBMS2005 asked departments to specify how many of their upper-division sections were taught by TTE faculty. In mathematics departments, about 78% of all upper-division mathematics and statistics courses were taught by TTE faculty. Looking at mathematics and statistics courses in these departments separately, one sees that TTE faculty taught about 84% of all upper-division mathematics courses offered in fall 2005 and about 59% of all upper-level statistics courses. In statistics departments, 74% of all upper-level courses were taught by TTE faculty in fall 2005. CBMS2005 has no data on who taught the remaining upper-division courses.

TABLE E.5 Percentage of sections, excluding distance learning, of mathematics, statistics, and computer science courses taught by tenured/tenure-eligible (TTE), other full-time (OFT), part-time (PT), graduate teaching assistants (GTAs), and unknown (Ukn) in mathematics departments and statistics departments by type of department in fall 2005, with fall 2000 figures in parentheses. (CBMS2000 data from Table E.12.)

	Percentage of mathematics sections taught by						Percentage of statistics sections taught by						Percentage of CS sections taught by					
	TTE %	OFT %	PT %	GTAs %	Ukn %	**No. of Math sections**	TTE %	OFT %	PT %	GTAs %	Ukn %	**No. of Stat sections**	TTE %	OFT %	PT %	GTAs %	Ukn %	**No. of CS sections**
Math Depts																		
Univ (PhD)	35	24	14	21	6	**17202**	39	44	7	9	2	**1498**	39	38	9	7	6	**236**
	(42)	(16)	(17)	(21)	(4)	**(15685)**	(63)	(9)	(11)	(14)	(3)	**(1407)**	(59)	(17)	(6)	(3)	(15)	**(214)**
Univ (MA)	45	20	22	8	6	**12303**	49	33	15	1	2	**1639**	43	8	18	0	30	**715**
	(48)	(19)	(22)	(5)	(6)	**(14765)**	(72)	(9)	(11)	(1)	(7)	**(1702)**	(47)	(11)	(35)	(0)	(7)	**(2545)**
Coll (BA)	54	20	23	1	3	**24652**	59	13	25	0	3	**3962**	80	9	9	0	1	**2811**
	(60)	(13)	(21)	(0)	(6)	**(23228)**	(59)	(13)	(22)	(0)	(6)	**(3100)**	(56)	(18)	(15)	(0)	(11)	**(4015)**
Total Math Depts	**46**	**21**	**20**	**9**	**5**	**54157**	**52**	**24**	**19**	**2**	**2**	**7099**	**70**	**11**	**11**	**0**	**7**	**3762**
	(52)	**(15)**	**(20)**	**(7)**	**(6)**	**(53678)**	**(63)**	**(11)**	**(17)**	**(4)**	**(5)**	**(6209)**	**(53)**	**(15)**	**(22)**	**(0)**	**(10)**	**(6774)**
Stat Depts																		
Univ (PhD)	Too few cases in the sample to make reliable estimates						41	22	7	14	15	**1195**	Too few cases in the sample to make reliable estimates					**13**
							(53)	(8)	(14)	(20)	(5)	**(1262)**						**(4)**
Univ (MA)							64	27	7	0	2	**342**						**36**
							(71)	(9)	(5)	(4)	(12)	**(245)**						**(22)**
Total Stat Depts							**46**	**23**	**7**	**11**	**12**	**1537**						**49**
							(56)	**(8)**	**(12)**	**(18)**	**(6)**	**(1507)**						**(26)**

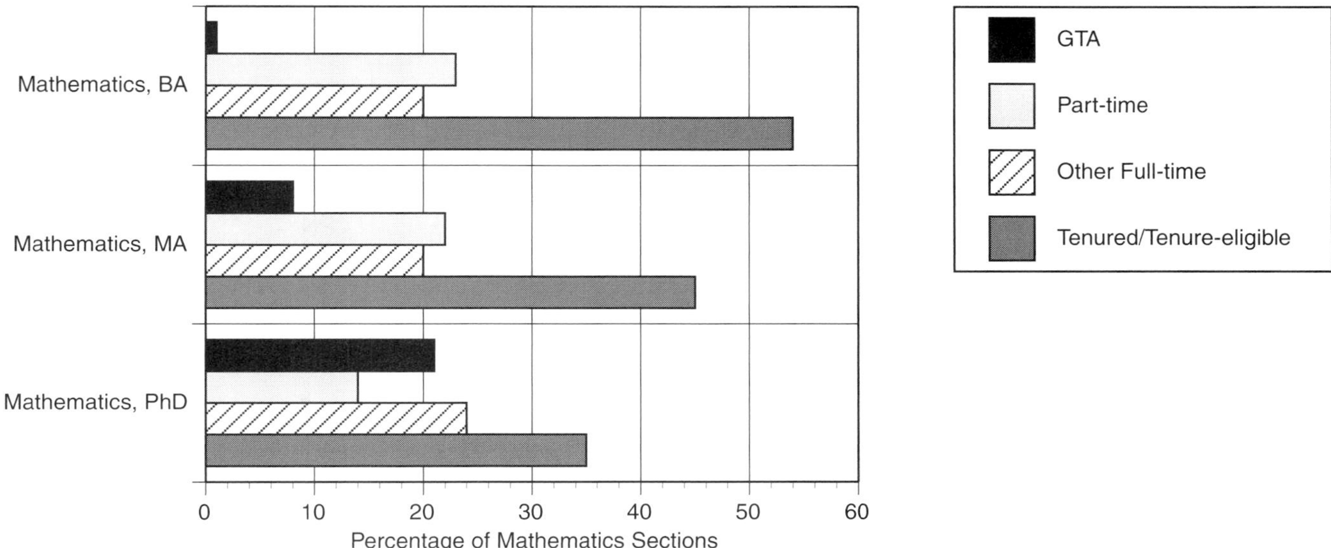

FIGURE E.5.1 Percentage of mathematics sections in mathematics departments whose instructors were tenured/tenure-eligible faculty, other full-time faculty, part-time faculty, and graduate teaching assistants (GTA), by type of department in fall 2005.

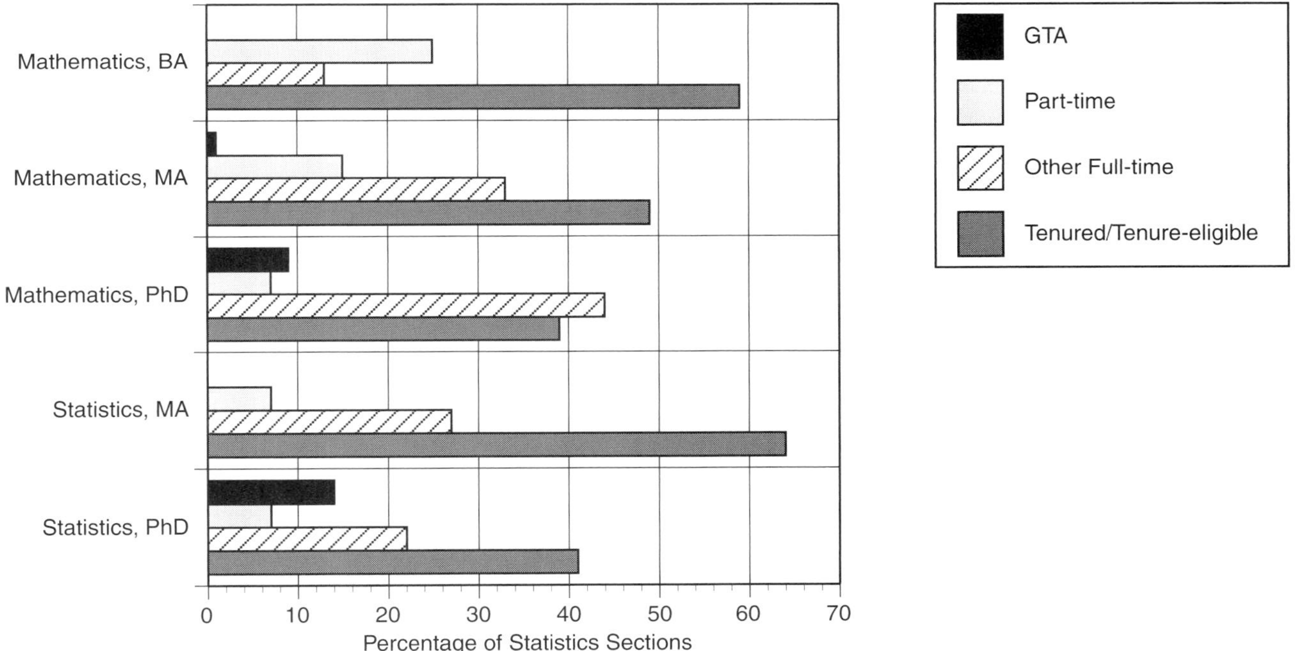

FIGURE E.5.2 Percentage of statistics sections whose instructors were tenured/tenure-eligible faculty, other full-time faculty, part-time faculty, and graduate teaching assistants (GTA), by type of mathematics or statistics department in fall 2005.

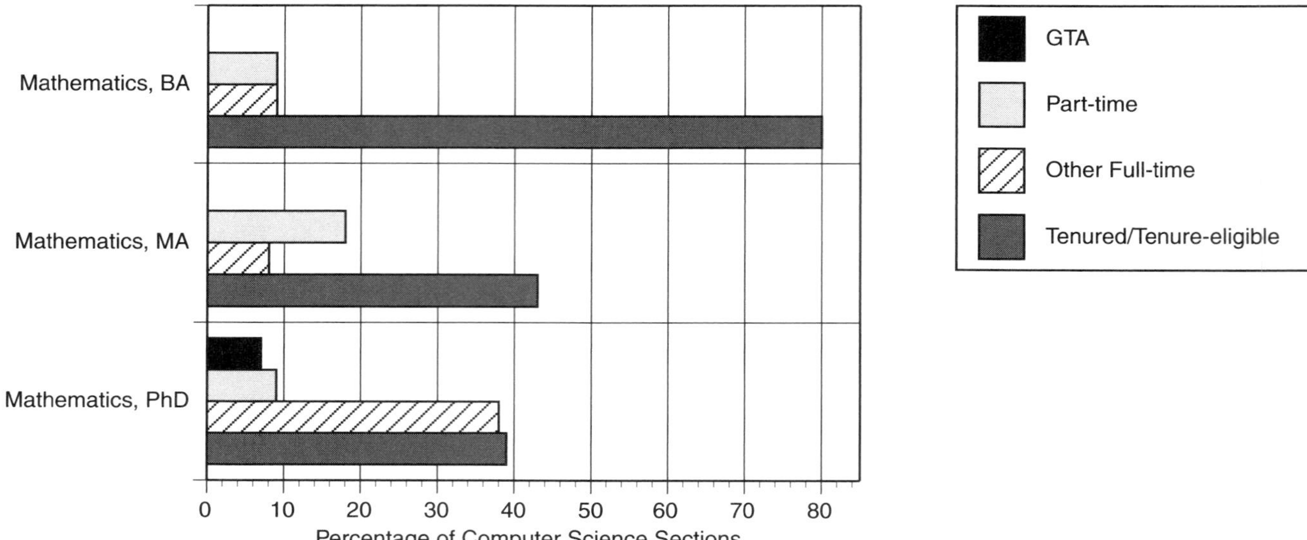

FIGURE E.5.3 Percentage of computer science sections taught in mathematics departments whose instructors were tenured/tenure-eligible faculty, other full-time faculty, part-time faculty, and graduate teaching assistants (GTA), by type of mathematics department in fall 2005. (Percentages do not sum to 100% due to "unknown" instructor percentages.)

TABLE E.6 Number of sections, not including distance learning, of precollege-level courses in mathematics departments taught by various types of instructor, by type of department in fall 2005, with fall 2000 figures in parentheses. (CBMS2000 data from Table E.13.)

	Number of precollege-level sections taught by						
	Tenured/ tenure-eligible	Other full-time (total)	Other full-time (doctoral)	Part-time	GTA	Ukn	**Total sections**
Mathematics Departments							
Univ (PhD)	29 (25)	312 (216)	34 (na)	579 (618)	376 (482)	66 (152)	**1363** **(1493)**
Univ (MA)	55 (120)	491 (475)	43 (na)	616 (807)	641 (221)	99 (149)	**1902** **(1772)**
Coll (BA)	576 (1387)	980 (698)	209 (na)	2091 (1829)	23 (26)	192 (448)	**3862** **(4388)**
Total	**660** **(1532)**	**1783** **(1389)**	**286** **(na)**	**3286** **(3254)**	**1040** **(729)**	**357** **(749)**	**7126** **(7653)**

Note: Round-off may make row and column sums seem inaccurate.

TABLE E.7 Number of sections (excluding distance learning) of introductory-level courses (including precalculus) in mathematics departments taught by various types of instructors, by type of department in fall 2005, with fall 2000 figures in parentheses. (CBMS2000 data from Table E.14.)

	Number of introductory-level sections taught by						
	Tenured/ tenure-eligible	Other full-time (total)	Other full-time (doctoral)	Part-time	GTA	Ukn	**Total sections**
Mathematics Departments							
Univ (PhD)	588 (683)	1457 (1159)	341 (na)	1176 (1261)	1902 (1714)	394 (215)	**5517** **(5032)**
Univ (MA)	1849 (2007)	1373 (1747)	197 (na)	1657 (1760)	295 (419)	369 (573)	**5543** **(6506)**
Coll (BA)	4079 (4397)	2385 (1407)	423 (na)	2998 (2676)	0 (0)	432 (507)	**9895** **(8987)**
Total	**6517** **(7087)**	**5215** **(4313)**	**960** **(na)**	**5831** **(5697)**	**2196** **(2133)**	**1196** **(1295)**	**20955** **(20525)**

Note: Round-off may make row and column sums seem inaccurate.

TABLE E.8 Number of sections (excluding distance learning) of calculus-level courses in mathematics departments taught by various types of instructor, by type of department in fall 2005, with fall 2000 figures in parentheses. (CBMS2000 data from Table E.15.)

	Number of calculus-level sections taught by						
	Tenured/ tenure-eligible	Other full-time (total)	Other full-time (doctoral)	Part-time	GTA	Ukn	**Total sections**
Mathematics Departments							
Univ (PhD)	3199 (3522)	1860 (1134)	1155 (na)	726 (762)	1261 (1087)	650 (263)	**7696** **(6768)**
Univ (MA)	2196 (3053)	375 (614)	159 (na)	402 (652)	16 (42)	249 (190)	**3237** **(4551)**
Coll (BA)	5754 (4854)	900 (820)	526 (na)	520 (409)	107 (0)	108 (355)	**7388** **(6438)**
Total	**11149** **(11429)**	**3135** **(2568)**	**1841** **(na)**	**1648** **(1823)**	**1384** **(1129)**	**1006** **(808)**	**18321** **(17757)**

TABLE E.9 Number of sections (excluding distance learning) of elementary level statistics taught in mathematics departments and statistics departments, by type of instructor and type of department in fall 2005 with fall 2000 figures in parentheses. (CBMS2000 data from Table E.16.)

Elementary Statistics	Number of elementary-level statistics sections taught by						**Total sections**
	Tenured/ tenure-eligible	Other full-time (total)	Other full-time (doctoral)	Part-time	GTA	Ukn	
Mathematics Departments							
Univ (PhD)	145 (307)	219 (130)	73 (na)	104 (157)	136 (198)	25 (35)	**629** **(827)**
Univ (MA)	441 (589)	185 (146)	34 (na)	250 (195)	15 (20)	34 (114)	**924** **(1064)**
Coll (BA)	1738 (1087)	366 (402)	90 (na)	987 (691)	0 (0)	100 (192)	**3191** **(2372)**
Total Math Depts	**2324** **(1983)**	**770** **(678)**	**197** **(na)**	**1341** **(1043)**	**151** **(218)**	**159** **(341)**	**4744** **(4263)**
Statistics Departments							
Univ (PhD)	144 (196)	111 (104)	60 (na)	88 (174)	172 (254)	180 (58)	**696** **(786)**
Univ (MA)	80 (51)	75 (23)	22 (na)	24 (9)	0 (11)	7 (29)	**186** **(123)**
Total Stat Depts	**224** **(247)**	**186** **(127)**	**82** **(na)**	**112** **(183)**	**172** **(265)**	**187** **(87)**	**882** **(909)**

Note: Round-off may make row and column sums seem inaccurate.

TABLE E.10 Number of sections (excluding distance learning) of lower-level computer science taught in mathematics departments, by type instructor and type of department in fall 2005, with fall 2000 figures in parentheses. (CBMS2000 data from Table E.17.)

	Number of lower-level computer science sections taught by						
	Tenured/ tenure-eligible	Other full-time (total)	Other full-time (doctoral)	Part-time	GTA	Ukn	**Total sections**
Mathematics Departments							
Univ (PhD)	31 (41)	44 (26)	24 (na)	10 (8)	14 (6)	15 (11)	**114** **(92)**
Univ (MA)	187 (559)	50 (204)	0 (na)	127 (677)	0 (0)	149 (113)	**512** **(1553)**
Coll (BA)	1199 (1162)	168 (549)	55 (na)	256 (504)	0 (12)	6 (330)	**1629** **(2557)**
Total Mathematics Depts	**1416** **(1762)**	**262** **(779)**	**79** **(na)**	**393** **(1189)**	**14** **(18)**	**169** **(454)**	**2254** **(4202)**

Note: Round-off may make row and column sums seem inaccurate.

TABLE E.11 Number of sections (excluding distance learning) of middle-level computer science taught in mathematics departments, by type of instructor and type of department in fall 2005, with fall 2000 figures in parentheses. (CBMS2000 data from Table E.18.)

	Number of middle-level computer science sections taught by						
	Tenured/ tenure-eligible	Other full-time (total)	Other full-time (doctoral)	Part-time	GTA	Ukn	**Total sections**
Mathematics Departments							
Univ (PhD)	19 (12)	36 (8)	19 (na)	3 (0)	3 (0)	0 (4)	**61** **(24)**
Univ (MA)	72 (286)	11 (27)	0 (na)	6 (106)	0 (0)	33 (46)	**121** **(465)**
Coll (BA)	613 (422)	98 (93)	70 (na)	6 (65)	0 (0)	22 (10)	**739** **(590)**
Total Math Depts	**703** **(720)**	**145** **(128)**	**89** **(na)**	**15** **(171)**	**3** **(0)**	**55** **(60)**	**921** **(1079)**

Note: Round-off may make row and column sums seem inaccurate.

TABLE E.12 Number of sections of advanced mathematics (including operations research) and statistics courses in mathematics departments, and number of sections of advanced statistics courses in statistics departments, taught by tenured and tenure-eligible (TTE) faculty, and total number of advanced level sections, by type of department in fall 2005. (Data for fall 2000 are not available.)

Mathematics Departments	Sections taught by TTE	Total sections	**Statistics Departments**	Sections taught by TTE	Total sections
Advanced mathematics courses					
Univ (PhD)	2184	2625			
Univ (MA)	1382	1622			
Coll (BA)	2941	3507			
Total advanced mathematics	6506	7754			
Advanced statistics courses			**Advanced statistics courses**		
Univ (PhD)	434	869	Univ (PhD)	343	499
Univ (MA)	359	714	Univ (MA)	140	156
Coll (BA)	604	771			
Total advanced statistics	1398	2354	Total advanced statistics	483	654
Total all advanced courses	**7904**	**10108**	**Total all advanced courses**	**483**	**654**

Note: Round-off may make row and column sums seem inaccurate.

Tables E.13 and E.14: Data on section sizes

Table E.13 summarizes data on average section sizes for a wide array of courses. Except in upper-level mathematics and statistics courses, average section size declined between fall 2000 and fall 2005. The Mathematical Association of America (MAA) has recommended 30 as the appropriate maximum class size in undergraduate mathematics [MAAGuidelines], and in fall 2005, national average section sizes were somewhat above that recommended limit. In particular, section sizes in doctoral departments often substantially exceeded that MAA guideline.

After the publication of CBMS2000, some doctoral department chairs asked for data on the average recitation size for calculus courses that are taught in lecture/recitation mode. CBMS2000 could provide only very rough estimates, but those estimates were good enough to convince several deans to add GTA slots to their doctoral mathematics departments. CBMS2005 collected better data on recitation sizes in various calculus courses and in elementary statistics courses, and these data are presented by type of department in Table E.13.

TABLE E.13 Average section size (excluding distance learning) for undergraduate mathematics, statistics, and computer science courses in mathematics and statistics departments, by level of course and type of department in fall 2005, with fall 2000 data in parentheses. Also, all departments' average section sizes from previous CBMS surveys. (CBMS2000 data from Table E.11.)

	Average section size Fall 2005 (2000)								
	Mathematics Depts			Statistics Depts					
	Univ (PhD)	Univ (MA)	Coll (BA)	Univ (PhD)	Univ (MA)	**All Depts 1990**	**All Depts 1995**	**All Depts 2000**	**All Depts 2005**
Mathematics courses									
Precollege	40 (39)	31 (33)	22 (23)			31	31	29	28
Introductory (incl. Precalc)	48 (51)	34 (35)	25 (26)			35	34	35	33
Calculus	45 (45)	27 (29)	21 (21)			35	31	32	32
Advanced Mathematics	20 (18)	15 (12)	10 (10)			16	12	13	14
Statistics courses									
Elementary Statistics	47 (46)	34 (33)	26 (27)	60 (58)	63 (65)	37	38	37	35
Upper Statistics	17 (21)	13 (19)	13 (15)	40 (36)	22 (25)	24	19	22	19
CS courses									
Lower CS	25 (50)	22 (21)	18 (20)	16 (13)	66 (58)	24	22	22	19
Middle CS	19 (39)	8 (16)	8 (16)	48 (na)	16 (90)	15	14	22	9
Upper CS	15 (21)	8 (12)	7 (10)	0 (na)	0 (30)	14	12	11	8

TABLE E.14 Average recitation size in Mainstream Calculus I and II and other Calculus I courses and in Elementary Statistics courses that are taught using lecture/recitation method, by type of department in fall 2005. Distance-learning sections are not included. (A calculus course is "mainstream" if it leads to the usual upper-division mathematical sciences courses.)

For Lecture/Recitation Courses	Average recitation section size		
	Univ (PhD)	Univ (MA)	College (BA)
Calculus Courses			
Mainstream Calculus I	28	19	21
Mainstream Calculus II	26	20	15
Other Calculus I	29	na	na
Elementary Statistics			
in Mathematics Depts	30	32	22
in Statistics Depts	32	19	na

Chapter 4

Faculty Demographics in Mathematical Sciences Departments of Four-Year Colleges and Universities

Introduction

In this chapter we consider data on the number, gender, age, and race/ethnicity of mathematics faculty in doctoral-level, masters-level, and bachelors-level mathematics departments, and also in doctoral-level statistics departments. The same topics were presented in Chapter 1 tables for the profession as a whole. In this chapter, we will show how faculty demographics differed among various types of departments, grouped by the highest degree offered by the department. So that the discussion can be relatively self-contained, we repeat some demographic data from Chapter 1.

- Table S.14 in Chapter 1 showed that there was an 11% increase in the total number of full-time faculty in mathematics departments (all levels combined) from 2000 to 2005. Table S.17 showed that the components of that increase were a 1% decrease in the total number of tenured faculty, coupled with a 33% increase in the number of tenure-eligible faculty, and a 31% increase in other full-time (OFT) faculty. The increase in OFT faculty was due in part to the increasing number of postdoctoral positions. In doctoral statistics departments, the total number of full-time faculty grew by 17%, the number of tenured faculty grew by 6%, the number of tenure-eligible faculty grew by 31%, and the number of OFT faculty expanded by 65%. In this chapter, Table F.1 breaks this data down by level of department.
- Table S.14 in Chapter 1 showed that the total number of part-time mathematics faculty in 2005 was about 10% below the high levels observed in fall 2000. Table F.1 shows that the decline was not uniform across all types of departments; declines of 25% and 20% in doctoral and masters-level departments, respectively, were coupled with a 1% increase in bachelors-level departments. In doctoral statistics departments there was a 10% increase in part-time faculty.
- Table S.17 in Chapter 1 showed that the percentage of women among all tenured faculty in four-year college and university mathematics departments rose three percentage points, from 15% in fall 2000 to 18% in fall 2005. Tables F.1, F.2, and F.3 give breakdowns in various categories of faculty in different types of departments. From these tables we see that the percentage of women among tenured faculty in doctoral-level mathematics departments rose from 7% to 9%, while the percentage of women among tenured faculty in bachelors-level departments rose from 20% to 24%. Doctoral statistics departments continued to show substantial growth in the numbers and percentages of women, especially in tenure-eligible positions.
- Table F.4 shows that the average ages of both tenured men and tenured women were up slightly in each type of mathematics department in fall 2005, compared to fall 2000, while Table S.19 shows that in doctoral statistics departments, the average age of tenured and tenure-eligible female faculty was down.
- Table F.5 shows that some increase in race/ethnicity diversity was observed from 2000 to 2005. In fall 2005, 80% of the total full-time mathematics faculty was classified as "White, non-Hispanic". That percentage varied by only a few points between mathematics departments of different types. Table F.6 shows the race/ethnicity breakdown of part-time faculty.

In the text that follows this introduction, differences in the trends in the various levels of departments will be explored in detail.

Data sources and notes on the tables

Each fall, the Joint Data Committee (JDC) of the AMS-ASA-IMS-MAA-SIAM conducts national surveys that include faculty demographic information. In previous CBMS survey years (2000, 1995, 1990, etc.) the CBMS survey has asked department chairs to provide essentially the same demographic information on the CBMS questionnaires. After the CBMS survey concluded in fall 2000, there were enough complaints about the multiple surveying that the JDC and the CBMS2005 committee agreed to use JDC data as the basis for faculty demographics tables in the CBMS2005 report. In addition to simplifying the CBMS questionnaires, this decision allows readers to compare fall 2005 data with annually published findings of the JDC. These JDC reports appear annually in the *Notices of the American Mathematical Society* and are available online at http://www.ams.org/employment/surveyreports.html.

The methodology of the JDC Annual Surveys differs from that of the CBMS surveys. In JDC surveys, all of the doctoral mathematics and statistics departments are surveyed, while in the CBMS surveys, the doctoral departments are part of a universe from which a random, stratified sample is drawn. Both the JDC's Annual Survey and the CBMS surveys use a stratified random sample of bachelors-level and masters-level institutions. The doctoral statistics departments surveyed by the JDC's Annual Survey include some departments that do not have undergraduate statistics programs, and such departments were removed from the analysis that appears in CBMS2005.

As noted in earlier chapters, there was a reclassification of certain masters-level mathematics departments by the AMS between the 2000 and 2005 surveys, with about 40 departments being reclassified as bachelors departments. Both the CBMS2005 survey and the JDC survey in fall 2005 used the new classification scheme when drawing their random samples of masters and bachelors mathematics departments, and this alone would account for some of the declines in enrollments, degrees granted, and faculty numbers that were detected among masters-level mathematics departments by the 2005 CBMS and JDC surveys, and for some of the corresponding growth among bachelors-level departments.

In each table in this chapter we have chosen the most appropriate comparison data for fall 2000. In most cases that data is the JDC's Annual Survey data from fall 2000, but in some cases it is CBMS2000 data. Sources of comparison data are clearly identified. Because the JDC's Annual Survey does not include masters-level statistics departments, data on faculty demographics in those departments (about 10 in number) do not appear in this CBMS2005 report even though such data did appear in CBMS2000. Consequently, we take special care to refer to "doctoral statistics departments" when reporting demographic data for fall 2005 in order to remind readers of that fact. This contrasts with the situation in other chapters of this CBMS2005 survey which include, for example, enrollment and degree-granted data for both masters- and doctoral-level statistics departments.

The JDC survey defined "full-time faculty" as "faculty who are full-time employees in the institution and at least half-time in the department" and then partitioned full-time faculty into four disjoint groups: tenured, tenure-eligible, postdoctoral (defined below in the section "Increases in numbers of other full-time faculty"), and other full-time. In order to make the classification of faculty used in Chapter 4 consistent with the terminology used in the remainder of this report and in previous CBMS reports, we have combined the two JDC questionnaire categories, "postdoctoral" and "other full-time", to make the CBMS2005 category "other full-time" (OFT). Consequently, *in this CBMS report, the term "other full-time faculty" means "all full-time faculty who are neither tenured nor tenure-eligible."* Therefore, when comparing the data in CBMS2005 to data in the JDC's Annual Survey publications, readers should keep in mind that beginning with the 2003 Annual Survey, the designation "OFT" in the JDC's Annual Survey does not include postdoctoral appointments, as it does in this, and in past, CBMS reports. In order to maintain comparability with previous CBMS surveys, and so that future CBMS reports can track changes in this growing subcategory of OFT faculty, in this chapter of the CBMS2005 report, the numbers of postdoctoral faculty are included in the OFT faculty column and also are broken out as separate columns.

Finally, a word of warning may be in order about the marginal totals in this chapter's tables. Table entries are rounded to the nearest integer, and the sum of rounded numbers is not always equal to the rounded sum.

Number of tenured and tenure-eligible faculty

From Tables S.14 and S.15, and Figure S.14.1, we see that the total number of full-time faculty in four-year college and university mathematics departments increased 11%, from 19,779 in 2000 to 21,885 in 2005. Table S.17 shows that across all types of departments, the total number of tenured full-time mathematics faculty decreased by 1%, the number of tenure-eligible full-time mathematics faculty increased by 33%, and the total number of tenured and tenure-eligible full time faculty, combined, increased by 6%. From Table F.1, where data are broken down by the level of the department, we see that most of this growth took place in bachelors-level departments, where the numbers of both tenured and tenure-eligible full-time faculty increased. In both doctoral-level and masters-level mathematics departments, the numbers of tenured faculty decreased, and the numbers of tenure-eligible faculty increased, with a net loss in the numbers of tenured and tenure-eligible faculty combined. In every category in Table F.1, the number of doctoral tenure-eligible faculty increased from 2000 to 2005.

In bachelors-level mathematics departments, the total number of tenured faculty rose 17%, from 4,817 in 2000 to 5,612 in 2005, and the total number of tenure-eligible faculty rose 52%, from 1,596 to 2,429. The AMS reclassification, mentioned above, that shifted some masters departments into the bachelors category would account for some of that increase in bachelors-level faculty numbers. However, with such a substantial change in the total number of faculty in bachelors-level mathematics departments, there is some concern that these estimates may be overestimates. Such concerns are based on the size of the standard error in the total number of full-time faculty in the fall 2005 survey (which was 595, more than double the standard error in the Third Report of the 2004 Annual Survey) and on what seem to be substantial differences between the 2005 survey estimates and the corresponding estimates from the five Annual Surveys between 2000 and 2004. For example, the JDC's 2005 Annual Survey estimated that there were 4,697 doctoral tenured faculty in bachelors-level mathematics departments, while the average number reported in the previous five annual JDC surveys was 4,053 (with a standard deviation of 102). Subsequent Annual Surveys should show whether the gains in bachelors-level departments in tenured and tenure-eligible faculty were as great as estimated in the 2005 Annual Survey.

In doctoral-level and masters-level mathematics departments, the number of tenured doctoral faculty decreased, and the number of tenure-eligible doctoral faculty increased. The total number of tenured faculty decreased 6% in doctoral-level mathematics departments, from 5,022 in 2000 to 4,719 in 2005, and it decreased 18% in masters-level mathematics departments, from 3,120 in 2000 to 2,544 in 2005. (Some of the decline at the masters level might be due to the reclassification mentioned above.) The number of tenure-eligible faculty increased 13% in doctoral-level mathematics departments, from 828 in 2000 to 933 in 2005, and it increased 18% in masters-level mathematics departments, from 863 in 2000 to 1,019 in 2005.

In doctoral statistics departments, the total full-time faculty increased 17%, from 808 in 2000 to 946 in 2005; both the number of tenured and the number of tenure-eligible doctoral full-time faculty increased in doctoral statistics departments from 2000 to 2005 (increases of 6% and 31%, respectively).

Increases in numbers of other full-time faculty

Table S.17 shows that the number of OFT faculty (defined as all full-time faculty who are neither tenured nor tenure-eligible) in four-year college and university mathematics departments rose 31%, from 3,533 in 2000 to 4,629 in 2005, and the finer breakdown of Table F.1 shows that the number of OFT faculty was up in 2005 over 2000 for every category of the table. In doctoral statistics departments, Tables S.17 and Table F.1 show that the number of OFT faculty increased 65%, from 99 in 2000 to 163 in 2005.

Nationally, there were many types of OFT appointments in fall 2005, some intended as research experiences and others carrying heavy teaching assignments. Starting in 2003, the JDC's Annual Survey has broken out the number of postdoctoral appointments (defined as "temporary positions primarily intended to provide an opportunity to extend graduate training or to further research experience") from the number of OFT faculty in its annual Third Report. These annual JDC reports show that there was an increase in the number of postdoctoral appointments from 2003 to 2005. When comparing the data in this CBMS report to that in the Annual JDC Survey, the reader is reminded that beginning with the 2003 Annual Survey, the designation "OFT" does not include postdoctoral appointments, while it does in this and other CBMS reports.

Numbers of part-time faculty

From Table S.14 we see that the total number of part-time faculty in four-year college and university mathematics departments in 2005 was 6,536, a 10% decrease from the 7,301 observed in 2000, but still above the 5,399 observed in 1995 (see Figures S.14.2 and S.14.3). Using Table F.1 to break down part-time faculty by type of department (doctoral-level, masters-level, and bachelors-level), and by doctoral and non-doctoral part-time faculty, we observe that the number of part-time faculty increased slightly in the bachelors-level group from 2000 to 2005, but decreased in the masters-level and doctoral-level groups (by 20% and 25%, respectively). The decrease in the number of part-time faculty in the doctoral-level groups was particularly large for non-doctoral part-time faculty (down 31%).

There was a different trend in the doctoral statistics departments (see Figure S.14.5). The number of part-time statistics faculty increased to 112 in 2005 from 102 in 2000; there were 125 part-time statistics faculty in 1995. Table F.1 shows that the increase in part-time faculty in doctoral statistics departments from 2000 to 2005 was due to an increase in the number of non-doctoral part-time faculty.

Non-doctoral faculty

The numbers of non-doctoral full-time faculty generally increased from 2000 to 2005 in four-year mathematics departments. In doctoral-level mathematics departments, the total number of non-doctoral full-time faculty increased 43%, from 484 in 2000 (7% of all full-time faculty) to 691 in 2005 (9% of all full-time faculty). In masters-level mathematics departments, the total number of non-doctoral faculty was up 9%, from 844 in 2000 to 921 in 2005. Were it not for the reclassification mentioned in an earlier section of this chapter, the numbers for masters-level departments might have been even higher. In bachelors-level mathematics departments, the number of non-doctoral faculty was up 22%, from 1,812 (24% of full-time faculty) in 2000 to 2,203 (23% of full-time faculty) in 2005. In doctoral-level statistics departments, non-doctoral faculty were almost exclusively in non-tenure-eligible positions, which increased from 12 in 2000 to 30 in 2005. While the increases in non-doctoral faculty are large in percentage terms, Table F.1 shows that in 2005 only about 17% of all full-time faculty in mathematics departments fell into the non-doctoral category, while only about 3% of full-time faculty in doctoral statistics departments failed to have doctoral degrees.

Gender

According to Joint Data Committee publications, between 2001 and 2005 women received about 30% of all mathematical sciences Ph.D. degrees each year, a percentage that is historically high and that is almost double the percentage of women among tenured mathematical sciences faculty in the U.S. Consequently it is no surprise that women continued to increase in numbers and percentages in most categories of faculty in four-year mathematics and statistics departments between 2000 and 2005. Table S.17 shows that the combined total number of female full-time mathematics faculty in four-year mathematics departments increased by about 30%, from 4,346 in 2000 to 5,641 in 2005. From 2000 to 2005 there were gains in the percentage of women in all faculty categories, except among tenure-eligible faculty, a category in which the percentage of women remained unchanged at 29%, essentially mirroring the percentage of women among new Ph.D. recipients. More specifically, in fall 2000, women comprised 22% of the full-time faculty, 15% of the tenured faculty, 29% of the tenure-eligible faculty, and 41% of the other full-time faculty. In fall 2005, women were 26% of the total full-time faculty, 18% of the tenured faculty, 29% of the tenure-eligible faculty, and 44% of the other full-time faculty. In fall 2005, 23% of the postdoctoral faculty in mathematics were women. Figure S.17.1 displays the percentages of tenured women and of tenure-eligible women in the combined four-year mathematics departments and in the doctoral statistics departments in 2000 and 2005.

Tables F.1 and F.2 and Figure F.3.1 provide data on the percentages of women in different types of departments, and we observe some differences among the percentages of women in doctoral-level, masters-level, and bachelors-level mathematics departments. In terms of both numbers of women and percentages of women, there are generally more women in bachelors-level departments, followed by masters-level departments, with the doctoral mathematics departments having the fewest women. In both doctoral-level and masters-level departments there was a decline in the number of all tenured positions from 2000 to 2005. At the same time, in the doctoral-level mathematics departments, the number of tenured women increased 18% from 2000 to 2005, while the number of tenured men decreased 8%; in masters-level mathematics departments, the numbers of tenured men and of tenured women both declined. The numbers of tenure-eligible women, and of other full-time women, increased from 2000 to 2005 in both the doctoral-level and masters-level departments; the number of tenure-eligible women increased 36% in the doctoral-level departments and 22% in the masters-level departments. In 2005 in the doctoral-level mathematics departments, women were 19% of the postdocs, and women postdocs were 20% of the women who held other full-time positions, while male postdocs were 47% of the men who held other full-time positions. Hence, in 2005, the other full-time women in doctoral departments were less likely to be in research-related temporary positions than the men. This difference also was due to the fact that in 2005 in the doctoral-level departments 60% of the non-doctoral other full-time positions were held by women. In bachelors-level departments, the number of women in each category increased from 2000 to 2005; for example, the number of tenured women increased 41%, from 972 in 2000 to 1,373 in 2005. In 2005, an astonishing 85% of the 48 postdoctoral positions in bachelors-level departments were held by women.

In fall 2005, women comprised a higher percentage of the part-time faculty than of the full-time faculty. In the four-year mathematics groups combined, women held 39% of the part-time positions. The percentage of women among part-time faculty was highest (41%) in the bachelors-level departments. For comparison, CBMS2000 shows that in fall 2000, women were 38% of the (larger) total part-time mathematics faculty.

Doctoral statistics departments continue to show impressive growth in numbers and percentages of women. From Table S.17 and Table F.3 we see that the total number of full-time women in doctoral statistics departments increased 51%, from 140 in 2000 to 211 in 2005. In 2005 women made up 22% of the total full-time doctoral statistics faculty, 13% of the tenured faculty, 37% of the tenure-eligible faculty, and 40% of the other full-time faculty; in 2000 these percentages were 17%, 9%, 34%, and 42%, respectively. In 2005 women were 29% of the part-time faculty (they were 28% of part-time faculty in 2000). The fact that women held 37% of the tenure-eligible positions in doctoral statistics departments is likely to lead to even greater numbers and percentages of tenured women in doctoral statistics departments in the future.

It is interesting to compare the percentages of women in doctoral statistics departments to those in doctoral mathematics departments. In doctoral-level mathematics departments in 2005, women comprised 18% of the total full-time faculty, 9% of the tenured faculty, 24% of the tenure-eligible faculty, and 19% of the postdocs; each of these percentages was lower than the corresponding percentages of women in doctoral statistics departments. The difference in the percentage of women among tenure-eligible faculty (37% in doctoral statistics departments and 24% in doctoral mathematics departments) is particularly striking. Indeed, as Figure F.3.1 demonstrates, the percentage of tenure-eligible women was greater in doctoral statistics departments than in any of the mathematics groups.

TABLE F.1 Number of faculty, and of female faculty (F), of various types in mathematics departments and PhD statistics departments, by highest degree and type of department, in fall 2005. (Fall 2000 figures are in parentheses, and postdocs are included in other full-time (OFT) faculty totals.)

	Univ (PhD)					Univ (MA)					Coll (BA)				
	Tenured	Tenure-eligible	OFT	Post-docs	Part-time	Tenured	Tenure-eligible	OFT	Post-docs	Part-time	Tenured	Tenure-eligible	OFT	Post-docs	Part-time
Mathematics Depts															
Doctoral faculty	4699 (4998)	930 (824)	1381 (993)	760 (na)	412 (483)	2412 (2851)	990 (819)	268 (262)	5 (na)	383 (445)	4697 (4129)	2179 (1357)	516 (407)	48 (na)	837 (800)
Doctoral (F)	420 (355)	218 (161)	336 (205)	147 (na)	95 (111)	480 (513)	319 (258)	97 (76)	2 (na)	102 (95)	1080 (799)	614 (428)	166 (123)	41 (na)	210 (167)
Non-doctoral faculty	20 (24)	3 (4)	668 (456)	4 (na)	634 (916)	132 (269)	29 (44)	760 (531)	2 (na)	1477 (1877)	915 (688)	251 (239)	1037 (885)	0 (na)	2793 (2780)
Non-doctoral (F)	7 (6)	2 (1)	399 (262)	1 (na)	291 (407)	52 (95)	18 (18)	435 (311)	0 (na)	588 (747)	293 (173)	79 (89)	626 (472)	0 (na)	1294 (1280)
Total Mathematics	4719 (5022)	933 (828)	2049 (1449)	764 (na)	1046 (1399)	2544 (3120)	1019 (863)	1027 (793)	7 (na)	1860 (2322)	5612 (4817)	2429 (1596)	1553 (1292)	48 (na)	3630 (3580)
Total Mathematics (F)	427 (361)	220 (162)	735 (467)	148 (na)	386 (518)	532 (608)	337 (276)	532 (387)	2 (na)	689 (842)	1373 (972)	693 (517)	792 (595)	41 (na)	1503 (1447)
PhD Statistics Depts															
Doctoral faculty	603 (571)	178 (136)	133 (87)	51 (na)	76 (81)										
Doctoral (F)	79 (51)	66 (46)	46 (36)	16 (na)	16 (17)										
Non-doctoral faculty	1 (1)	1 (1)	30 (12)	0 (na)	36 (21)										
Non-doctoral (F)	0 (0)	0 (1)	20 (6)	0 (na)	17 (12)										
Total PhD Statistics	604 (572)	179 (137)	163 (99)	51 (na)	112 (102)										
Total PhD Statistics (F)	79 (51)	66 (47)	66 (42)	16 (na)	33 (29)										

TABLE F.2 Number of tenured, tenure-eligible, postdoctoral, and other full-time faculty in mathematics departments of four-year colleges and universities by gender and type of department in fall 2005 and 2000. (Note: Postdoctoral faculty are included in Other full-time totals.)

	Univ (PhD)				Univ (MA)				Coll (BA)				**Total**			
	Tenured	Tenure-eligible	Other full-time[1]	Post-docs[1]	Tenured	Tenure-eligible	Other full-time[1]	Post-docs[1]	Tenured	Tenure-eligible	Other full-time[1]	Post-docs[1]	Tenured	Tenure-eligible	Other full-time[1]	Post-docs[1]
Men, 2005	4292	713	1314	616	2011	682	495	4	4239	1737	761	8	10542	3132	2570	628
Women, 2005	427	220	735	148	532	337	532	2	1373	693	792	41	2332	1250	2059	191
Total, 2005	**4719**	**933**	**2049**	**764**	**2544**	**1019**	**1027**	**7**	**5612**	**2429**	**1553**	**48**	**12874**	**4382**	**4629**	**819**
Men, 2000	4661	667	982	na	2512	587	405	na	3845	1079	697	na	11018	2333	2084	na
Women, 2000	361	162	467	na	608	276	388	na	972	517	595	na	1941	955	1450	na
Total, 2000	**5022**	**828**	**1449**	**na**	**3120**	**863**	**793**	**na**	**4817**	**1596**	**1292**	**na**	**12959**	**3287**	**3533**	**na**

[1] A postdoctoral appointment is a temporary position primarily intended to provide an opportunity to extend graduate training or to further research experience. Postdoctoral faculty are included in the other-full-time-faculty totals throughout CBMS2005. This contrasts with publications of the Joint Data Committee since 2003, which list postdoctoral faculty as a category separate from other-full-time faculty. Before 2003, JDC data did not collect separate counts of postdoctoral faculty.

Note: Round-off may make marginal totals seem inaccurate.

TABLE F.3 Number of tenured, tenure-eligible, other full-time, and postdoctoral faculty in doctoral statistics departments, by gender, in fall 2005 and 2000. (Postdoctoral faculty are included in Other full-time faculty totals.)

	Doctoral Statistics Departments			
	Tenured	Tenure-eligible	Other full-time	Postdocs[1]
Men, 2005	525	113	97	35
Women, 2005	79	66	66	16
Total, 2005	**604**	**179**	**163**	**51**
Men, 2000	521	90	57	na
Women, 2000	51	47	42	na
Total, 2000	**572**	**137**	**99**	**na**

[1] A postdoctoral appointment is a temporary position primarily intended to provide an opportunity to extend graduate training or to further research experience. Throughout CBMS2005, postdoctoral faculty are included in other full-time faculty totals. This contrasts with publications of the Joint Data Committee since 2003, which list postdoctoral faculty as a category separate from other full-time faculty. Before 2003, JDC data did not collect separate counts of postdoctoral faculty.

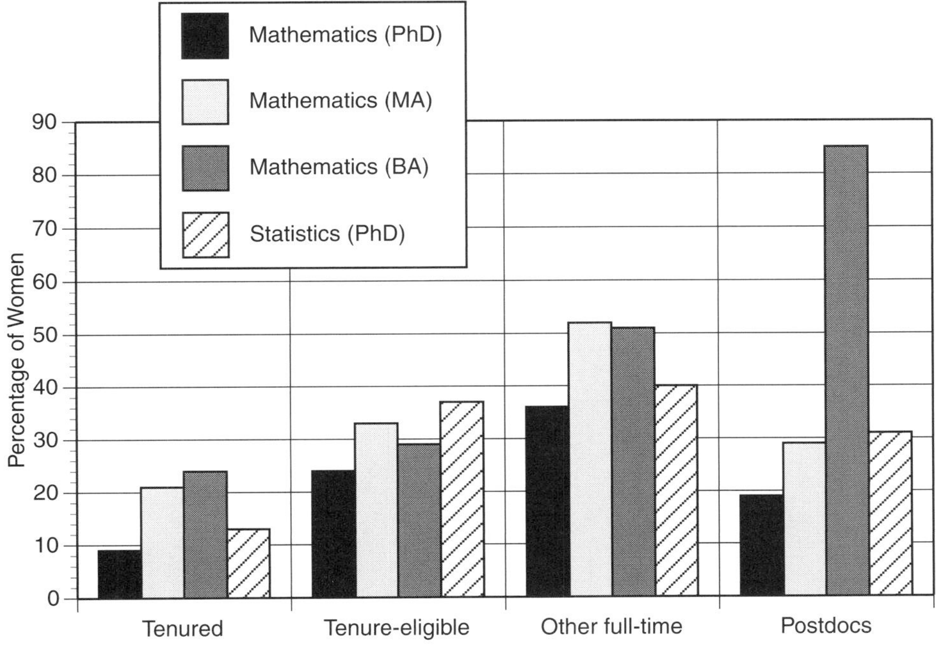

FIGURE F.3.1 Percentage of women in various faculty categories, by type of department, in fall 2005.

Age distribution

Table S.18 and Figure S.18.1 in Chapter 1 present the age distribution of tenured and tenure-eligible men and women in all four-year mathematics departments in fall 2005, and Table F.4 and Figures F.4.1, F.4.2, and F.4.3 display the finer breakdown of faculty ages by level of mathematics department. Table S.19 and Figure S.19.1 in Chapter 1 present the same information for doctoral statistics departments. The tables also show average ages within each type of department, and the percentages within each type of department total 100%, except for possible round-off errors.

Table F.4 can be used to compare the average ages of mathematics faculty in 2000 and 2005 for various categories of full-time faculty and different types of departments. The average age of both tenured men and tenured women was higher in 2005 than 2000 in each type of mathematics department. The age of tenure-eligible men and women was up noticeably in the bachelors-level departments (in 2000, men averaged 35.8 years and women averaged 36.8 years, while in 2005, men averaged 40.2 years and women averaged 38.9 years). Table S.19 shows that the average ages of men in doctoral statistics departments were about the same in 2005 as in 2000, but the average ages of women were lower: in 2000, tenured women averaged 48.3 and tenure-eligible women averaged 38.0, while in 2005, tenured women averaged 45.6 and tenure-eligible women averaged 33.2. Indeed, as Figures S.18.1 and S.19.1 show, the distribution of women was much more skewed toward younger women in doctoral statistics departments than in all four-year mathematics departments combined.

TABLE F.4 Percentage of tenured and tenure-eligible mathematics department faculty at four-year colleges and universities belonging to various age groups by type of department and gender in fall 2005.

	<30 %	30-34 %	35-39 %	40-44 %	45-49 %	50-54 %	55-59 %	60-64 %	65-69 %	>69 %	Average age 2000	Average age 2005
Univ (PhD)												
Tenured men	0	1	4	9	12	13	12	13	8	4	52.1	54.4
Tenured women	0	0	1	1	1	1	1	1	0	0	49.6	50.0
Tenure-eligible men	1	5	4	2	1	0	0	0	0	0	36.6	36.3
Tenure-eligible women	0	1	1	1	0	0	0	0	0	0	37.8	37.3
Total Univ (PhD)	**1**	**8**	**10**	**13**	**14**	**15**	**14**	**13**	**8**	**4**		
Univ (MA)												
Tenured men	0	0	4	6	11	10	9	10	4	2	53.1	53.8
Tenured women	0	0	2	1	3	2	2	1	1	1	49.2	52.1
Tenure-eligible men	2	6	7	3	1	1	1	0	0	0	37.5	38.3
Tenure-eligible women	1	3	3	2	1	1	0	0	0	0	38.8	38.7
Total Univ (MA)	**3**	**9**	**16**	**12**	**15**	**13**	**12**	**13**	**5**	**3**		
Coll (BA)												
Tenured men	0	1	4	8	7	8	10	10	3	1	52.7	52.9
Tenured women	0	1	2	4	2	4	3	2	0	0	47.3	49.6
Tenure-eligible men	1	6	6	3	2	1	1	0	0	0	35.8	40.2
Tenure-eligible women	1	3	2	1	1	1	0	0	0	0	36.8	38.9
Total Coll (BA)	**2**	**10**	**13**	**16**	**13**	**13**	**15**	**12**	**4**	**1**		

Note: 0 means less than half of 1%.

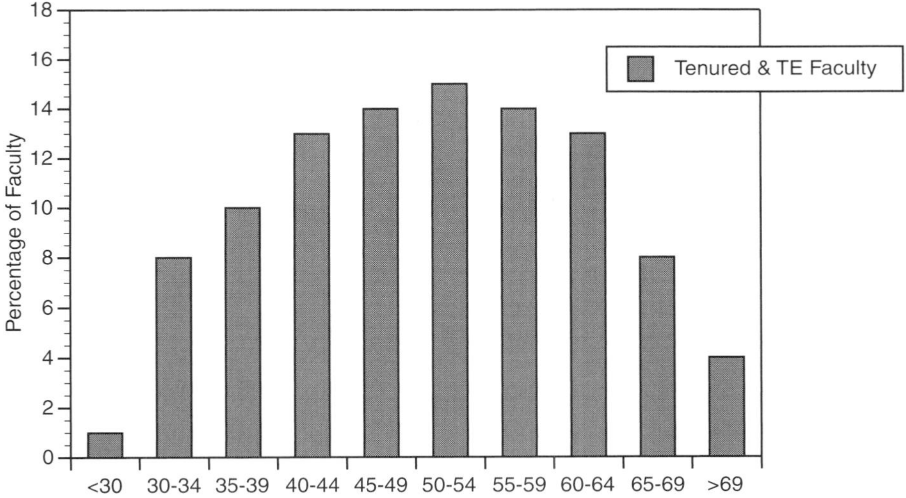

FIGURE F.4.1 Percentage of tenured and tenure-eligible faculty in doctoral mathematics departments in various age groups in fall 2005.

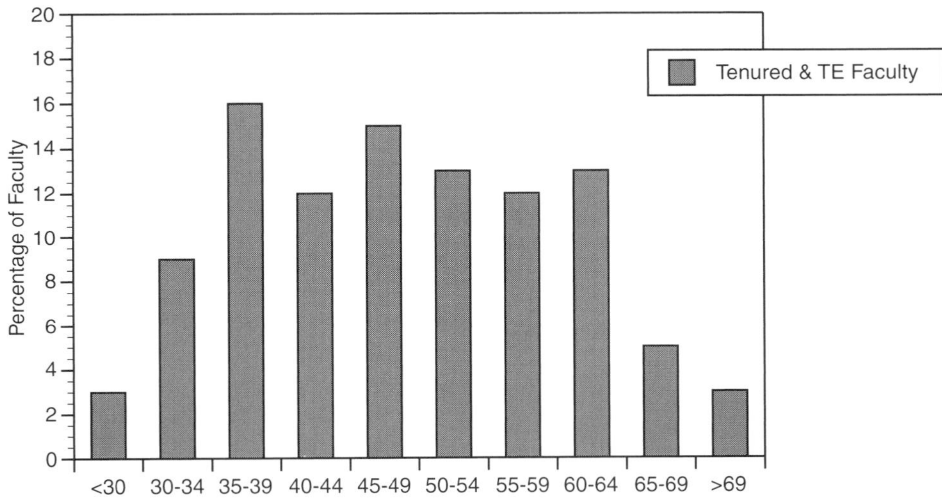

FIGURE F.4.2 Percentage of tenured and tenure-eligible faculty in masters-level mathematics departments belonging to various age groups in fall 2005.

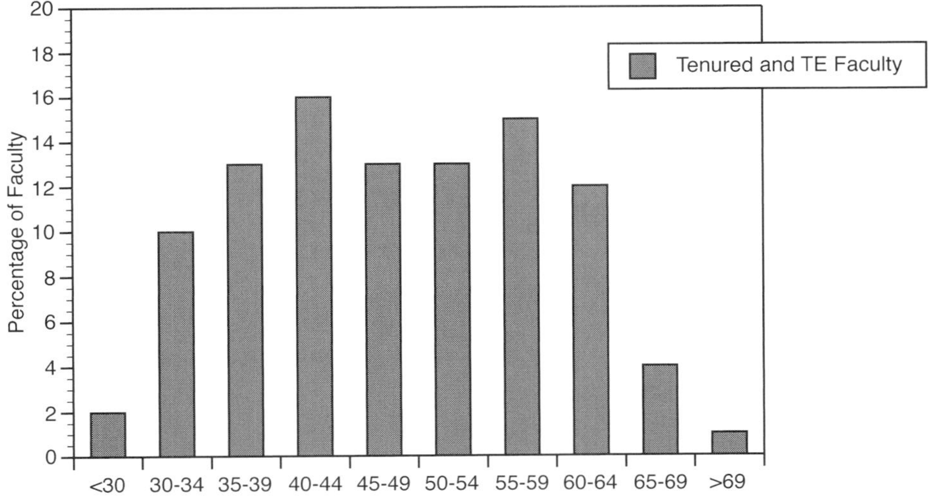

FIGURE F.4.3 Percentage of tenured and tenure-eligible faculty in bachelors-level mathematics departments belonging to various age groups in fall 2005.

Race, ethnicity, and gender

Table S.20 gives the percentages of faculty in fall 2005, by gender and in various racial/ethnic groups, for tenured, tenure-eligible, postdoctoral, and other full-time mathematics faculty in all types of mathematics departments combined. The comparison table for fall 2000 is Table SF.11 in CBMS2000.

Joint Data Committee surveys follow the federal pattern for racial and ethnic classification of faculty. However, in the text of this report, some of the more cumbersome federal classifications will be shortened. For example, "Mexican-American/Puerto Rican/other Hispanic" will be abbreviated to "Hispanic." Similarly, the federal classifications "Black, not Hispanic" and "White, not Hispanic" will be shortened to "Black" and "White" respectively, and "Asian/Native Hawaiian/ Pacific Islander" will be shortened to "Asian."

Generally, there was an increase in diversity in the racial/ethnic composition of mathematical sciences faculty between 2000 and 2005. Percentages of White faculty declined, and percentages of some other racial/ethnic groups increased slightly. Table S.20 shows that the overall percentages of full-time, Asian male and female mathematics faculty were up in 2005 compared to 2000, as was the percentage of Black female mathematics faculty. Percentages of White full-time mathematics faculty were all the same or lower in 2005 compared with 2000 except tenure-eligible men, which rose from 9% to 11%; the percentage of total White, male, full-time mathematics faculty was down from 63% in 2000 to 59% in 2005.

Table F.5 gives the finer breakdown of the racial, ethnic, and gender composition of the mathematics full-time faculty by type of department; it can be compared to Table F.6 of CBMS2000. For example, Table F.5 shows that in bachelors- and masters-level mathematics departments, the percentage of Asian full-time faculty rose between fall 2000 and fall 2005, and that in doctoral-level mathematics departments, the percentage of Asian, male, full-time faculty declined slightly. The percentage of Hispanic full-time mathematics faculty was up in 2005 over 2000, except in masters-level departments where the percentage of men decreased, while the percentage of women was unchanged from fall 2000 levels. The percentages of White, full-time faculty were down in 2005 from 2000 except in the doctoral-level mathematics departments, where the percentage of White, female faculty rose from 13% to 14%.

Table S.21 in Chapter 1 gives the analogous breakdown for full-time faculty in doctoral-level statistics departments in 2005; it may be compared to Table F.7 in CBMS2000. In doctoral-level statistics departments, the percentage of Asian full-time faculty was either down or the same from 2000 to 2005, with the percentage of all male, Asian, full-time faculty in doctoral-level statistics departments rising from 17% in 2000 to 18% in 2005. The percentage of Black faculty in doctoral statistics departments increased for both male and female faculty, and the same was true for male Hispanic faculty. The percentage of White, female faculty in doctoral-level statistics departments increased from 12% in 2000 to 16% in 2005, consistent with the growth in numbers of women in the doctoral-level statistics departments that was noted earlier in the chapter.

Table F.6 gives the fall 2005 percentages of faculty in various racial/ethnic groups for part-time faculty, broken down by gender, in each type of mathematics department and for doctoral-level statistics departments. The comparison table from CBMS2000 is Table F.8. From fall 2000 to fall 2005, there were decreasing percentages of White part-time faculty, both men and women, in all types of mathematics departments and in doctoral-level statistics departments, except for an increase in the percentage of White, female, part-time faculty in masters-level mathematics departments. The percentage of Black, part-time, female faculty was down in doctoral-level mathematics departments, but otherwise the percentages of Black faculty were up or unchanged from 2000 to 2005. Percentages of Hispanic part-time faculty were generally down in 2005 from 2000, except for increases in these percentages for bachelors-level mathematics part-time female faculty, and for doctoral-level statistics male part-time faculty. The percentage of Asian part-time faculty increased among men and women in doctoral-level and masters-level mathematics departments, increased among men in bachelors-level mathematics departments, and decreased among both men and women in doctoral statistics departments.

For a small percentage of the faculty, race and ethnicity data were listed as "unknown" by responding departments, and these faculty are listed as "unknown" in Tables F.5 and F.6.

TABLE F.5 Percentages of full-time faculty belonging to various ethnic groups, by gender and type of department, in fall 2005. Except for round-off, the percentages within each departmental type sum to 100%.

	Percentage of Full-time Faculty				
	Asian %	Black, not Hispanic %	Mexican American/ Puerto Rican/ other Hispanic %	White, not Hispanic %	Other/Unknown %
PhD Mathematics Departments					
All full-time men	12	1	2	66	1
All full-time women	3	0	1	14	0
MA Mathematics Departments					
All full-time men	10	3	2	54	2
All full-time women	4	1	2	22	1
BA Mathematics Departments					
All full-time men	6	2	2	57	3
All full-time women	3	1	1	25	2
PhD Statistics Departments					
All full-time men	18	1	1	55	2
All full-time women	7	1	0	16	1

Note: Zero means less than one-half of one percent.

Note: The column "Other/Unknown" includes the federal categories Native American/Alaskan Native and Native Hawaiian/Other Pacific Islander.

TABLE F.6 Percentages of part-time faculty belonging to various ethnic groups, by gender and type of department, in fall 2005. Except for round-off, the percentages within each departmental type sum to 100%.

	Percentage of Part-time Faculty				
	Asian %	Black, not Hispanic %	Mexican American/ Puerto Rican/ other Hispanic %	White, not Hispanic %	Other/ Unknown %
PhD Mathematics Departments					
All part-time men	4	2	0	50	6
All part-time women	3	0	0	31	2
MA Mathematics Departments					
All part-time men	3	2	2	46	7
All part-time women	2	3	1	33	3
BA Mathematics Departments					
All part-time men	3	3	2	44	7
All part-time women	1	2	1	31	6
PhD Statistics Departments					
All part-time men	11	2	1	44	12
All part-time women	1	0	0	23	5

Note: Zero means less than one-half of 1%.

Note: The column "Other/Unknown" includes the federal categories Native American/Alaskan Native and Native Hawaiian/Other Pacific Islander.

Chapter 5

First-Year Courses in Four-Year Colleges and Universities

Tables in this chapter further explore topics from Tables S.7 to S.13 in Chapter 1 and Tables E.2 to E.9 of Chapter 3, presenting details by type of department on certain first-year mathematics courses in four-year colleges and universities—their enrollments, their teachers, and how they were taught. Courses studied include a spectrum of introductory-level courses, several first-year calculus courses, and elementary statistics courses. Among introductory-level mathematics courses, the chapter focuses on:

a) two general education courses (with names such as Finite Mathematics and Mathematics for Liberal Arts) that are specifically designed for students fulfilling a general education requirement,
b) courses for pre-service elementary education teachers, and
c) the cluster of precalculus courses with names such as College Algebra, Trigonometry, Algebra and Trigonometry, and Elementary Functions.

First-year calculus courses are divided into "mainstream" and "non-mainstream" courses, where a calculus course is classified as "mainstream" if it typically leads to upper-division mathematical sciences courses. That definition has been used in almost all CBMS surveys, and before 2005, it was roughly true to say that mainstream calculus courses were typically designed for mathematics, engineering, and physical sciences majors. By fall 2005, that rough characterization was less and less accurate. With the increasing national emphasis on mathematical biology, there was a growing body of calculus courses specifically designed for students with biological interests that could fall into the "mainstream" classification. Whether a particular calculus course was classified as mainstream or non-mainstream was left up to responding departments, and based on calls and e-mails to the project directors in fall 2005, responding departments had few doubts about which calculus courses were mainstream and which were not. The final group of courses studied in this chapter are the elementary statistics courses, where the term "elementary" refers only to the fact that such courses do not have a calculus prerequisite. Most of these courses are also part of the curriculum of two-year colleges, and details about the courses in the two-year-college setting appear in Chapter 6.

Enrollments (Tables FY.2, FY.4, FY.6, FY.8, and FY.10 and Appendix I Tables A.1 and A.2)

- Table A.1 in Appendix I shows that combined enrollments in Finite Mathematics and Liberal Arts Mathematics, two general education courses, increased markedly between fall 1995 and fall 2005, growing from 133,000 in 1995 to 168,000 in fall 2000 and finally to 217,000 in fall 2005. That is a 63% increase over ten years, and in fall 2005 combined enrollment in these two general education courses exceeded the total enrollment in Mainstream Calculus I.
- Enrollments in first-year courses designed for pre-service elementary teachers rose between fall 1995 and fall 2000 and rose again by fall 2005. Table FY.2 shows an increase from roughly 59,000 in fall 1995 to about 72,000 in fall 2005, a 22% increase.
- Enrollments in the cluster of four precalculus courses listed in c) above were roughly 368,000 in fall 1995, grew to about 386,000 in fall 2000, and declined to 352,000 in fall 2005, ending the decade more than 9% below 1995 levels. See Table FY.2.
- Table A.2 in Appendix I shows that the combined enrollment in the Elementary Statistics course in mathematics and statistics departments (including distance-learning enrollments) grew from 132,000 in fall 1995 to 155,000 in fall 2000 and to 167,000 in fall 2005, an increase of about 27% between 1995 and 2005, with the rate of enrollment growth appearing to slow in the last five years of the decade. Mathematics departments taught almost three-quarters of the nation's Elementary Statistics. Tables FY.8 and FY.10 display the non-distance-learning enrollments in this course in fall 2005.

Who taught first-year courses? (Tables FY.1, FY.3, FY.5, FY.7, and FY.9)

CBMS1995 and CBMS2000 presented data on the type of instructors assigned to teach first-year courses in terms of *percentages of enrollments*, but those enrollment estimates relied on certain assumptions that made standard errors difficult to calculate. To allow standard error calculations in this report, CBMS2005 expresses its conclusions in terms of *percentages of sections*. Consequently, direct numerical comparisons between CBMS2005 and earlier CBMS studies are problematic. Even if one assumes that percentage of sections converts linearly into percentage of enrollments, a conservative approach to making comparisons suggests drawing only tentative conclusions.

In Chapter 5, as in previous CBMS surveys, tenured and tenure-eligible (TTE) faculty were combined into a single category. All other full-time faculty were put into the class called other full-time (OFT) faculty. To get a better picture of the mathematical qualifications of teachers in first-year courses, CBMS2005 subdivided the OFT faculty into those with doctoral degrees (OFT-doctoral) and those without doctorates. This was a new feature of CBMS2005. In order to maintain some degree of comparability with CBMS1995 and CBMS2000, tables in this chapter contain a column called "OFT (total)" as well as the column called "OFT (doctoral)."

- In fall 2005, about forty percent of introductory-level courses in bachelors- and masters-level departments were taught by TTE or OFT-doctoral faculty, compared to about 17% in doctoral departments. Doctoral departments assigned about a third of introductory-level courses to graduate teaching assistants (GTAs), meaning that the GTAs were the instructors of record in those courses. See Table FY.1.
- Doctoral departments used a combination of TTE and OFT-doctoral faculty to teach about half of their Mainstream Calculus I sections. In masters-level departments, the combined percentage was closer to 75%, and in bachelors-level departments it was about 85%.
- Table FY.1 of CBMS2000 shows that doctoral mathematics departments taught 62% of their Mainstream Calculus I enrollment using TTE faculty in fall 1995, and 50% in fall 2000. Table FY.3 in CBMS2005 shows that in fall 2005, doctoral mathematics departments used TTE faculty to teach 36% of their Mainstream Calculus I sections. With the usual caveat about comparing percentages of enrollment from CBMS2000 with percentages of sections in CBMS2005, Tables FY.1 in CBMS2000 and FY.3 in CBMS2005 suggest a marked trend in doctoral mathematics departments away from using TTE faculty in Calculus I.
- The percentage of Mainstream Calculus I sections taught by graduate teaching assistants (GTAs) in fall 2005 was only slightly lower than the percentage of enrollments in Mainstream Calculus I taught by GTAs in fall 2000, and this suggests that there was not much change in the use of GTAs to teach Mainstream Calculus I between 2000 and 2005. See Table FY.1 in CBMS2000 and CBMS2005.
- There appears to be a continuing trend among mathematics departments to shift the teaching of the Elementary Statistics course from TTE faculty to OFT faculty. In mathematics departments, the percentage of Elementary Statistics sections taught by TTE faculty was below the percentage of enrollment taught by TTE faculty in 1995. At the same time, among bachelors- and masters-level mathematics departments, the percentage of Elementary Statistics sections taught by OFT faculty in fall 2005 was more than double the percentage of enrollment in the same course taught by OFT faculty in fall 1995. Among doctoral mathematics departments, the fall 2005 percentage of sections taught by OFT faculty was almost four times as large as was the percentage of enrollment taught by OFT faculty in 1995. See Table FY.6 in CBMS2000 and Table FY.7 of this chapter.

How are first-year courses taught? (Tables FY.2, FY.4, FY.6, FY.8, and FY.10)

The CBMS1995 survey asked departments about the impact of the calculus reform movement on the way that their calculus courses were taught. In fall 1995, a meaningful question was "What percentage of your calculus sections are taught using a reform text?" By fall 2000, that question was no longer meaningful, with almost every publisher claiming to have incorporated calculus reform into every calculus text. To trace the continuing impact of calculus reform in fall 2000, the CBMS2000 survey focused attention on a spectrum of pedagogical methods that had come to be thought of as "reform methods". These were of two general types—those related to technology (the use of graphing calculators and computers), and those that were sometimes described as "humanistic pedagogies," e.g., the use of writing assignments and group projects. Tables FY.2, FY.4, FY.6, FY.8, and FY.10 continue that study and suggest some conclusions about the spread of reform pedagogies during the 1995–2005 decade, once again subject to the caveat that comparing percentages of enrollment in CBMS1995 and CBMS2000 with percentages of sections in CBMS2005 leads to tentative conclusions at best.

- In fall 2005, none of the four reform pedagogies were universal in Calculus I (whether the mainstream version, or non-mainstream). Graphing calculators were the most widely used reform pedagogy in Calculus I courses and were used about twice as widely in Calculus I as computer assignments. See Table FY.4.

- The percentage of Calculus I sections taught using writing assignments and group projects was generally below 20%, and they were mostly in the single-digit range among doctoral-level departments. This is consistent with findings of CBMS2000. See Table FY.4.
- In contrast to the situation in Calculus I, a markedly larger percentage of Elementary Statistics sections used computer assignments compared to graphing calculators. In addition, while the use of writing assignments and group projects seems to have declined among Elementary Statistics sections taught in mathematics departments, it apparently increased markedly in Elementary Statistics sections taught in doctoral statistics departments. See Tables FY.8 and FY.10.

Earlier CBMS studies did not examine the pedagogical methods used in introductory-level courses (such as College Algebra and Precalculus), so it is not possible to trace the spread of reform pedagogies over time in courses of that type. However, Table FY.2 does allow some comparisons between introductory-level and other first-year courses in fall 2005.

- The cluster of precalculus courses (namely College Algebra, Trigonometry, Algebra & Trigonometry (combined course), and Precalculus) resembled Mainstream Calculus I in pedagogical pattern, with graphing calculators being twice as commonly used as computer assignments, and with writing assignments and group projects trailing far behind.
- Writing assignments and group projects were used much more extensively in Mathematics for Elementary Teachers than in any other introductory-level course, while graphing calculators were used less.

A new question in CBMS2005 asked departments about the extent to which they used online resource systems in their first-year courses. The CBMS2005 questionnaires described these systems as online packages for generating and grading homework. In four-year colleges and universities, the percentage of first-year sections (i.e., introductory-level courses, Calculus I, or Elementary Statistics) using such systems was typically in the single digits in mathematics departments. By contrast, it was closer to twenty percent in Elementary Statistics courses taught in doctoral statistics departments.

In fall 2005, reform pedagogies had been more widely adopted in two-year college courses than in the same courses at four-year colleges and universities, often by wide margins. See Table TYE.10 of Chapter 6 for details about the use of reform pedagogies and online resource systems in courses taught in two-year colleges.

Special Note on Chapter 5 Estimates: As can be seen from the Appendix on standard errors, many of the estimates in Chapter 5 had large standard error values so that the values in the entire population might be quite different from the estimates given in Chapter 5 tables.

TABLE FY.1 Percentage of sections (excluding distance-learning sections) of certain introductory-level courses taught by various types of instructors in mathematics departments in fall 2005, by type of department. Also average section sizes.

	Percentage of sections taught by																		Average section size		
	Tenured/ tenure-eligible %			Other full-time (total) %			Other full-time (doctoral) %			Part-time %			Graduate teaching assistants %			Unknown %					
Course & Department Type	PhD	MA	BA	PhD	MA	BA	PhD	MA	BA	PhD	MA	BA	PhD	MA	BA	PhD	MA	BA	PhD	MA	BA
Mathematics for Liberal Arts	18	36	43	19	13	16	5	4	4	28	38	32	25	3	0	11	10	9	46	34	25
Finite Mathematics	17	49	31	32	28	14	7	4	4	12	17	55	23	0	0	16	6	0	74	34	23
Business Math (non-calculus)	14	30	36	20	23	30	9	5	11	21	41	32	43	2	0	2	3	3	47	34	26
Math for Elem Sch Teachers	19	45	59	38	24	24	10	2	3	22	24	12	14	1	0	6	6	6	29	27	22
College Algebra	4	24	34	25	36	31	3	5	3	21	26	29	44	6	0	6	7	5	46	41	27
Trigonometry	10	31	30	26	36	32	3	0	2	19	19	39	43	0	0	2	14	0	37	31	27
College Alg & Trig (combined)	6	26	61	45	8	29	10	2	8	19	36	11	29	30	0	1	0	0	57	28	25
Elem Functions, Precalculus	7	32	43	22	21	22	8	3	0	24	33	35	40	10	0	7	4	0	48	31	25
Intro to Math Modeling	25	36	11	75	14	78	38	0	22	0	50	11	0	0	0	0	0	0	81	31	20
Total All Intro Level Courses	**11**	**33**	**41**	**26**	**25**	**24**	**6**	**4**	**4**	**21**	**30**	**30**	**34**	**5**	**0**	**7**	**7**	**4**	**48**	**34**	**25**

Note: 0 means less than one half of 1%.

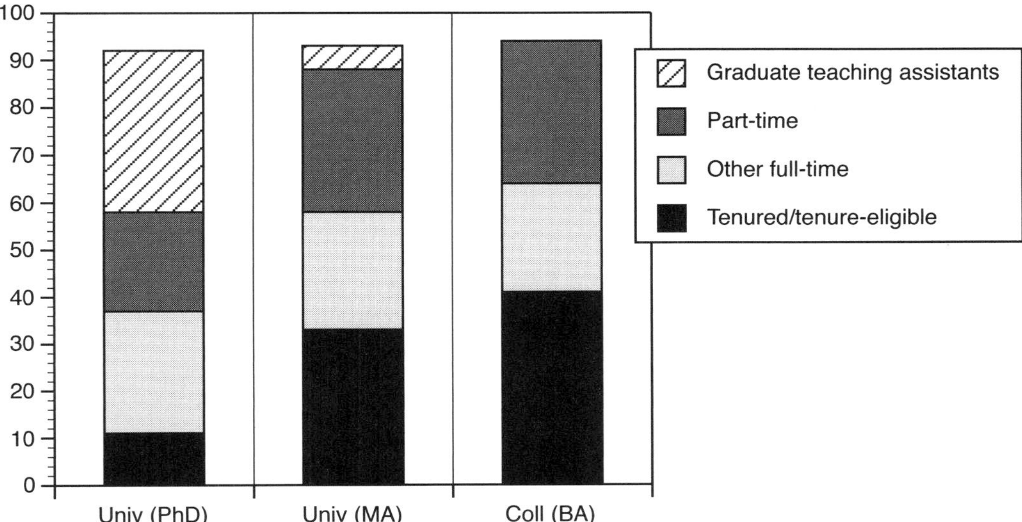

FIGURE FY.1.1 Percentage of sections (excluding distance-learning sections) in introductory-level mathematics courses (including College Algebra and Precalculus) taught in mathematics departments by various kinds of instructors in fall 2005, by type of department. (Deficits from 100% represent unknown instructors.)

TABLE FY.2 Percentage of sections (excluding distance-learning sections) in certain introductory-level courses taught using various reform methods in mathematics departments in fall 2005, by type of department. Also total enrollments (in 1000s) and average section size.

	Percentage of sections in certain Introductory Level courses taught using																				
	Graphing calculators %			Writing assignments %			Computer assignments %			On-line resource systems %			Group projects %			Enrollment in 1000s			Average section size		
Course & Department Type	PhD	MA	BA	PhD	MA	BA	PhD	MA	BA	PhD	MA	BA	PhD	MA	BA	PhD	MA	BA	PhD	MA	BA
Mathematics for Elem School Teachers	14	38	14	36	58	55	10	13	20	3	2	2	25	31	43	15	20	37	29	27	22
College Algebra	47	41	47	4	13	3	18	3	5	18	6	7	4	3	3	71	63	62	46	41	27
Trigonometry	31	51	70	1	18	5	12	0	5	15	0	5	1	7	5	17	6	7	37	31	27
College Algebra & Trig (combined)	32	57	19	4	4	0	2	0	0	12	0	0	0	4	0	18	7	9	57	28	25
Elementary Functions, Precalculus	47	50	77	2	6	13	6	2	11	17	2	4	2	7	9	47	20	25	48	31	25
Intro to Mathematical Modeling	25	59	48	25	59	44	0	0	59	0	0	4	13	0	56	1	4	3	81	31	20
All courses in FY.2	39	44	42	7	23	21	12	4	12	10	3	4	6	10	17	169	120	143	44	34	25

Note: 0 means less than one half of 1% in columns 1-15, and less than 500 in the Enrollment columns.

TABLE FY.3 Percentage of sections (excluding distance-learning sections) in Mainstream Calculus I and Mainstream Calculus II taught by various types of instructors in four-year mathematics departments in fall 2005, by size of sections and type of department. Also average section sizes.

	Percentage of sections taught by																		Average section size		
	Tenured/ tenure-eligible %			Other full-time (total) %			Other full-time (doctoral) %			Part-time %			Graduate teaching assistants %			Unknown %					
Course & Department Type	PhD	MA	BA	PhD	MA	BA	PhD	MA	BA	PhD	MA	BA	PhD	MA	BA	PhD	MA	BA	PhD	MA	BA
Mainstream Calculus I																					
Lecture/ recitation	42	72	62	31	16	24	19	3	17	6	2	14	9	0	0	11	11	0	65	29	23
Regular section <31	42	78	83	19	5	9	10	1	5	5	5	5	32	4	0	2	7	2	25	24	21
Regular section >30	28	71	94	21	16	0	14	6	0	12	8	6	29	0	0	11	5	0	37	34	33
Total Mainstream Calculus I	**36**	**73**	**79**	**25**	**12**	**12**	**15**	**4**	**7**	**8**	**6**	**7**	**22**	**1**	**0**	**9**	**7**	**2**	**46**	**29**	**22**
Mainstream Calculus II																					
Lecture/ recitation	51	63	79	29	0	18	20	0	4	4	21	0	7	0	0	8	16	4	64	23	19
Regular section <31	38	70	96	20	7	4	14	4	4	6	13	0	36	0	0	1	9	0	26	22	20
Regular section >30	34	78	100	25	12	0	13	12	0	14	4	0	18	0	0	9	6	0	38	31	35
Total Mainstream Calculus II	**42**	**73**	**94**	**26**	**8**	**6**	**16**	**7**	**3**	**8**	**10**	**0**	**17**	**0**	**0**	**7**	**9**	**1**	**47**	**25**	**20**
Total Mainstream Calculus I & II	**38**	**73**	**83**	**25**	**11**	**10**	**15**	**5**	**6**	**8**	**7**	**5**	**20**	**1**	**0**	**9**	**7**	**1**	**46**	**28**	**22**

Note: 0 means less than one half of 1% in columns 1 through 18.

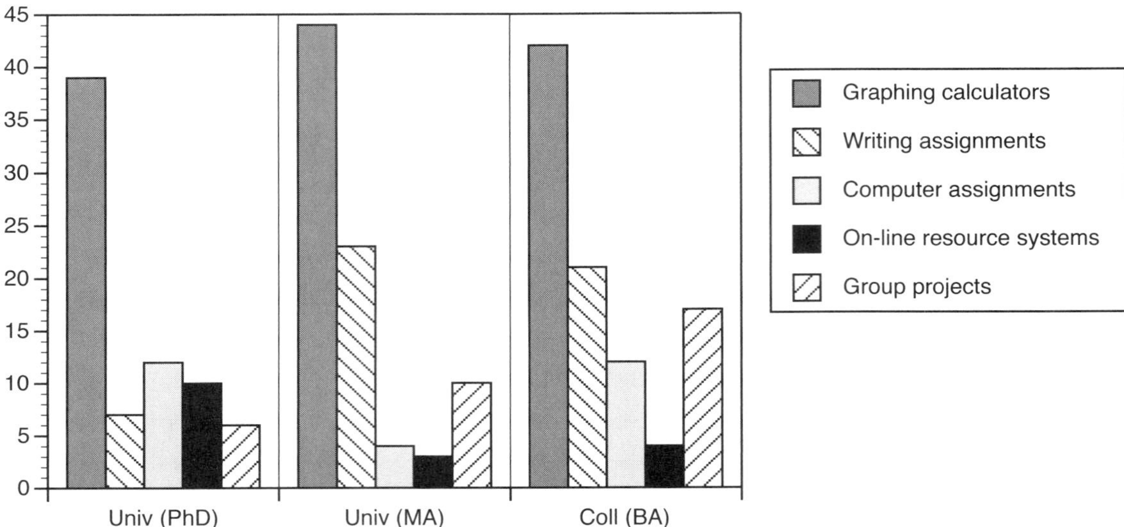

FIGURE FY.2.1 Percentage of sections (excluding distance enrollment) in introductory-level mathematics courses in Table FY.2 (including College Algebra and Precalculus) taught in mathematics departments using various reform methods in fall 2005, by type of department.

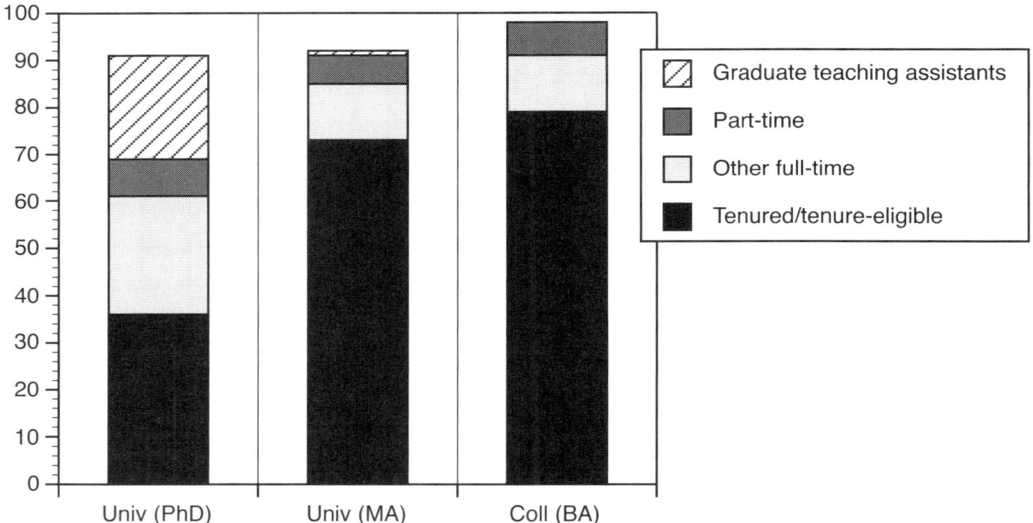

FIGURE FY.3.1 Percentage of sections (excluding distance learning) in Mainstream Calculus I in four-year mathematics departments by type of instructor and type of department in fall 2005. (Deficits from 100% represent unknown instructors.)

TABLE FY.4 Percentage of sections (excluding distance-learning sections) in Mainstream Calculus I & II taught using various reform methods in mathematics departments by type of section and type of department in fall 2005. Also total enrollments (in 1000s) and average section size.

	Percentage of Mainstream Calculus I & II sections taught using																				
	Graphing calculators %			Writing assignments %			Computer assignments %			On-line resource systems %			Group projects %			Enrollment in 1000s			Average section size		
Course & Department Type	PhD	MA	BA	PhD	MA	BA	PhD	MA	BA	PhD	MA	BA	PhD	MA	BA	PhD	MA	BA	PhD	MA	BA
Mainstream Calculus I																					
Lecture/recitation	37	69	57	5	9	25	14	39	33	10	6	0	4	0	27	60	5	14	65	29	23
Regular section <31	44	66	59	2	27	16	9	10	25	4	1	2	5	19	6	11	8	44	25	24	21
Regular section >30	42	36	65	5	18	14	26	4	32	11	0	0	11	10	32	34	17	7	37	34	33
Total Mainstream Calculus I	**40**	**52**	**59**	**5**	**20**	**18**	**18**	**12**	**27**	**9**	**2**	**2**	**7**	**11**	**12**	**105**	**30**	**65**	**46**	**29**	**22**
Mainstream Calculus II																					
Lecture/recitation	23	75	64	4	0	25	8	46	43	6	0	0	1	0	28	31	2	3	64	23	19
Regular section <31	42	54	47	6	12	15	17	6	31	3	0	2	3	12	4	6	4	15	26	22	20
Regular section >30	37	44	86	1	8	28	15	16	57	8	0	0	2	8	28	16	6	1	38	31	35
Total Mainstream Calculus II	**32**	**53**	**52**	**3**	**8**	**17**	**13**	**16**	**34**	**6**	**0**	**2**	**2**	**8**	**9**	**54**	**12**	**19**	**47**	**25**	**20**
Total Mainstream Calculus I&II	**38**	**52**	**57**	**4**	**16**	**18**	**16**	**14**	**29**	**8**	**1**	**2**	**5**	**10**	**11**	**159**	**42**	**84**	**46**	**28**	**22**

Note: 0 means less than one half of 1% in columns 1 through 15, and less than 500 in the Enrollment columns.

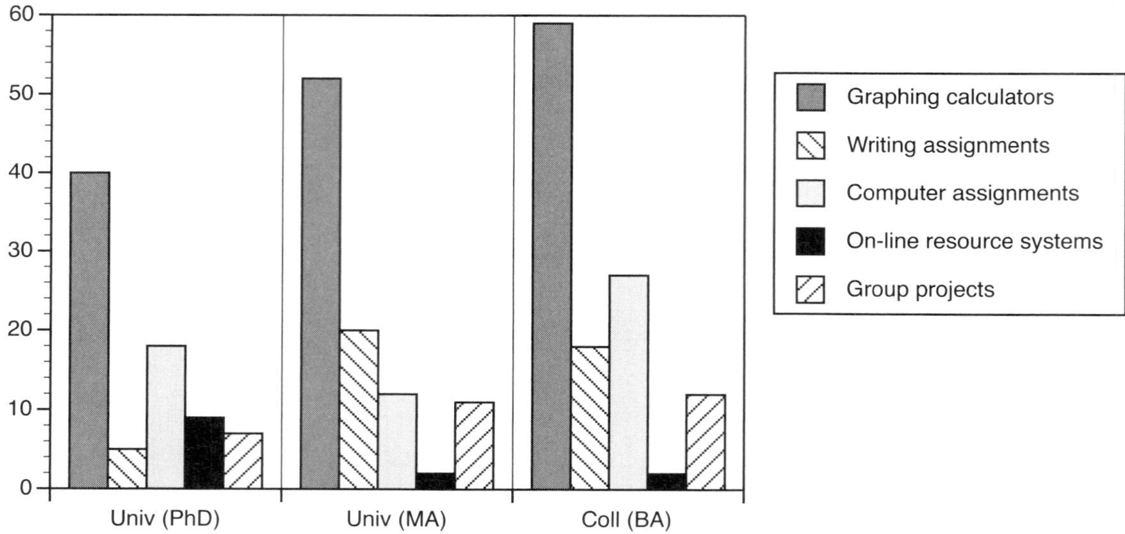

FIGURE FY.4.1 Percentage of sections (excluding distance-learning sections) in Mainstream Calculus I taught using various reform methods in four-year mathematics departments by type of department in fall 2005.

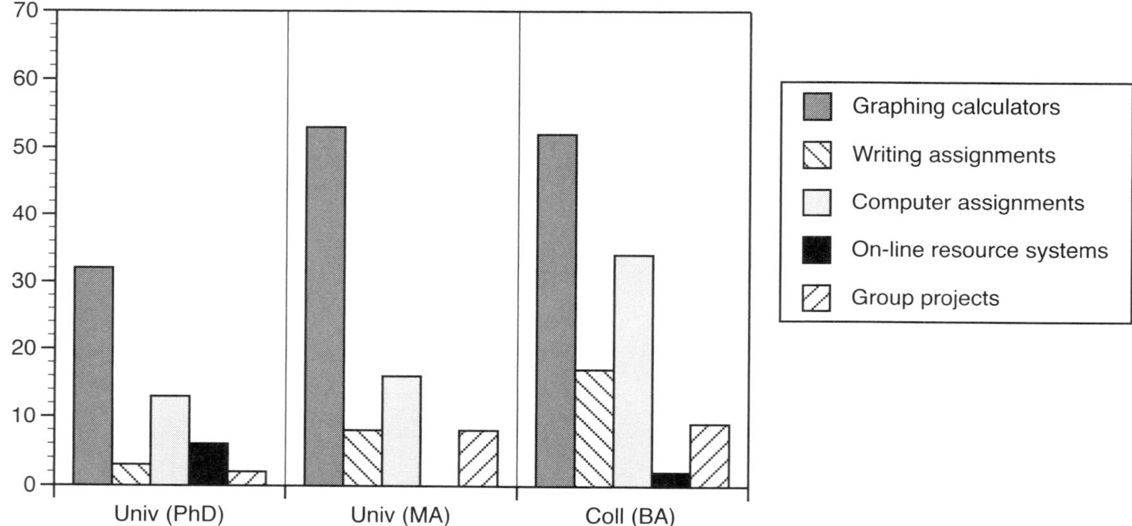

FIGURE FY.4.2 Percentage of sections (excluding distance-learning sections) in Mainstream Calculus II taught using various reform methods in four-year mathematics departments by type of department in fall 2005.

TABLE FY.5 Percentage of sections (excluding distance-learning sections) in Non-Mainstream Calculus I and II taught by various types of instructors in mathematics departments in fall 2005, by size of sections and type of department. Also average section size.

Course & Department Type	Percentage of sections taught by: Tenured/ tenure-eligible %			Other full-time (total) %			Other full-time (doctoral) %			Part-time %			Graduate teaching assistants %			Unknown %			Average section size		
	PhD	MA	BA	PhD	MA	BA	PhD	MA	BA	PhD	MA	BA	PhD	MA	BA	PhD	MA	BA	PhD	MA	BA
Non-Mainstream Calculus I																					
Lecture/ recitation	16	27[2]	40	33	9[2]	60	13	9[2]	0	11	0[2]	0	11	0[2]	0	29	64[2]	0	72	28[2]	22
Reg. section <31	7	46	47	24	7	20	4	1	5	12	27	20	36	0	13[1]	20	20	0	26	23	24
Reg. section >30	21	40	75	27	27	6	11	8	3	24	31	19	27	0	0	1	2	0	53	39	28
Total Non-Mnstrm Calculus I	**17**	**42**	**51**	**28**	**18**	**19**	**10**	**5**	**4**	**17**	**28**	**19**	**24**	**0**	**10[1]**	**14**	**12**	**0**	**52**	**33**	**25**
Total Non-Mnstrm Calculus II	**25**	**47**	**100**	**31**	**13**	**0**	**9**	**0**	**0**	**20**	**40**	**0**	**22**	**0**	**0**	**1**	**0**	**0**	**56**	**18**	**14**
Total Non-Mnstrm Calculus I & II	**18**	**42**	**52**	**29**	**18**	**19**	**10**	**5**	**4**	**18**	**28**	**19**	**23**	**0**	**10**	**12**	**12**	**0**	**53**	**32**	**25**

Note: 0 means less than one half of 1% in columns 1 through 18.

[1] See discussion of this percentage in the text of report.

[2] The sample size for this cell was very small.

Special Note on Table FY.5: Table FY.5 asserts that thirteen percent of smaller sections of the Non-mainstream Calculus I course taught in bachelors-level mathematics departments were taught by graduate teaching assistants (GTAs), and that seems anomalous. Part of that thirteen percent figure can be accounted for by the fact that some bachelors-level departments borrow GTAs from graduate science departments at their universities and assign the borrowed GTAs to teach mathematics courses. However, follow-up calls revealed that the bulk of that figure was caused by the inclusion of some M.A.T. programs in the bachelors-level universe of the CBMS2005 study. Such departments assigned M.A.T. students to teach some of their calculus courses, and the statistical calculations used this raw data to make the national projection of thirteen percent.

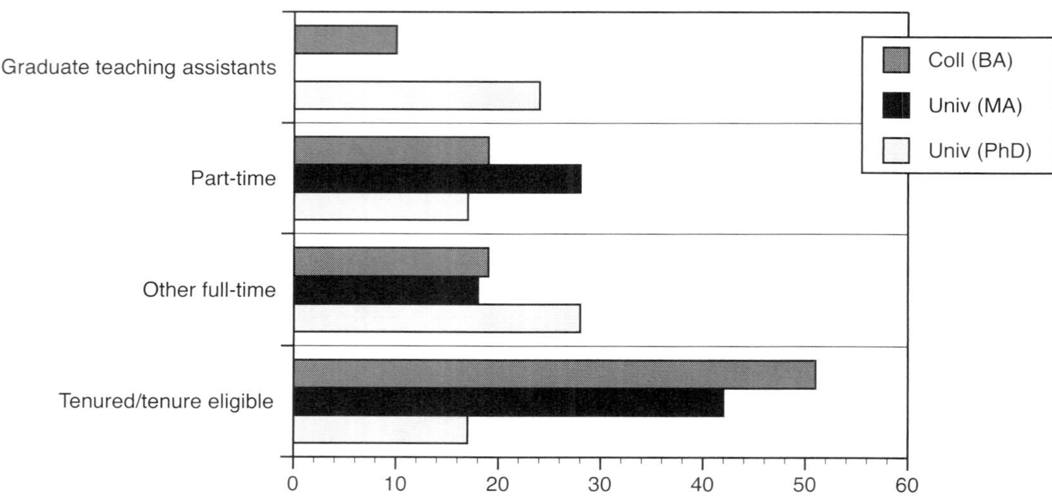

FIGURE FY.5.1 Percentage of sections (excluding distance-learning sections) in Non-mainstream Calculus I in four-year mathematics departments taught by various kinds of instructors, by type of department in fall 2005. (See the text of the report for discussion of the use of GTAs in bachelors-only departments.)

TABLE FY.6 Percentage of sections (excluding distance-learning sections) in Non-mainstream Calculus I taught using various reform methods in four-year mathematics departments in fall 2005, by type of section and type of department. Also total enrollments (in 1000s) and average section size.

Course & Department Type	Percentage of Non-mainstream Calculus I sections taught using															Enrollment in 1000s			Average section size		
	Graphing calculators %			Writing assignments %			Computer assignments %			On-line resource systems %			Group projects %								
	PhD	MA	BA	PhD	MA	BA	PhD	MA	BA	PhD	MA	BA	PhD	MA	BA	PhD	MA	BA	PhD	MA	BA
Non-mainstream Calculus I																					
Lecture/recitation	60	36	80	4	0	60	10	0	0	8	0	0	5	0	0	26	1	1	72	28	22
Regular section <31	45	44	75	1	2	0	1	0	7	1	0	5	1	0	1	5	5	20	26	23	24
Regular section >30	31	47	35	6	9	6	7	0	0	6	0	13	4	7	6	30	15	5	53	39	28
Total Non-mainstream Calculus I	**43**	**45**	**68**	**4**	**6**	**3**	**7**	**0**	**6**	**6**	**0**	**6**	**4**	**4**	**2**	**61**	**21**	**26**	**52**	**33**	**25**

Note: 0 means less than one half of 1% in columns 1 through 15, and less than 500 in the Enrollment columns.

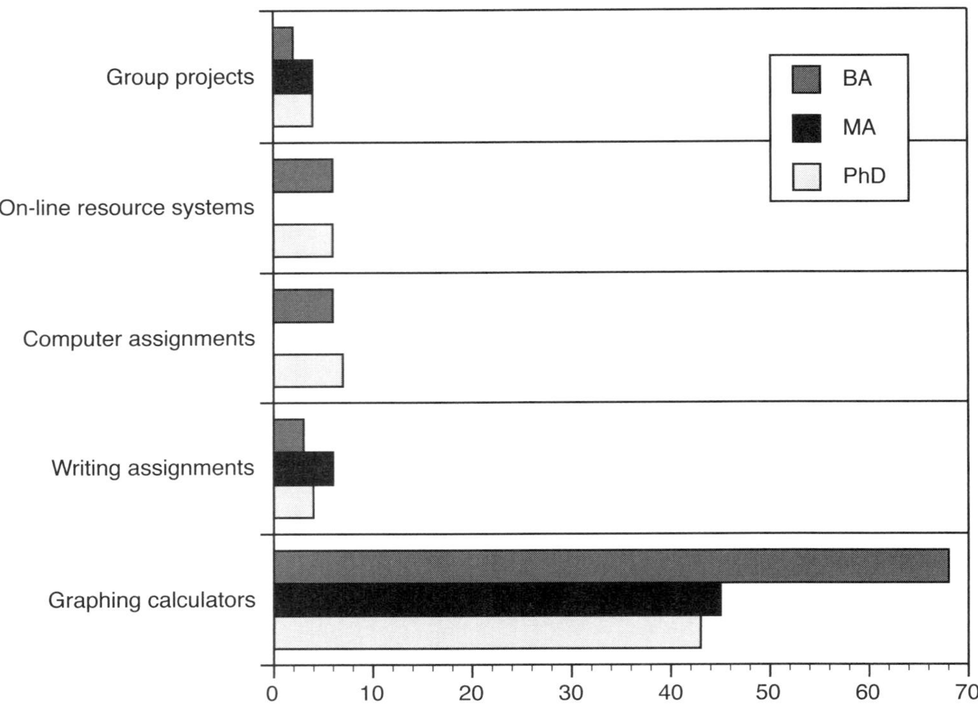

FIGURE FY.6.1 Percentage of sections (excluding distance-learning sections) in Non-mainstream Calculus I taught using various reform methods in four-year mathematics departments by type of department in fall 2005.

TABLE FY.7 Percentage of sections (excluding distance-learning sections) in Elementary Statistics (non-Calculus) and Probability and Statistics (non-Calculus) taught by various types of instructors in mathematics departments in fall 2005, by size of sections and type of department. Also average section size.

	Percentage of sections taught by																				
	Tenured/ tenure-eligible %			Other full-time (total) %			Other full-time (doctoral) %			Part-time %			Graduate teaching assistants %			Unknown %			Average section size		
Mathematics Departments	PhD	MA	BA	PhD	MA	BA	PhD	MA	BA	PhD	MA	BA	PhD	MA	BA	PhD	MA	BA	PhD	MA	BA
Elementary Statistics (non-Calculus)																					
Lecture/ recitation	15	13	41	58	14	17	32	9	0	14	63	34	9	0	0	4	9	8	70	37	22
Regular section <31	1	35	61	51	28	8	22	4	3	14	31	29	33	6	0	0	0	2	24	26	24
Regular section >30	31	53	54	25	20	13	5	2	5	12	22	27	26	1	0	6	3	6	48	41	36
Total Elementary Statistics	**21**	**45**	**57**	**38**	**21**	**10**	**14**	**3**	**3**	**13**	**28**	**29**	**24**	**2**	**0**	**4**	**3**	**4**	**46**	**37**	**27**
Total Probability & Statistics (non-Calculus)	25	53	15	29	17	27	2	5	4	37	25	58	10	0	0	0	6	0	49	33	23
Total both courses	**21**	**47**	**53**	**37**	**20**	**12**	**13**	**4**	**3**	**17**	**27**	**32**	**22**	**2**	**0**	**3**	**4**	**3**	**47**	**36**	**26**

Note: 0 means less than one half of 1% in columns 1 through 18.

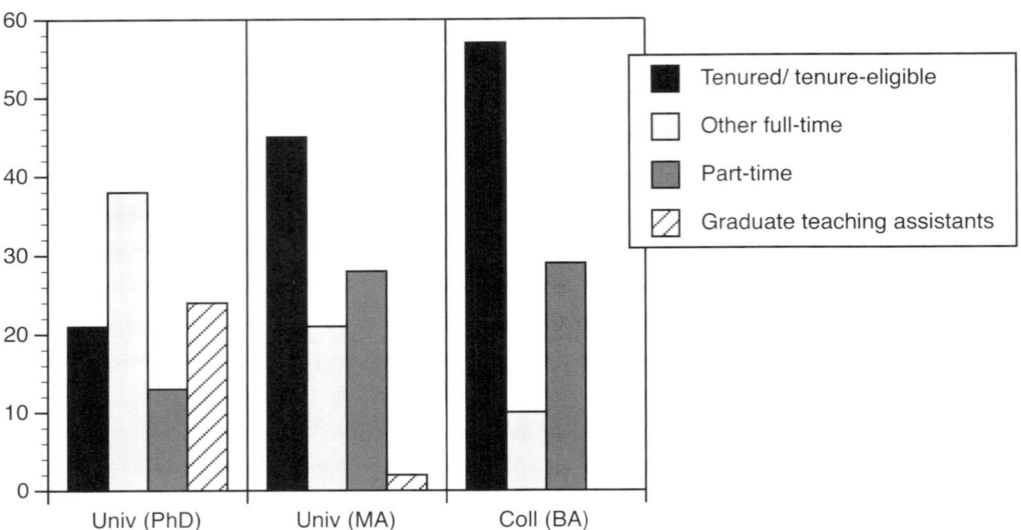

FIGURE FY.7.1 Percentage of sections (excluding distance-learning sections) in Elementary Statistics (non-Calculus) in four-year mathematics departments, by type of instructor and type of department in fall 2005.

TABLE FY.8 Percentage of sections (excluding distance-learning sections) in Elementary Statistics (non-Calculus) and Probability & Statistics (non-Calculus) taught using various reform methods in four-year mathematics departments in fall 2005, by type of section and type of department. Also total enrollments (in 1000s) and average section size.

	Percentage of Statistics & Probability (non-Calculus) sections taught using																				
	Graphing calculators %			Writing assignments %			Computer assignments %			On-line resource systems %			Group projects %			Enrollment in 1000s			Average section size		
Mathematics Departments	PhD	MA	BA	PhD	MA	BA	PhD	MA	BA	PhD	MA	BA	PhD	MA	BA	PhD	MA	BA	PhD	MA	BA
Elementary Statistics (non-Calculus)																					
Lecture/recitation	0	33	62	0	67	62	69	67	92	0	0	0	0	0	65	7	1	5	70	37	22
Regular section <31	0	59	29	3	27	31	57	35	58	0	7	4	0	25	20	3	4	47	24	26	24
Regular section > 30	36	39	52	24	12	26	17	40	64	0	1	3	12	6	2	14	20	23	48	41	36
Total Elementary Statistics	**21**	**43**	**37**	**14**	**21**	**33**	**36**	**41**	**62**	**0**	**2**	**4**	**7**	**10**	**20**	**23**	**24**	**74**	**46**	**35**	**27**
Total Probability & Statistics (non-Calculus)	19	3	35	8	0	79	85	13	61	0	0	0	19	0	53	4	7	7	49	33	23
Total both courses	**21**	**34**	**37**	**13**	**16**	**37**	**43**	**35**	**62**	**0**	**2**	**3**	**8**	**7**	**23**	**27**	**31**	**81**	**43**	**32**	**26**

Note: 0 means less than one half of 1% in columns 1 through 15.

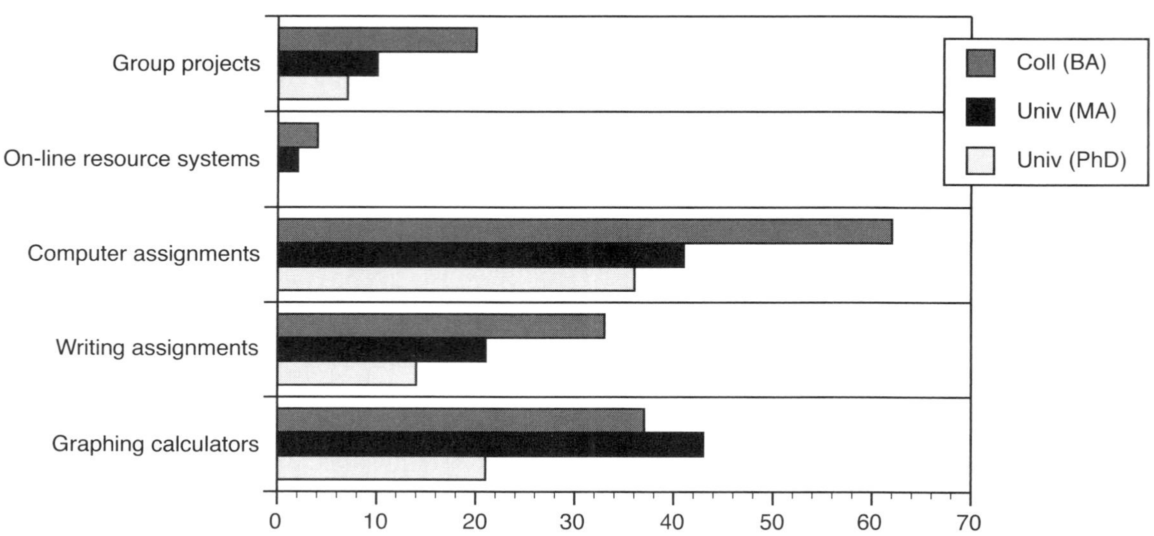

FIGURE FY.8.1 Percentage of sections (excluding distance-learning sections) in Elementary Statistics (non-Calculus) taught using various reform methods in four-year mathematics departments by type of department in fall 2005.

TABLE FY.9 Percentage of sections (excluding distance-learning sections) in Elementary Statistics (non-Calculus) and Probability and Statistics (non-Calculus) taught by instructors of various types in statistics departments in fall 2005, by size of sections and type of department. Also average section size.

	Percentage of sections taught by													
	Tenured/ tenure-eligible %		Other full-time (total) %		Other full-time (doctoral) %		Part-time %		Graduate teaching assistants %		Unknown %		Average section size	
Statistics Departments	PhD	MA	PhD	MA	PhD	MA	PhD	MA	PhD	MA	PhD	MA	PhD	MA
Elementary Statistics (non-Calculus)														
Lecture/recitation	18	26	21	63	8	0	16	11	20	0	25	0	75	121
Regular section <31	31	40	8	60	8	60	8	0	28	0	24	0	21	29
Regular section >30	18	58	11	20	10	4	18	17	48	0	5	5	58	38
Total Elementary Statistics	**19**	**46**	**17**	**37**	**9**	**6**	**16**	**14**	**30**	**0**	**18**	**3**	**67**	**66**
Probability & Statistics (non-Calculus)	**41**	**25**	**19**	**63**	**13**	**63**	**0**	**0**	**28**	**0**	**13**	**13**	**95**	**30**
Total Elementary Statistics and Probability & Statistics	**20**	**44**	**17**	**39**	**9**	**11**	**16**	**13**	**29**	**0**	**18**	**4**	**64**	**62**
Statistics Literacy	**13**	**0**	**22**	**67**	**12**	**33**	**10**	**33**	**20**	**0**	**35**	**0**	**61**	**94**
Total all courses in Table FY.9	**19**	**43**	**17**	**40**	**9**	**12**	**14**	**13**	**27**	**0**	**23**	**4**	**68**	**63**

Note: In the first 12 columns, 0 means less than one half of 1%.

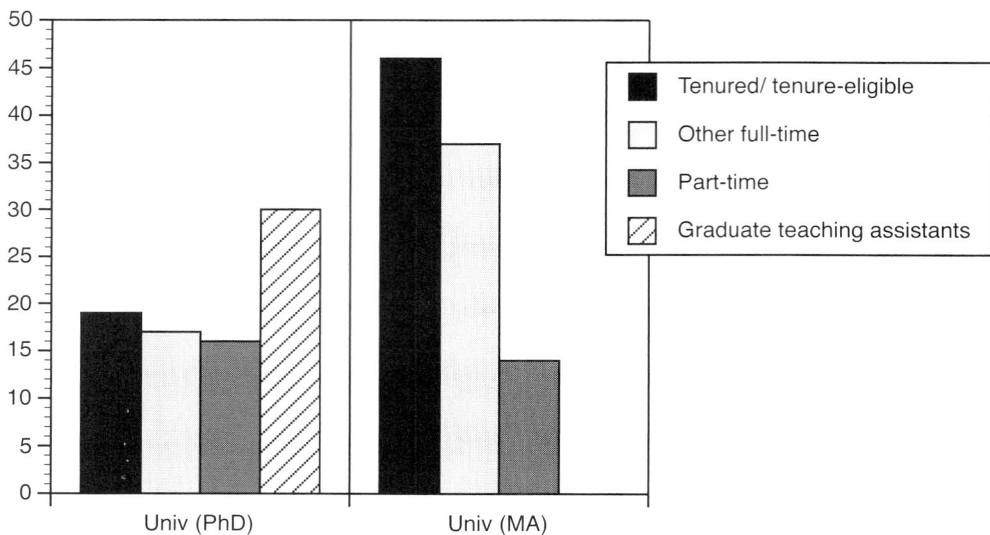

FIGURE FY.9.1 Percentage of sections (excluding distance-learning sections) in Elementary Statistics (non-calculus) taught in statistics departments in fall 2005, by type of instructor and type of department. (Deficits from 100% represent unknown instructors.)

TABLE FY.10 Percentage of sections (excluding distance-learning sections) in Elementary Statistics (non-Calculus) taught using various reform methods in statistics departments in fall 2005, by type of section and type of department. Also total enrollments (in 1000s) and average section size.

	Percentage of Elementary Statistics (non-Calculus) sections taught using													
	Graphing calculators %		Writing assignments %		Computer assignments %		On-line resource systems %		Group projects %		Enrollment in 1000s		Average section size	
Statistics Departments	PhD	MA	PhD	MA	PhD	MA	PhD	MA	PhD	MA	PhD	MA	PhD	MA
Elementary Statistics (non-Calculus)														
Lecture/recitation	10	0	37	74	56	74	28	15	29	41	22	7	75	121
Regular section <31	2	0	24	0	82	100	20	80	20	0	0	0	21	29
Regular section >30	2	0	62	48	43	67	0	2	6	48	9	4	58	38
Total Elementary Statistics	**7**	**0**	**44**	**54**	**54**	**71**	**18**	**11**	**20**	**43**	**31**	**11**	**67**	**66**

Note: 0 means less than one half of 1% in columns 1-12 and less than 500 in the Enrollment columns.

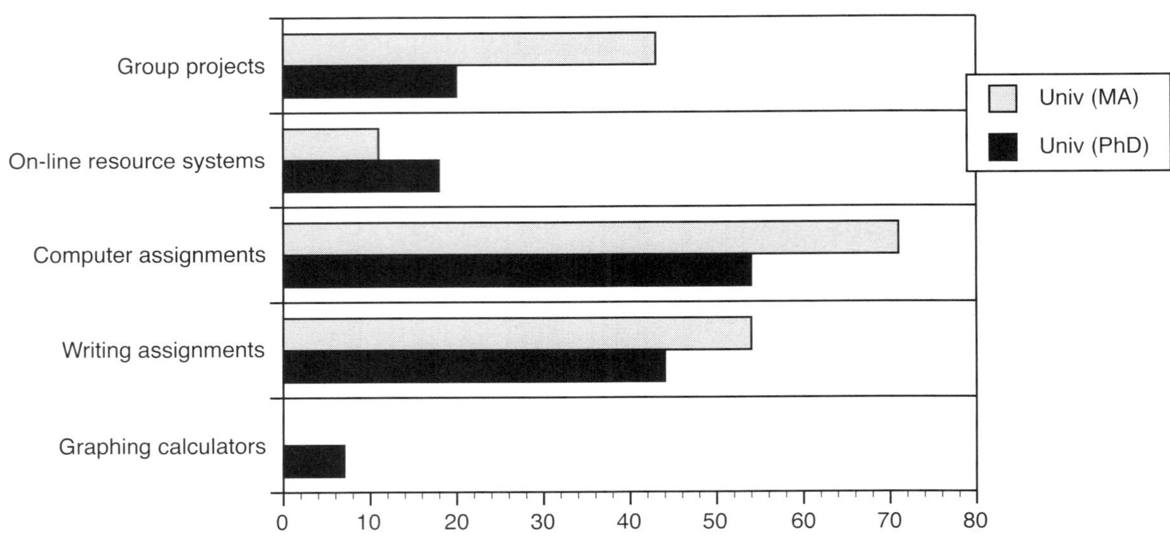

FIGURE FY.10.1 Percentage of sections (excluding distance-learning sections) in Elementary Statistics (non-Calculus) taught using various reform methods in statistics departments, by type of department in fall 2005.

Chapter 6

Enrollment, Course Offerings, and Instructional Practices in Mathematics Programs at Two-Year Colleges

This chapter reports enrollment and instructional practices in fall 2005 in mathematics and statistics courses at public two-year colleges in the United States. Also included are total enrollment for these two-year colleges, average mathematics class size, trends in availability of mathematics courses, enrollment in mathematics courses offered outside of the mathematics programs, and services available to mathematics students. Many tables contain data from previous CBMS surveys (1975, 1980, 1985, 1990, 1995, and 2000) and hence allow for historical comparisons. Further analysis of many of the items discussed in this chapter can be found in Chapter 1, where they are discussed from a comprehensive point of view in comparison to similar data for four-year colleges and universities.

In the 1990 and earlier CBMS surveys, computer courses taught outside the mathematics department, and the faculty who taught them, were considered part of the "mathematics program." By 1995, computer science and data processing programs at two-year colleges for the most part were organized separately from the mathematics program. Hence, in 1995, 2000, and again in this 2005 report, such outside computer science courses and their faculty are not included in mathematics program data. In 1995 and 2000, enrollment data were collected about computer courses taught within the mathematics program and can be found in those reports. Because such courses had become rare, the 2005 survey contains no specific data about even these "inside mathematics program" computer courses, though some, no doubt, were reported by mathematics programs under the Other Courses category. Furthermore, the enrollment tables that follow have been adjusted to eliminate all specific computer science enrollments that appeared in previous CBMS reports. (See, for example, TYE.3 and TYE.4.) This adjustment allows for a more accurate comparison of mathematics program enrollments over time.

Because of the small number of non-public two-year colleges, in contrast to previous surveys, CBMS2005 included only public two-year colleges. Historically, impact on two-year data by non-public colleges has been small. As regards enrollment comparisons with previous surveys, see the explanatory text accompanying Table S.1 in Chapter 1. The two-year college data in this report were projected from a stratified random sample of 241 such institutions chosen from a sample frame of 975 colleges. Survey forms were returned by 130 colleges (54% of the sample). The return rate for all institutions, two-year and four-year, in CBMS2005 was 58% (345 of 600). For comparison purposes, we note that in 2000 the survey return rate for two-year colleges was 60% (179 of 300 colleges), and in 1995 the return rate was 65% (163 of 250). All three two-year rates (1995, 2000, and 2005) are dramatically higher than two-year college return rates had been prior to 1995, reflecting a decade in which two-year college mathematics faculty greatly broadened their professional involvement and in which more intense follow-up efforts were exerted in collecting survey data. For more information on the sampling and projection procedures used in this survey, see Appendix II. A copy of the two-year college survey questionnaire for CBMS2005 may be found in Appendix V.

The terms "permanent full-time" and "temporary full-time" faculty occasionally are used in this chapter. For a detailed explanation of what these terms mean, see the introductory notes in Chapter 7.

The Table display code in this chapter is TYE, for "two-year enrollment," since the chapter deals mostly with issues related to enrollment.

Highlights of Chapter 6

- When all students were counted, including dual-enrollment students at local high schools, in fall 2005 enrollment in mathematics and statistics courses in mathematics programs at public two-year colleges reached an historic high of 1,739,014 students. When about 42,000 dual-enrolled students were omitted, the number is about 1,697,000, still an historic high. See Table S.1 in Chapter 1, Table SP.16 in Chapter 2, and Table TYE.2.
- Using the 1,697,000 figure above, in fall 2005 two-year colleges enrolled about 48% of all undergraduate mathematics students in U.S. colleges and universities. Two-year colleges accounted for about 44% of all collegiate undergraduate enrollments.
- Depending on what comparison is made, the enrollment growth in two-year college mathematics programs from 2000 to 2005 was between 27%

and 30%. For details, see the discussion before Table TYE.2.

- The mathematics and statistics enrollment increase from 2000 to 2005 described above more than doubled the 12% overall enrollment increase at public two-year colleges in the same period. For details, see the discussion before and after Table TYE.1.
- Two-year college enrollment growth in mathematics from 2000 to 2005 was in dramatic contrast to what occurred in the nation's four-year colleges and universities, where for the same time period, enrollment in mathematics declined slightly and lagged far behind total enrollment growth. See Table S.1 in Chapter 1.
- About 57% of the two-year college mathematics and statistics enrollment in fall 2005 was in precollege (formerly called remedial) courses. This was almost identically the percentage in 2000. See Table TYE.4.
- The total number of precollege (remedial) enrollments in mathematics programs at two-year colleges dropped by 5% from 1995 to 2000 but jumped 26% from 2000 to 2005 to end the decade 21% higher than 1995, a pattern very similar to that for overall mathematics program enrollment. This contrasts with four-year colleges (see Table S.2) in which precollege enrollments dropped by 8% between 2000 and 2005. See Table TYE.4.
- Within the cohort of precollege courses, Arithmetic/Basic Skills showed a 15% drop in enrollment even though the whole precollege group had a 26% enrollment increase. The movement was toward pre-algebra courses, which experienced a 57% increase in enrollment. See Table TYE.3.
- Enrollment in the precalculus course group grew about 17% from 2000 to 2005, generally reflecting the large overall increase in mathematics enrollment. See Tables TYE.3 and TYE.4.
- Enrollment in calculus-level courses, which made up 9% of overall enrollment in 1995 and 8% of enrollment in 2000, continued to slide with only 6% of enrollment in 2005 and showed only a slight total headcount increase from 2000, in spite of the large overall mathematics enrollment increase. However, there was a 31% surge in Non-mainstream Calculus I, perhaps reflecting a growth in calculus enrollment by biology and life-science majors. See Tables TYE.3 and TYE.4.
- Enrollment was level or up for every course type except Arithmetic and Basic Mathematics, combined College Algebra/Trigonometry, Mainstream Calculus I and II, Differential Equations, Discrete Mathematics, and calculus-based Technical Mathematics. See Table TYE.3.
- Among the usual college-level, transferable mathematics and statistics courses, the largest enrollment increases in percentage order were as follows: Mathematics for Elementary School Teachers (11,000 increase; 61%), Elementary Statistics (40,000 increase; 56%), Mathematics for Liberal Arts (16,000 increase; 37%), and College Algebra (33,000 increase; 19%). See Table TYE.3.
- The fall 2005 survey indicated the following reductions (in comparison to fall 2000) in the percentage of colleges offering various advanced courses over a two-year window: Mainstream Calculus I, down 7 percentage points to 87%; Mainstream Calculus II, down 10 percentage points to 78%; Differential Equations, down 1 percentage point to 58%. See Table TYE.5.
- Compared directly to fall 2000, fall 2005 saw the following notable increases in the percentage of two-year colleges offering various courses required for baccalaureate degrees: Mathematics for Liberal Arts, up 6 percentage points to 56% and Mathematics for Elementary School Teachers, up 10 percentage points to 59%. See Table TYE.6.
- In fall 2005, average size of on-campus classes decreased by about two students to 23, with only 21% of class sections above 30, the class size recommended by the Mathematical Association of America (MAA). See Tables TYE.7 and TYE.8. For comparable four-year data, see Tables E.13 and E.14 in Chapter 3.
- The percentage of class sections taught by part-time faculty in fall 2005 was 44%, a two-percentage-point drop from 2000, reversing the direction of the eight-percentage-point increase that had occurred from 1995 to 2000. Once again, the percentage of sections taught by part-time faculty varied significantly by course type, with part-time faculty teaching 56% of precollege courses but only 12% of mainstream calculus courses. See Table TYE.9.
- For easy reference concerning part-time faculty, we note here that part-time faculty (including those paid by third parties such as school districts) constituted about 68% of the total faculty in mathematics programs at public two-year colleges in fall 2005, up two points from 2000. If 1,915 part-time faculty members paid by a third party are excluded, in 2005 the part-time percentage of the total faculty was 66%. In 2000, the comparable figure was 65%. Information on faculty size is given in Table TFY.1 in Chapter 7.
- The predominant instructional modality continued to be the standard lecture method, with this reported as the preferred methodology for all but two courses by percentages that ranged as high as 93%. In Mainstream Calculus I, the use of writing,

computer assignments, and group projects dropped 10 to 15 percentage points. For details, see Tables TYE.10, TYE.11 and the surrounding discussion.

- Perhaps surprisingly, the use of on-line resource systems for homework, tutoring, and testing was low, at 14% and 11% of course sections for Arithmetic and each of Elementary/Intermediate Algebra, and 10% for statistics. Use was half this percentage in most other courses. Data about on-line resource use were collected for the first time in CBMS2005, replacing a question about weekly use of computer labs. See Table TYE.10.
- About 5% of mathematics program enrollment at two-year colleges in fall 2005 was in distance learning, defined as an instructional format in which at least half the students received the majority of instruction using methods where the instructor is not physically present. Most courses showed less than 5% enrollment in this format. Some courses, such as Geometry, Mathematics for Elementary School Teachers, and Elementary Statistics, however, had distance enrollment near or over 10%. See Table TYE.12.
- Virtually all two-year college mathematics programs made diagnostic or placement testing available, with 97% requiring placement testing of first-time enrollees. Discussion of scores with advisors was required by 90% of colleges, and 88% of colleges used placement tests as part of mandatory placement. See Table SP.11 in Chapter 2.
- About 95% of two-year colleges had a mathematics lab or tutorial center. There was about a ten-percentage-point increase in the number of colleges whose students participated in mathematics contests and a similar increase in the number of colleges with special programs to encourage minority students in mathematics. See Table TYE.13.
- After a 42% decline in 2000, the collection of precollege (remedial) courses taught "outside" the mathematics program (e.g., in developmental studies divisions) experienced an 89% rise in 2005, almost triple the enrollment increase within mathematics programs. These "outside" enrollments, offered at 31% of colleges, are not included in Table TYE. 1. If they were, total mathematics enrollment in fall 2005 at public two-year colleges would exceed 1,900,000. See the discussion before Tables TYE.3 and TYE.5 and especially the discussion before Tables TYE.15 and TYE.16.

Enrollment, Class Size, and Course Offerings In Mathematics Programs

Number of two-year-college students

About 6,389,000 students were enrolled in public two-year colleges in fall 2005. This estimate is based on a mid-range overall 2005 enrollment projection for public two-year colleges by the National Center for Educational Statistics (NCES). Enrollment in two-year colleges in fall 2005 constituted about 44% of the total undergraduate enrollment in the United States. See Table S.1 in Chapter 1.

In CBMS surveys prior to 2005, mathematics enrollment was collected from both public and private two-year colleges. The reader should note that, in contrast to Table S.1, the data in Table TYE.1 include actual (not projected) overall NCES enrollment for both public and private two-year colleges, with 2004 being the last year for which the actual NCES data is available. The data in TYE.1 allows readers to compare mathematics enrollment to overall enrollment for years 2000 and earlier. See Table S.1 for 2005 data on public colleges only.

TABLE TYE.1 Total enrollment (all disciplines) and percentage of part-time enrollments in public and private two-year colleges, in fall 1975, 1980, 1985, 1990, 1995, 2000, and 2004.

	1975	1980	1985	1990	1995	2000	2004
Number of students	3,970,119	4,526,287	4,531,077	5,240,083	5,492,529	5,948,431	6,545,863
Percentage part-time	56	61	63	64	64	63	59

Sources: Table 177, National Center for Educational Statistics, 2005 and NCES IPEDS Table 1. In Table 177, 2004 was the latest year for which data, rather than projections, were available.

Note: Table TYE.1 differs from Table S.1 of Chapter 1 because Table S.1 includes public two-year colleges only.

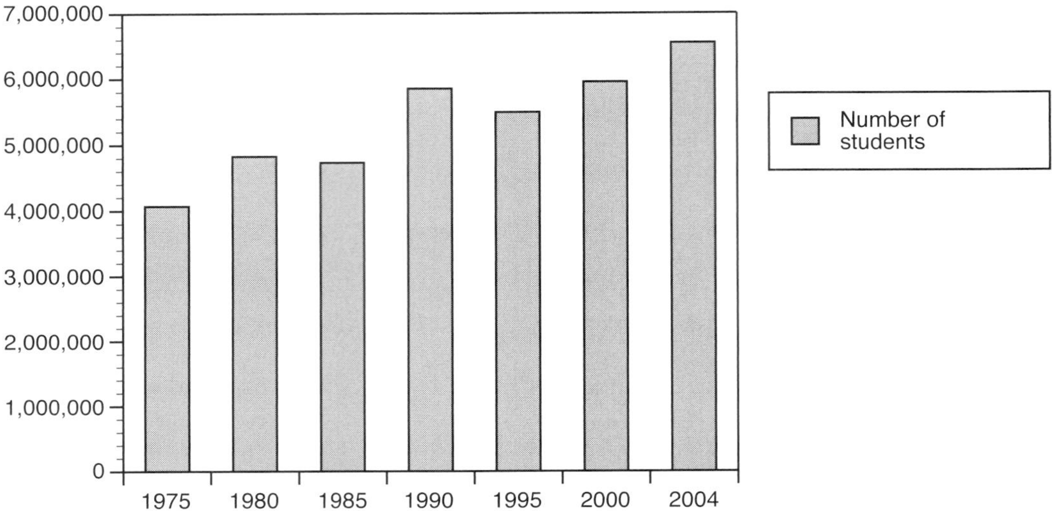

FIGURE TYE.1.1 Total enrollments (all disciplines) in public and private two-year colleges in fall 1975, 1980, 1985, 1990, 1995, 2000, and 2004, from NCES data.

Enrollment trends in mathematics programs

As in CBMS1995 and 2000, Table TYE.2 for 2005 does not include any computer science enrollments. Moreover, enrollment totals in Table TYE.2 reported from CBMS surveys prior to CBMS1995 have been adjusted to remove all computer science enrollments. For more detail on this reporting issue, see the second paragraph above at the start of this chapter.

When dual-enrollment students are included—about 42,000 high school students who took courses taught by high school teachers on a high school campus and received course credit at both the high school and at the two-year college—fall 2005 enrollment in mathematics and statistics courses in mathematics programs at public two-year colleges reached an all-time high of 1,739,014 students. In comparison to 2000, this was an enrollment increase of at least 29%. It sharply reversed the 7.5% decrease in mathematics program enrollment that had occurred between 1995 and 2000. See Tables SP.16 in Chapter 2 as well as Table TYE.2 below.

However, in fall 2005, the growth at public two-year colleges actually was slightly larger than 29%. The 2000 entry in Table TYE.2, the base for comparison, includes private two-year college enrollments. Data from the National Center for Educational Statistics (NCES) indicated about 99% of overall two-year college enrollment in 2002 was at public institutions. Assuming the 99% was valid in 2000 also, the enrollment growth in mathematics programs at public colleges from 2000 to 2005 exceeded 30%.

Dual-enrollment students, numbering about 42,000, were one reason for the mathematics program growth that appeared in 2005, but they accounted for only about 3% of the growth. When these students are excluded, mathematics programs at public two-year colleges still had an historically high enrollment of 1,697,000. Again using the 99% adjustment described in the previous paragraph, without dual enrollments the increase from 2000 to 2005 was 27%. See Table TYE.2 below as well as Table S.1 in Chapter 1 and Table SP.16 in Chapter 2.

A 29% enrollment increase in mathematics and statistics courses from 2000 to 2005 more than doubled the 12% overall enrollment increase at public two-year colleges in the same period. The overall enrollment increase is reported in Table S.1 and above in Table TYE.1. The percentage is based on a mid-range NCES overall enrollment projection of 6,389,000 students at public two-year colleges in 2005. The reader is reminded that the data in Table TYE.1 includes actual (not projected) enrollment for both public and private two-year colleges for the years indicated, with 2004 the last year for which actual NCES data is available.

Two-year college mathematics growth from 2000 to 2005 also contrasted sharply with the pattern in the nation's four-year colleges and universities. During the same time period, at four-year institutions, mathematics enrollment declined slightly and lagged far behind total enrollment growth. See Table S.1 in Chapter 1. This decline created yet another alternation in an interesting interlocking of collegiate mathematics enrollment patterns that first emerged over the decade from 1990 to 2000. Both two-year and four-year colleges came to the millennium with mathematics enrollment at about the same level each had reported in 1990, but they had followed very different paths in reaching that point. Four-year enrollments fell from 1990 to 1995 and rebounded in 2000 to earlier levels. By contrast, two-year enrollments rose sharply from 1990 to 1995 but by 2000 had fallen to 1990 levels. In 2005, when two-year enrollments were exploding, the enrollment in mathematics at four-year institutions declined slightly.

In addition to the tables that follow, the reader should consult Chapter 1 of the current report. Chapter 1 contains a detailed analysis of mathematics department enrollments at both two-year and four-year colleges over the decade 1995 to 2005 and also contains additional enrollment comparisons between two-year and four-year colleges.

The 2005 survey confirmed that the typical two-year college mathematics program principally offered courses for remedial or general education and in support of disciplinary majors other than mathematics. This observation is consistent with past CBMS surveys that have suggested few two-year college students intended to transfer to a four-year college or university to study mathematics as a major.

TABLE TYE.2 Enrollments in mathematics and statistics (no computer science) courses in mathematics programs at two-year colleges in fall 1975, 1980, 1985, 1990, 1995, 2000, and 2005. (Total for fall 2005 includes only public two-year colleges, and includes dual enrollments.)

	1975	1980	1985	1990	1995	2000	2005
Mathematics & Statistics enrollments in TYCs	864,000	953,000	936,000	1,295,000	1,456,000	1,347,000	1,739,000[1]

[1] Data for 2005 include only public two-year colleges and include 42,000 dual enrollments from Table SP.16.

Note: Data for 1990, 1995, and 2000 in Table TYE.2 differ from corresponding data in Table S.1 of Chapter 1 because the totals in TYE.2 do not include any computer science courses, while the totals in S.1 do.

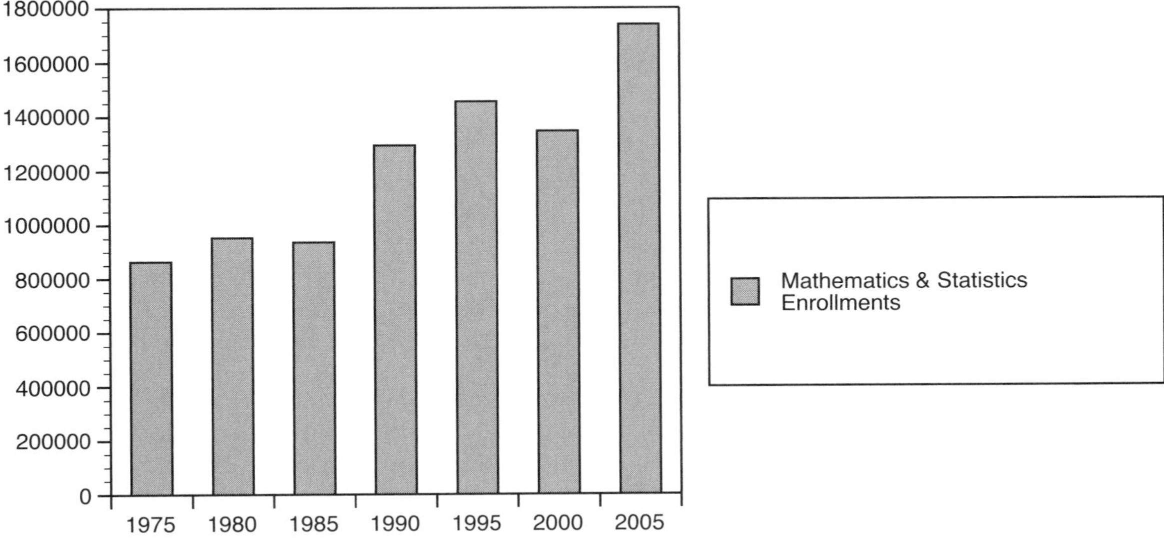

FIGURE TYE.2.1 Enrollments in mathematics and statistics courses (no computer science) in mathematics programs in two-year colleges in fall 1975, 1980, 1985, 1990, 1995, 2000, and 2005. (Data for 2005 include only public two-year colleges and include dual enrollments from Table SP.16.)

Enrollment trends in course groups and in specific courses

Table TYE.3 reports enrollment in individual mathematics courses. Table TYE.4 reports enrollment for categories of courses. Table TYE.4 is constructed from Table TYE.3 and reports headcounts and percentages from 1990 through 2005 for the following course groupings: precollege, precalculus, calculus, and statistics. Each category consists of five or more specific courses from Table TYE.3. Percentages in Table TYE.4 will differ slightly from the corresponding percentages in the CBMS2000 report because of the computer science enrollment adjustment discussed in the introduction to this chapter.

In fall 2005, precollege courses (formerly called remedial) comprised over half (57%) of mathematics program enrollment. The percentage of precollege enrollments in the overall mathematics program enrollment also was 57% in fall 2000. In fact, this percentage has been essentially stable at 57% since 1990, a very long run without significant change.

The total size of the precollege course enrollment has varied over time as follows: down by 5% from 1995 to 2000 but up 26% from 2000 to 2005, to end the decade in 2005 at 21% higher than 1995. Interestingly, these swings in the number of precollege enrollments have almost exactly paralleled the rises and falls in the total mathematics program enrollment at two-year colleges during these years: down 7% from 1995 to 2000 but up 29% from 2000 to 2005, for a decade-long change of plus 19%. These percentages are calculated from Table TYE.4, which does not include 42,000 dual enrollments used in other calculations.

Additionally, more than 30% of two-year colleges conducted all or part of their precollege (remedial) mathematics program outside of the mathematics program in an alternate structure like a developmental studies division or learning laboratory. These enrollments are not included in Tables TYE.3 and TYE.4. These "outside" precollege enrollments also grew substantially from 2000 to 2005 (by 89%), reflecting a continued difference in strategy at two-year colleges about how best to supervise precollege mathematics students. For more information on these "outside" precollege courses, see the discussion for Tables TYE.15 and TYE.16 later in this chapter.

Precalculus-level courses accounted for 19% of 2005 enrollment, almost identical to the 20% reported in 2000. Together with precollege courses, these two categories of preparatory courses below calculus accounted for 76% of mathematics and statistics enrollment at public two-year colleges in fall 2005.

Calculus-level courses continued a ten-year decline in which they progressively accounted for smaller proportions of the overall mathematics program enrollment. They made up 9% of overall mathematics program enrollment in 1995 and 8% of enrollment in 2000 but only 6% of enrollment in 2005. The total headcount in calculus-level courses in 2005 was only very slightly larger than the headcount in these courses in 2000, in spite of the very large increase in overall mathematics program enrollment in 2005. However, there was a 31% enrollment increase in the special non-mainstream calculus course. The distinction between "mainstream" and "non-mainstream" calculus is discussed below.

In contrast to what happened from 1995 to 2000, between fall 2000 and fall 2005 enrollments increased in every major mathematics course category. See Table TYE.4. The increases within these course categories were precollege (remedial) 26%; precalculus 17%; calculus 1%; and statistics 59%.

Refer to Table TYE.3 for enrollment in individual courses. In dramatic contrast to the five-year period 1995–2000, 21 of the 28 courses surveyed remained level or increased in enrollment between 2000 and 2005. The seven exceptions were Arithmetic and Basic Mathematics, combined College Algebra/Trigonometry, Mainstream Calculus I and II, Differential Equations, Discrete Mathematics, and calculus-based Technical Mathematics. From 1995 to 2000, the only courses that had shown enrollment gain were Elementary Statistics (3%), Mathematics for Elementary School Teachers (12.5%), and Mathematics for Liberal Arts (13%). These three courses once again led the enrollment gain from 2000 to 2005 with increases respectively of 56%, 61%, and 37%.

As reported in Table TYE.3, business mathematics enrollment increased 73% from 2000 to 2005, thereby returning to its 1995 level, but this enrollment number is an amalgam of transferable and non-transferable courses. The fact that in fall 2005 there was an eight-point increase in the number of colleges offering the non-transferable business mathematics course at least once during a two-year cycle and a decrease in the number of programs offering the transferable course suggests that the 73% enrollment increase was skewed toward lower-level business courses.

In reading the enrollment tables, the reader is reminded that mainstream calculus consists of those calculus courses that lead to more advanced mathematics courses and usually is required of majors in mathematics, the physical sciences, and engineering. Non-mainstream calculus includes the calculus courses most often taught for biology, behavioral science, and business majors. Additionally, refer to the comments at the start of this chapter about adjustments made in the tables because of computer science enrollments that were included in previous CBMS surveys. Finally, note that additional enrollment data and analysis can be found in Chapter 1.

TABLE TYE.3 Enrollment in thousands in mathematics and statistics courses (not including dual enrollments) in mathematics programs at two-year colleges in fall 1990, 1995, 2000, and 2005. (This table does not include any computer science enrollments appearing in previous CBMS reports. Also, 2005 data include only public two-year colleges.)

Course Number	Type of course	1990	1995	2000	2005
	Precollege level				
1	Arithmetic & Basic Mathematics	147	134	122	104
2	Pre-algebra	45	91	87	137
3	Elementary Algebra (HS level)	262	304	292	380
4	Intermediate Algebra (HS level)	261	263	255	336
5	Geometry (HS level)	9	7	7	7
	Precalculus level				
6	College Algebra (above Intermed Algebra)	153	186	173	206
7	Trigonometry	39	43	30	36
8	College Algebra & Trig (combined)	18	17	16	14
9	Intro to Mathematical Modeling	na	na	7	7
10	Precalc/ Elem Fnctns/ Analyt Geom	35	50	48	58
	Calculus level[1]				
11	Mainstream Calculus I	53	58	53	51
12	Mainstream Calculus II	23	23	20	19
13	Mainstream Calculus III	14	14	11	11
14	Non-mainstream Calculus I	31	26	16	21
15	Non-mainstream Calculus II	3	1	1	1
16	Differential Equations	4	6	5	4
	Other mathematics courses				
17	Linear Algebra	3	5	3	3
18	Discrete Mathematics	1	3	3	2
19	Elem Statistics (with or w/o Probability)	47	69	71	111
20	Probability (with or w/o Statistics)	7	3	3	7
21	Finite Mathematics	29	24	19	22
22	Mathematics for Liberal Arts	35	38	43	59
23	Math for Elementary School Teachers	9	16	18	29
24 & 25	Business Mathematics	26	25	15	26
26	Technical Math (non-calculus)	17	17	13	16
27	Technical Math (calculus-based)	1	2	2	1
28	Other mathematics courses	0	0	14	28
	Total all TYC math courses	1272	1425	1347	1696

Note: 0 means fewer than 500 enrollments and na means not available. Round-off may make column sums seem inaccurate.

[1] Mainstream calculus is for mathematics, physics, science & engineering. Non-mainstream calculus is for biological, social, and management sciences.

TABLE TYE.4 Enrollment in 1000s (not including dual enrollments) and percentages of total enrollment in mathematics and statistics courses by type of course in mathematics programs at two-year colleges, in fall 1990, 1995, 2000, and 2005. (This table does not include any computer science enrollments appearing in previous CBMS reports. Also, 2005 data include only public two-year colleges.)

Course numbers	Type of course	1990	1995	2000	2005
1–5	Precollege	724 (57%)	800 (56%)	763 (57%)	964 (57%)
6–10	Precalculus	245 (19%)	295 (21%)	274 (20%)	321 (19%)
11–16	Calculus	128 (10%)	129 (9%)	106 (8%)	107 (6%)
19–20	Statistics	54 (4%)	72 (5%)	74 (5%)	118 (7%)
17,18, & 21–28	Other	121 (10%)	130 (9%)	130 (10%)	186 (11%)
1–28	Total all courses	1272 (100%)	1426 (100%)	1347 (100%)	1696 (100%)

Note: This table was constructed using Table TYR.3. Notice that the breakdown into type of course is different from that in Chapter 1 Table S.2 and Appendix I for four-year colleges and universities. Data from CBMS reports before 2005 have been modified to remove all computer science enrollments.

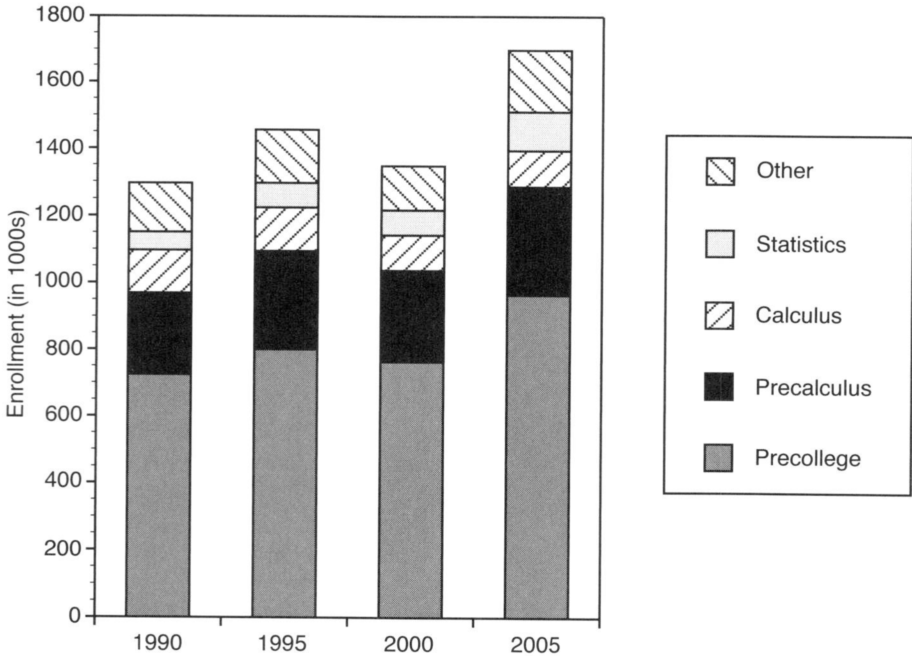

FIGURE TYE.4.1 Enrollment in 1000s (not including dual enrollments) in mathematics and statistics courses by type of course in mathematics programs at two-year colleges in fall 1990, 1995, 2000, and 2005. Totals do not include any computer science enrollments and data for 2005 include only public two-year colleges.

Trends in availability of courses in mathematics programs

Tables TYE.5 and TYE.6 should be considered together. The first shows the percentage of public two-year colleges offering a course within the mathematics program at least once in a two-year academic period. The second shows the percentage of colleges offering certain courses specifically during fall 2005. The availability of some of these courses (such as differential equations and linear algebra) over a two-year period is considerably higher than availability during a single fall semester.

The reader should also note that 31% of two-year colleges in fall 2005 reported that some or all of the precollege (remedial) mathematics courses at the college were organized separately from the mathematics department. This was up slightly from the 29% reported in both 1995 and 2000. See Table TYE.17. These "outside" courses are not included below in Tables TYE.5 and TYE.6 in reporting the availability of particular courses. The "outside" headcount enrollment is estimated in Tables TYE.15 and TYE.16. Also see the last highlight bullet at the start of this chapter.

Table TYE.5 reports that the percentage of two-year college mathematics programs offering a separately titled arithmetic/basic mathematics course continued a steep decline from 70% in 1995 to 56% in 2000 and finally to 48% in 2005. This does not mean that arithmetic material was not part of the department's or the college's overall curriculum, only that a stand-alone course called "arithmetic" continued to become less available within the mathematics program. At the same time, from 2000 to 2005, the percentage of mathematics programs offering a pre-algebra course, which almost certainly included arithmetic skills, rose six percentage points to 47% (Table TYE.5), and enrollment in these pre-algebra courses rose 57% (TYE.3). Also simultaneously, combined arithmetic/pre-algebra enrollment grew by 39% (Table TYE.15) in courses outside the mathematics program.

Intermediate Algebra, which is roughly equivalent to the second year of high school algebra, was offered in 88% of colleges in fall 2005, down slightly since 2000. Historically, Intermediate Algebra has been the bridge between a developmental studies division and a mathematics program. Within a mathematics program, Intermediate Algebra often is the preparatory course for transferable college-credit mathematics. The wide availability of the course in fall 2005 suggests Intermediate Algebra continued to play these roles. The availability of Elementary Algebra within mathematics programs grew in 2005 to 80%. The discussion below about mathematics courses taught "outside" the mathematics program also is relevant here. Table TYE.17 suggests that, historically, two to three times as many two-year colleges find a home for Elementary Algebra outside the mathematics program as those who do the same for Intermediate Algebra.

A surprising result in CBMS2005 was the sharp increase from 14% in fall 2000 to 24% in fall 2005 in the percentage of two-year colleges offering high-school-level geometry courses, though the overall geometry enrollment remained constant.

Here is availability data for courses directly preparatory for calculus, using a two-year window and compared to 2000. See Table TYE.5. The percentage of colleges that offered a separate College Algebra course decreased by four points to 79% and returned to its 1995 level. The percentage of colleges offering a separate Trigonometry course also dropped slightly, by 3 points to 63%. It had been 71% in 1995. The combined course College Algebra/Trigonometry had seen a sharp rise in availability from 1995 to 2000 but in 2005 had an identical drop in availability. Precalculus/Elementary Functions, which had a 19-percentage-point increase in availability from 1995 to 2000, dropped off five points to 60% in 2005.

When considered over the same two-year window, the percentage of colleges offering the first semester of mainstream calculus fell back to 87%. This number had been 94% in 2000 and 83% in 1995. In fall 2005 alone (Table TYE.6), 82% of colleges offered Mainstream Calculus I, and enrollment was down slightly from 2000 (Table TYE.3). The availability of Mainstream Calculus II over a two-year period was down 10 percentage points, but that of Non-mainstream Calculus I was up six points to 46%, moving back toward its 1995 level of 52%. One explanation for the rise in the latter percentage in 2005 might be an increase in the number of students pursuing transferable biology-oriented degrees in which some calculus, but not mainstream, is required. The percentage of colleges offering the second semester non-mainstream calculus remained constant at 6%.

Introductory Mathematical Modeling was a new course first surveyed in 2000. In that year, 12% of colleges reported offering the course. In 2005, this percentage had dropped to 7%. The drop may be explained in part by the fact that curriculum reform within the traditional College Algebra course was very active between 2000 and 2005, lessening the demand for newly-created modeling courses.

The CBMS1995 survey noted that many students at two-year colleges could not complete lower-division mathematics requirements in certain majors because essential courses such as Linear Algebra, Mathematics for Liberal Arts, and Mathematics for Elementary School Teachers were offered at fewer than half of two-year college mathematics programs, even over a two-year window. Using this window (Table TYE.5), CBMS2000 noted an important increase in availability for all three of these baccalaureate-essential courses. In 2005, the availability of all three jumped again.

Using data from CBMS2000, the pattern of these gains in availability (using a two-year window) over the ten-year period 1995 to 2005 is as follows: Linear Algebra, 30% to 39% to 41%; Liberal Arts, 46% to 50% to 65%; and Elementary Education, 43% to 49% to 66%. The same decade-long pattern for Differential Equations is 53% to 59% to 58%. For Mainstream Calculus I, it is 83% to 94% to 87%, and for Mainstream Calculus II, it is 79% to 88% to 78%.

Availability of other courses important to baccalaureate degrees in science, technology, engineering, mathematics, and computer science—such as Differential Equations, Discrete Mathematics, Elementary Statistics, and Finite Mathematics—had small gains or losses in 2005 but overall remained nearly constant from 2000. Overall, the continued availability of baccalaureate-transfer courses in what the National Science Foundation calls STEM degrees (science, technology, engineering, and mathematics) indicates that two-year college mathematics programs continue to support the important national effort to have more students pass through two-year college mathematics programs on their way to STEM baccalaureate degrees, though declines in availability or in the rate of enrollment growth in these courses need continual monitoring.

TABLE TYE.5 Percentage of two-year college mathematics programs teaching selected mathematics courses at least once in either 1999–2000 or 2000–2001, and at least once in either 2004-2005 or 2005–2006. (Data for 2005 include only public two-year colleges.)

Course number	Type of course	2000	2005
1	Arithmetic/Basic Mathematics	56	48
2	Pre-algebra	41	47
3	Elementary Algebra (HS level)	78	80
4	Intermediate Algebra (HS level)	90	88
5	Geometry	14	24
6	College Algebra	83	79
7	Trigonometry	66	63
8	College Algebra & Trigonometry	32	17
9	Introductory Mathematical Modeling	12	7
10	Precalculus/ Elem Functions/ Analytic Geometry	65	60
11	Mainstream Calculus I	94	87
12	Mainstream Calculus II	88	78
13	Mainstream Calculus III	67	70
14	Non-mainstream Calculus I	40	46
15	Non-mainstream Calculus II	6	6
16	Differential Equations	59	58
17	Linear Algebra	39	41
18	Discrete Mathematics	19	22
19	Elementary Statistics	83	80
20	Probability	4	8
21	Finite Mathematics	32	35
22	Mathematics for Liberal Arts	50	65
23	Mathematics for Elementary School Teachers	49	66
24	Business Mathematics (not transferable) [1]	14	22
25	Business Mathematics (transferable) [2]	19	17
26	Technical Mathematics (non-calculus)	36	36
27	Technical Mathematics (calculus-based)	9	7

[1] Not transferable for credit toward a bachelors degree.
[2] Transferable for credit toward a bachelors degree.

TABLE TYE.6 Percentage of two-year college mathematics programs teaching selected mathematics courses in the fall term of 1990, 1995, 2000, and 2005. (Data for 2005 include only public two-year colleges.)

Course number	Type of course	Percentage of two-year colleges teaching course			
		1990	1995	2000	2005
11	Mainstream Calculus I	na	83	94	82
16	Differential Equations	53	53	59	25
17	Linear Algebra	34	30	39	19
18	Discrete Mathematics	21	12	19	12
19	Elementary Statistics	69	80	83	78
21	Finite Mathematics	46	31	32	28
22	Mathematics for Liberal Arts	35	46	50	56
23	Mathematics for Elementary School Teachers	32	43	49	59
26	Technical Mathematics (non-calculus based)	36	33	36	35
27	Technical Mathematics (calculus based)	6	11	9	5

Trends in average section size

In fall 2005, the average number of students per class section in two-year college mathematics courses continued a downward trend begun in 1990. As the footnote in Table TYE.7 explains, when computer science classes taught in the mathematics department are excluded, the average class size in fall 2000 was 24.8 students. In fall 2005, this size was 23 students. Refer to the general comments at the beginning of this chapter for more detail on the exclusion of computer science courses.

The precollege (remedial) and precalculus course strata each had average class size almost exactly 23, the average for all courses. Calculus classes were about 3 persons below the average while statistics classes were a little above the average of all classes.

For a closer examination of individual course average section sizes, see Table TYE.8. As one would expect, except for some specialized courses, the smallest class sizes were among advanced courses at the two-year college such as Mainstream Calculus III, Differential Equations, and Linear Algebra.

Table TYE.7 reports that 21% of all class sections in fall 2005 had size greater than 30. There is no comparable figure for 2000 since in CBMS2000 the comparison size for two-year colleges was 35 students per class section. In 2000, 10% of class sections were over 35 students.

In 2005, the lower cut-off of 30 students per class was chosen to make data for two-year colleges directly comparable to that collected for four-year institutions and to coincide with the recommendation from the Mathematical Association of America that undergraduate class size not exceed 30 students. At two-year colleges, 79% of all class sections in fall 2005 met the MAA goal. At four-year institutions, the average class size for freshman/sophomore-level courses through calculus ranged from 28 students to 33 students, depending on course type. At Ph.D.-granting institutions, these numbers ranged from 40 to 48. See Tables E.13 and E.14 in Chapter 3 for four-year institutional data.

TABLE TYE.7 Average on-campus-section size by type of course in mathematics programs at two-year colleges, in fall 2000 and 2005. Also percentage of sections with enrollment above 30 in fall 2005. (Data for 2005 include only public two-year colleges.)

Course number [1]	Type of course	2000 average section size	2005 average section size	Percentage of 2005 sections with size > 30
1–5	Precollege	24.5	23.9	21%
6–10	Precalculus	24.8	23.6	23%
11–16	Calculus	20.8	20.0	16%
19–20	Statistics	25.2	25.9	33%
1–28	Total all courses	24.8 [2]	23.0	21%

[1] For names of specific courses see Table TYR.3.

[2] The average section size of 23.7 reported in CBMS2000 included computer science courses taught in mathematics programs. Combining data from Tables TYR.4 and TYR.9 of CBMS2000 gives an estimate of 24.8 for the average section size of non-computer-science courses (numbered 1-28) in fall 2000.

TABLE TYE.8 Average on-campus section size for public two-year college mathematics program courses, in fall 2005.

Course number	Type of course	Average section size	Course number	Type of course	Average section size
1	Arithmetic & Basic Math	22.7	16	Differential Equations	14.2
2	Pre-algebra	22.3	17	Linear Algebra	16.3
3	Elem Algebra (HS level)	24	18	Discrete Mathematics	14.3
4	Intermed Algebra (HS level)	25.1	19	Elementary Statistics	26.1
5	Geometry (HS level)	17.8	20	Probability	22.6
6	College Algebra	24.7	21	Finite Mathematics	25.3
7	Trigonometry	22.5	22	Math for Liberal Arts	24
8	College Alg & Trig. (combined)	21.7	23	Math for Elem Teachers	15.4
9	Intro to Math Modeling	24.6	24	Business Math (not transferable)	21.1
10	Precalculus [1]	21.2	25	Business Math (transferable)	8.6
11	Mainstream Calculus I	21.9	26	Technical Math (non-calculus)	18.7
12	Mainstream Calculus II	18.2	27	Technical Math (calculus-based)	18.1
13	Mainstream Calculus III	15.6	28	Other mathematics	22
14	Non-mainstream Calculus I	22.9			
15	Non-mainstream Calculus II	20.8			

[1] Includes Precalculus, Elementary Functions, and Analytic Geometry.

Trends in the use of part-time faculty

In fall 2005, part-time faculty made up a slightly larger part of the overall mathematics faculty at two-year colleges than they did in 2000. However, this statement requires some explanation. The relevant issue, as the faculty data in Table TYF.1 in Chapter 7 reflect, is who is included in the various categories. When faculty of every sort are included, such as part-time faculty paid by third parties and also temporary full-time faculty, part-time faculty in fall 2005 made up about 68% of the total faculty. The comparable figure in 2000 was 66%. If the 1,915 third-party-payee part-time faculty members are excluded, in fall 2005 about 66% of the faculty had part-time status. The comparable figure for 2000 was 65%.

Though making up about two-thirds of the faculty by headcount, part-time faculty taught only about 44% of mathematics program class sections in fall 2005. This occurred because most institutions impose a limit on the maximum number of credits a part-time faculty member can teach in comparison to the 15 contact hours weekly most full-time faculty teach. Again, see Chapter 7 for details. In fall 2000, 46% of class sections were taught by part-time faculty. In fall 1995, this figure was 38%.

Concerning the important instructional issue of which types of courses are taught most often by part-time faculty, the pattern in fall 2005 did not change from fall 2000. Once again in fall 2005, it was more likely that a part-time faculty member was teaching a course below calculus than a calculus course. It was most likely of all that the part-time faculty member was teaching a precollege (remedial) course. Table TYE.9 contains the relevant percentages.

TABLE TYE.9 Number of sections and number and percentage of sections taught by part-time faculty in mathematics programs at public two-year colleges by type of course, in fall 2005.

Course number[1]	Type of course	Number of sections	Number of sections taught by part-time faculty	Percentage of sections taught by part-time faculty
1–5	Precollege level	38814	21696	56%
6–10	Precalculus level	12898	3914	30%
11–13	Mainstream Calculus	3973	493	12%
14–15	Non-mainstream Calculus	923	254	28%
16–18	Advanced level	617	58	9%
19–20	Statistics	4142	1452	35%
21–25	Service courses	6710	1913	29%
26–27	Technical mathematics	927	339	37%
28	Other mathematics	1193	552	46%
1–28	Total all courses	70197	30671	44%

[1] For names of specific courses see Table TYR.3.

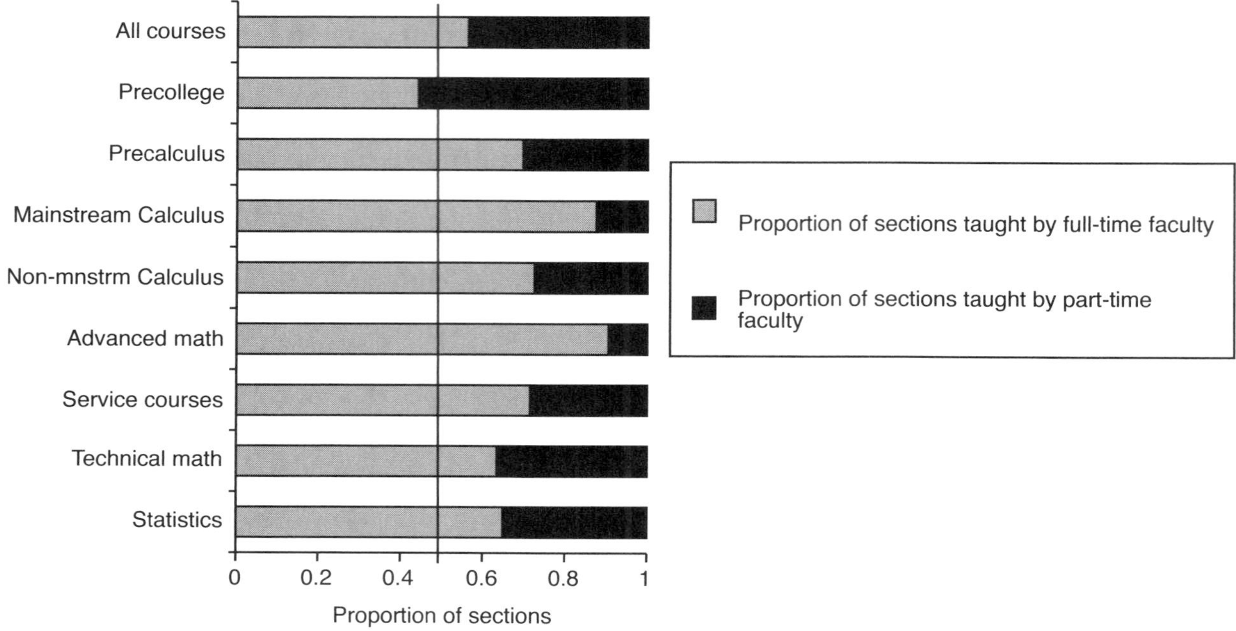

FIGURE TYE.9.1 Proportion of sections of mathematics and statistics courses taught by full-time and part-time faculty in mathematics programs at public two-year colleges by type of course in fall 2005.

Instructional Practices In Mathematics Programs

Table TYE.10 presents the percentage of class sections in mathematics courses at public two-year colleges that used various instructional practices in fall 2005. The predominant instructional method was the standard lecture format, with percentage of use in an individual course ranging from 93% in Differential Equations and 81% in Mainstream Calculus I to 74% in each of College Algebra and Elementary Algebra. The only exceptions to the predominance of the lecture method were Mathematics for Elementary School Teachers and certain business mathematics courses. CBMS2000 reported that 78% of all class sections used the lecture method. This last percentage was 77% in 1995.

Data and analysis on how first-year courses were taught at four-year institutions can be found in Chapter 5 of this report in Tables FY.2 through FY.10. For comparative data about four-year and two-year institutions, see Chapter 1, Tables S.11 through S.13.

Instructional methods in precalculus and calculus courses

In fall 2005 there also were clear patterns among various types of courses regarding the four instructional techniques included in the survey (use of a graphing calculator, inclusion of a writing component, computer assignments, and the use of group projects). For all calculus courses (both mainstream and non-mainstream) and for precalculus courses, the graphing calculator was used more frequently than any other technique. The percentage of sections using graphing calculators in calculus and precalculus courses ranged from 74% to 81%, very similar to the range in 2000 of 69% to 83%. Only Non-mainstream Calculus II had a distinctly lower use (40%), and this may well be attributed to its extremely low reported enrollment.

Table TYE.11 gives an historical perspective over ten years on the use of writing assignments and group projects in various types of calculus courses. This table reflects monitoring by the CBMS survey of the overall effect of the calculus reform movement on calculus instruction. In earlier years, use of these methods was associated closely with adoption of "calculus reform" either by entire departments or by individual faculty members, but by 2005, the best aspects of the 1990s movement for calculus instructional and content reform had settled into almost every available calculus textbook, making it hard to classify any mathematics program as reformist or non-reformist based on the use of such instructional techniques.

For a broader perspective than Tables TYE.10 and TYE.11 can give, the following display adds computer assignments to the overall picture, as well as the percentage use of all three techniques in the

Precalculus course. This layout focuses on what happened in these areas from 2000 to 2005. As noted above, during this period there was a slight increase in the already high percentage usage of graphing calculators in all these courses. But in almost every course and for almost every one of the three techniques, the percentage of use declined over this five-year period, sometimes substantially. The three exceptions were under group projects. Only one of the three exceptions had a significant percentage increase, and this increase was in the low-enrollment Non-mainstream Calculus II course for which data were less reliable.

	Writing Assignments		Computer Assignments		Group Projects	
	2000	2005	2000	2005	2000	2005
Precalculus	22%	14%	16%	9%	20%	21%
Main Cal I	31%	19%	35%	20%	27%	19%
Main Cal II	25%	18%	37%	30%	25%	25%
Main Cal III	21%	16%	35%	28%	23%	20%
Non-M Cal I	20%	14%	15%	9%	20%	14%
Non-M Cal II	39%	21%	24%	0%	8%	27%

Calculus data for four-year institutions can be found in Tables S.11 and S.12 in Chapter 1, broken down by the size of the lecture section used by the institution.

On-line resource systems

CBMS2005 added a new survey question related to the use of on-line resource systems in instruction. These systems, which have been vigorously promoted by publishers as supplements to textbooks and sometimes as stand-alone instructional systems, can involve a wide variety of teaching aids such as automated outside-of-class practice, automated graded homework assignments, and automated testing. As Table TYE.10 reports, these systems were used in only a small percentage of precalculus and calculus classes at two-year colleges. Their proportion of use was about the same in four-year institutions (S.11 and S.12). Only in arithmetic courses, algebra courses of all kinds, and statistics courses did their use reach 10% of sections.

Instructional methods in courses other than precalculus and calculus

Graphing calculator usage in courses other than Precalculus and the various levels of calculus held steady or grew modestly between 2000 and 2005. However, the use of graphing calculators in sections of College Algebra showed a 14-point drop to 60%. In sections of the combined College Algebra/Trigonometry course, which also had a large decline in availability, calculator usage dropped 33 points to 53%. Courses reporting an especially large growth in percentage of sections using graphing calculators were Differential Equations, up 29 points to 81%; Probability, up 27 points to 83%; Statistics, up 14 points to 73%; and Mathematics for Liberal Arts, up 13 points to 33%.

For writing assignments, there was an almost across-the-board decline in use between 2000 and 2005 in courses other than Precalculus and the various levels of calculus. In most cases, the decline was small, in the range of five percentage points, but a few cases stand out. Geometry, which was being offered at notably more colleges in 2005, reported use of writing in 25% of sections, up 21 points from 2000. Writing was down 35 points to 38% in Introduction to Mathematical Modeling and was down 14 points to 52% in courses for future elementary school teachers. Use of writing in courses for liberal arts students was down five points to 36%, but still maintained their standing in the top six of courses that used writing.

Changes in the percentage of sections using computer assignments between 2000 and 2005 varied greatly. Geometry was up 20 points to 23%. Combined College Algebra/Trigonometry was up 14 points to 25%. Discrete Mathematics and Finite Mathematics were up 10 and 11 points to 33% and 19%, respectively. On the other hand, Linear Algebra dropped 11 points to 29%. Probability dropped 10 points to 49%. Introduction to Mathematical Modeling dropped seven points to 17%. Mathematics for Liberal Arts and Mathematics for Elementary School Teachers each dropped eight points to 7% and 13%, respectively.

TABLE TYE.10 Percentage of on-campus sections using different instructional methods by course in mathematics programs at public two-year colleges, in fall 2005.

		Percentage of sections taught using						
	Type of Course	Graphing calculators %	Writing assignments %	Computer assignments %	Group projects %	On-line resource systems %	Standard lecture method %	Number of sections
1	Arithmetic	2	3	13	9	14	64	4,400
2	Pre-algebra	5	9	18	9	7	74	5,954
3	Elementary Algebra (HS)	17	7	14	8	11	74	15,331
4	Intermed Algebra (HS)	32	8	13	9	11	77	12,773
5	Geometry (HS)	33	25	23	15	0	68	356
6	College Algebra	60	17	8	14	14	74	7,866
7	Trigonometry	67	14	3	16	7	81	1,529
8	College Algebra & Trig	53	8	25	10	13	78	654
9	Intro Math Modeling	80	38	17	59	6	64	248
10	Precalculus [1]	75	14	9	21	6	76	2,601
11	Mnstrm Calculus I	79	19	20	19	5	81	2,226
12	Mnstrm Calculus II	81	18	30	25	7	86	1,054
13	Mnstrm Calculus III	74	16	28	20	4	83	693
14	Non-mstrm Calculus I	77	14	9	14	3	76	883
15	Non-mstrm Calculus II	40	21	0	27	0	89	40
16	Differential Equations	81	11	27	21	5	93	290
17	Linear Algebra	60	18	29	14	0	68	204
18	Discrete Mathematics	47	39	33	23	0	82	123
19	Elementary Statistics	73	44	45	24	10	85	3,872
20	Probability	83	55	49	50	0	68	270
21	Finite Mathematics	55	17	19	11	3	68	844
22	Math for Liberal Arts	33	36	7	25	6	79	2,232
23	Math for Elem Tchrs	21	52	13	48	3	48	1,665
24	Business Math [2]	6	2	18	1	0	87	539
25	Business Math [3]	18	7	7	6	2	24	1,430
26	Tech Math (non-calc)	39	4	5	5	5	72	863
27	Tech Math (calc)	63	17	21	30	0	83	64
28	Other math	27	10	5	7	6	63	1,193

[1] Includes precalculus, elementary functions, and analytic geometry.
[2] Not transferable for credit toward a bachelors degree.
[3] Transferable for credit toward a bachelors degree.

TABLE TYE.11 Percentage and number of calculus sections in mathematics programs at two-year colleges that assign group projects and that have a writing component, in fall 1995, 2000, and 2005. (Data for 2005 include only public two-year colleges.)

		Percentage of sections with group projects			Percentage of sections with a writing component			Number of sections		
Course number	Type of course	1995	2000	2005	1995	2000	2005	1995	2000	2005
11	Mainstream Calculus I	22	27	19	20	31	19	2325	2298	2226
12	Mainstream Calculus II	18	25	25	13	25	18	1008	957	1054
13	Mainstream Calculus III	22	23	20	16	21	16	733	686	693
14	Non-mstrm Calculus I	20	20	14	17	20	14	1010	728	883
15	Non-mstrm Calculus II	22	8	27	16	39	21	75	57	40

Distance learning

The comments that precede Table E.4 in Chapter 3 explain why the survey question in CBMS2005 about "distance learning" was phrased in terms of course enrollment, rather than the number of class sections, for both four-year and two-year colleges.

In the 1995 CBMS survey, two-year colleges were asked about course sections taught using television. Technology rapidly made this question obsolete. The 2000 survey inquired about the number of course sections taught via "distance learning," which was described as a course structure in which at least half the students in the section received the majority of instruction in a format where the instructor was not physically present. CBMS2005 asked the same question of two-year colleges as was asked in 2000, but CBMS2005 asked in terms of course enrollment because distance-learning sections are not bound by room-size limits and tend to vary dramatically in enrollment depending on local administrative practice.

Looking back over ten years, less than 1% of mathematics class sections at two-year colleges were offered via television in 1995, and only 2.5% of sections in 2000 were described as using distance learning. Among high-enrollment courses in 2000, College Algebra had 6.7% of sections offered via distance learning, and Elementary Statistics had 5.8%.

For fall 2005 in two-year colleges, the relevant data are in Table TYE.12. The rounded-by-course enrollment figures given in that table exclude dual enrollments and total 1,670,000 students. When per-course distance enrollment is calculated, using the percentages in Table TYE.12, almost 81,000 students are reported in some form of distance education in fall 2005, about 5% of the mathematics program enrollment at two-year colleges.

At four-year institutions in fall 2005, "distance learning" was used sparingly, with only one of the course groupings in Table E.4 showing more than 2% of total enrollment in a distance format. By contrast, in two-year colleges (again, see Table TYE.12), only six of the 27 individual courses listed show a distance enrollment of less than 2%. At two-year colleges, the percentage of distance enrollment was quite high in some courses such as Geometry (12%), Business Mathematics (11%), Introduction to Mathematical Modeling (11%), and Mathematics for Elementary School Teachers (10%). In Elementary Statistics the percentage was 9%.

TABLE TYE.12 Percentage of distance-learning enrollments (= where at least half of the students receive the majority of their instruction using a method where the instructor is not physically present) among all enrollments (excluding dual enrollments) in certain courses in mathematics programs at public two-year colleges in fall 2005, and total enrollments (in 1000s) in those courses.

	Type of Course	Total Enrollment[4] (1000s)	Percentage Distance Enrollments
1	Arithmetic	104	4%
2	Pre-algebra	137	3%
3	Elementary Algebra (HS)	380	4%
4	Intermed Algebra (HS)	336	5%
5	Geometry (HS)	7	12%
6	College Algebra	206	6%
7	Trigonometry	36	4%
8	College Algebra & Trig.	14	1%
9	Intro Math Modeling	7	11%
10	Precalculus	58	4%
11	Mainstream Calculus I[1]	51	5%
12	Mainstream Calculus II	19	1%
13	Mainstream Calculus III	11	2%
14	Non-mstrm Calculus I	21	5%
15	Non-mstrm Calculus II	1	0%
16	Differential Equations	4	0%
17	Linear Algebra	3	2%
18	Discrete Mathematics	2	2%
19	Elementary Statistics	111	9%
20	Probability	7	7%
21	Finite Mathematics	22	5%
22	Mathematics for Liberal Arts	59	8%
23	Math for Elem Teachers	29	10%
24	Business Mathematics[2]	13	9%
25	Business Mathematics[3]	14	11%
26	Tech Math (non-calculus)	16	1%
27	Tech Math (calculus)	1	0%

Note: 0% means less than one-half of one percent.

[1] Mainstream calculus courses are typically for mathematics, physics, and engineering majors.
[2] Not transferable for credit toward a bachelors degree.
[3] Transferable for credit toward a bachelors degree.
[4] Does not include dual enrollments.

Services Available to Students

Chapter 2 of this report contains a comparison of academic services and other resources available to four-year college students and to two-year college students in fall 2005. See Tables SP.11 through SP.15 in that chapter. Table TYE.13 gives the percentage of mathematics programs at two-year colleges that offered various services to students in fall 2005.

Placement testing, tutorial laboratories, outreach projects, independent study, honors programs, programs for minorities, and programs for women

Table TYE.13 reports that diagnostic or placement testing was almost universally available in two-year colleges (97%). SP.11 reports that 97% of these colleges made such testing mandatory for first-time students, 90% of colleges required that the student discuss the placement scores with an advisor, and 88% used this score as part of a mandatory course placement program.

SP.11 also reports the source of placement tests used by two-year colleges. The decrease in locally produced tests was dramatic, from 99% to 11%. About one-third of colleges reported using commercial tests from American College Testing (ACT), and one-third reported using tests from Educational Testing Service (ETS). About 25% used other test providers. This almost-universal movement to commercial test providers likely is related to the transfer of many advising responsibilities, as discussed below, to centralized advising centers.

Mathematics tutorial centers or labs were available at almost all colleges (95%).

Two new items associated with the mathematics program had been included for the first time in the 2000 survey: outreach projects to K–12 schools and opportunities for independent study. In 2005, both had grown in availability at two-year colleges, from 20% to 25% and from 25% to 38%, respectively. By contrast (see SP.14 in Chapter 2), opportunities for involvement with K–12 schools dropped in four-year colleges from 47% to 34%, though many other opportunities at four-year colleges were more broadly available.

Special programs to encourage minorities in mathematics were reported in 15% of two-year colleges, up from 4% in 2000 and surpassing the 11% reported in 1995. Over ten years, honors sections in mathematics programs continued to grow, from 17% in 1995 to 20% in 2000 and to 24% in 2005. Participation in mathematics contests was reported by 37% of colleges.

Faculty advisors and advising

The period from 1995 to 2000 witnessed a 50% drop (down 32 percentage points) in colleges that offered mathematics advising to students by members of the mathematics faculty. By 2005, this pattern had partly reversed itself with 40% of colleges, up from 33%, reporting that advising was available from mathematics faculty.

CBMS2000 attributed these numbers to a systematic move among two-year colleges over the previous decade to locate academic advising within a student services unit where generalists offered academic counseling in all subject areas. The motivation for such a move offered in the CBMS2000 report remained valid in 2005. Two-thirds of the mathematics faculty are part-time, many of whom do not assist with advising. Hence, the full-time faculty is stretched thin to cover this duty. The student body itself is very fluid—part-time, drop-in/drop-out, night-only, weekend, working, non-residential—and not readily available on campus when the relatively few permanent full-time faculty members are present. Hence, offering advising through a student services unit, where it can be tied directly to diagnostic and placement testing, makes advising accessible to more students.

Anecdotally, mathematics faculty members complain about the accuracy of the advice students receive from non-mathematicians working in multi-disciplinary advising units. This might in part explain the increase in faculty involvement in advising that appeared in fall 2005.

The 2006 Community College Survey of Student Engagement (CCSSE), conducted under the auspices of the Community College Leadership Program at The University of Texas at Austin, reported that the majority of community college students felt academic advising was the most important support service their colleges provided, even more important than financial aid. Yet in that survey 29% of part-time students and 16% of full-time students (23% of all students) reported that they did not use advising services. Among remedial students, 26% reported that they rarely or never participated in academic advising. This last percentage was an extremely large 41% for students taking college-level courses.

The largest student group (43%) in the CCSSE survey reported that the best source of academic advising was a faculty member. Friends, family, or other students were listed as the best advising source by 26%. Only 10% of students indicated that the best academic advice came from a non-faculty-member academic advisor, and only 7% said that the best advice was on-line or obtained via computer. A companion survey, the 2006 Community College Faculty Survey of Student Engagement, indicated that about 90% of full-time faculty and 60% of part-time faculty spent some time advising students during a typical week, though CCSSE reported this fact negatively, namely, that 10% of full-time faculty and 40% of part-time faculty reported spending zero hours weekly advising students.

The CCSSE survey, based on data from 2004, 2005, and 2006, included 249,548 community college students at 447 colleges in 46 states. The survey can be downloaded at http://www.ccsse.org. A news release about the survey is at http://www.edb.utexas.edu/education/news/2006/CCSSE_06.php. Highlights are given at http://www.edb.utexas.edu/education/news/2006/CCSSE_highlights06.php. The survey is reported in the December 1, 2006 issue of *The Chronicle of Higher Education*.

In light of the CCSSE data about faculty involvement in advising and the increase in mathematics faculty advising reported in CBMS2005, there is evidence that many students seek and get mathematical advising from faculty members. This occurs in spite of the apparent systematic institutional shift of advising to generic advising centers suggested in earlier CBMS surveys. The CCSSE survey strongly suggests that faculty advising is what students prefer.

TABLE TYE.13 Percentage of two-year colleges offering various opportunities and services to mathematics students, in fall 2000 and 2005. (Data for 2005 include only public two-year colleges.)

Opportunity/Service	2000	2005
Diagnostic or placement testing	98	97
Mathematics lab or tutorial center	98	95
Advising by a member of the mathematics faculty	33	40
Opportunities to compete in mathematics contests	28	37
Honors sections	20	24
Mathematics club	14	22
Special mathematics programs to encourage minorities	4	15
Lectures/colloquia for students, not part of math club	9	6
Special mathematics programs to encourage women	4	7
K-12 outreach opportunities	20	25
Undergraduate research opportunities	4	9
Independent mathematics studies	25	38
Other	4	4

Mathematics labs and tutoring centers

In fall 2005, as noted above, 95% of mathematics programs at two-year colleges reported a mathematics lab or tutorial center. Table TYE.14 shows the various services available in these centers. Almost all labs (94%) offered tutoring by students. Media-oriented tools such as videotapes, computer-aided instruction, computer software, and internet access were common in labs, as reported by three-quarters of the colleges. The involvement of full-time faculty in tutoring labs was reported by 50% of colleges, up 10 points from 2000, with part-time-faculty involvement about the same. Paraprofessionals were part of the personnel in two-thirds of the labs. These latter are non-faculty staff who may not hold any collegiate degrees or no collegiate degrees beyond the bachelors.

TABLE TYE.14 Percentage of two-year colleges with a mathematics lab or tutorial center that offer various services to students in fall 1995, 2000, and 2005. (Data for 2005 include only public two-year colleges.)

	Percentage of two-year colleges with math lab/tutorial center that offer various services to students		
Services offered in mathematics lab or tutorial center	1995	2000	2005
Computer-aided instruction	69	68	75
Computer software such as computer algebra systems or statistical packages	65	69	72
Internet resources	na	53	77
Media such as videotapes	70	74	68
Organized small-group study sessions	na	na	62
Tutoring by students	84	96	94
Tutoring by paraprofessionals	58	68	67
Tutoring by part-time mathematics faculty	39	48	48
Tutoring by full-time mathematics faculty	38	42	51

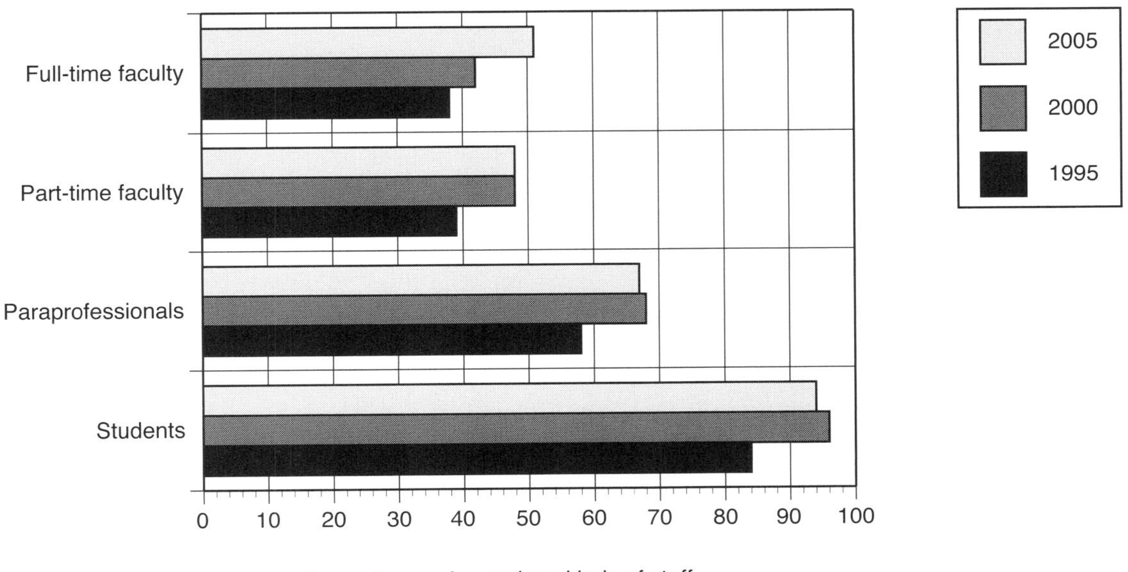

FIGURE TYE.14.1 Percentage of two-year colleges using various sources of personnel for staffing mathematics labs or tutoring centers in fall 1995, 2000, and 2005. (Data for 2005 include only public two-year colleges.)

Mathematics Courses Taught Outside of the Mathematics Programs

Not unlike their four-year counterparts, two-year colleges have a long history of offering mathematics courses in instructional units outside of the mathematics program. Tables TYE.15, TYE.16, and TYE.17 give the enrollment in mathematics courses offered outside of mathematics programs. These enrollments were estimated by mathematics program heads. Thus, they may not be as accurate as the numbers given for enrollment within mathematics programs.

In fall 2005, 80% of the outside enrollment was in precollege (remedial) courses taught either in a learning lab or in another unit such as a developmental studies division. The remainder of this outside enrollment was concentrated in business mathematics taught in a business division, statistics and probability also mostly taught in a business division, and technical mathematics taught in occupational training programs.

Percentage of precollege mathematics taught outside of the mathematics program

The largest and most important component of this "outside" mathematics enrollment is precollege developmental courses. The structure of precollege course offerings within a particular college is affected by the institution's philosophy concerning developmental education. Two views predominate. Either a student takes all developmental courses (mathematics, reading, and writing) in a self-contained unit devoted to developmental studies, or developmental courses are offered as part of the disciplinary curriculum.

The earliest CBMS survey for which "outside" precollege mathematics enrollment data are available on a course-by-course basis was in 1990. The following percentages are obtained by using Table TYE.3 and Table TYE.15. They trace the pattern of enrollment outside the mathematics program from 1990 to 2005 in Arithmetic, Elementary Algebra and Intermediate Algebra as a percentage of total enrollments in the course or the course group.

	1990	1995	2000	2005
Arithmetic/Prealgebra	18%	19%	17%	20%
Elementary Algebra	13%	12%	12%	15%
Intermediate Algebra	9%	4%	4%	7%

These "outside of mathematics program" precollege-level courses experienced a 42% drop in enrollment from 1995 to 2000 but rebounded with an 89% enrollment increase from 2000 to 2005. Though built on a much smaller base, nonetheless this percentage increase was about three times the percentage enrollment increase from 2000 to 2005 within the mathematics program itself.

Organization of mathematics courses outside of the mathematics program

Table TYE.17 shows 31% of colleges indicated that some part of their developmental mathematics program was administered separately from the mathematics program. This percentage was 29% in both 2000 and 1995. Almost all of the precollege enrollment outside of the mathematics program likely was in a learning center or some form of a developmental education division within the college.

The "shift to outside enrollment" for precollege mathematics courses that shows up in CBMS2005 is too small to harbinger a return to the large, independent developmental mathematics divisions of the 1970s, but it is a statistic that is interesting to watch.

TABLE TYE.15 Estimated enrollment (in 1000s) in mathematics and statistics courses taught outside of mathematics programs at two-year colleges, in fall 1990, 1995, 2000, and 2005. (Data for 2005 include only public two-year colleges.)

	Enrollment (in 1000s)			
Type of course	1990	1995	2000	2005
Arithmetic/Pre-algebra	42	54	43	60
Elementary Algebra (HS level)	38	41	27	65
Intermediate Algebra (HS level)	27	10	10	26
Business Mathematics	32	26	18	15
Statistics & Probability	15	9	7	12
Technical Mathematics	10	8	5	10
Total	164	148	110	188

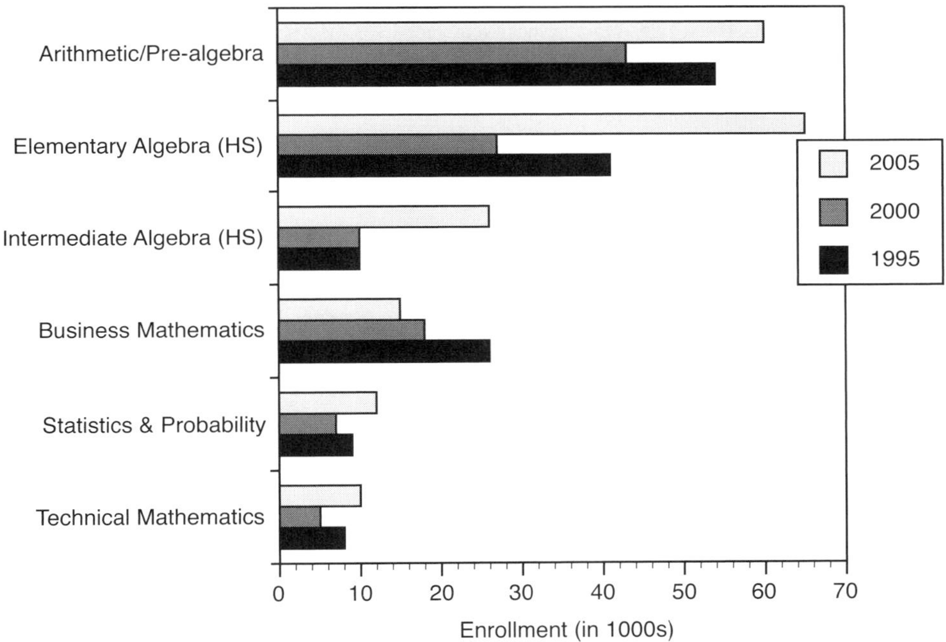

FIGURE TYE.15.1 Estimated enrollment (in 1000s) in mathematics and statistics courses taught outside of mathematics programs at two-year colleges in fall 1995, 2000, and 2005.

TABLE TYE.16 Estimated enrollment (in 1000s) in mathematics courses taught outside of mathematics programs at public two-year colleges, by division where taught, in fall 2005.

	Mathematics Enrollment (in 1000s) in Other Programs			
Course	Occupational Programs	Business	Learning Center	Other Depts/ Divisions[1]
Arithmetic/Pre-algebra	1	1	9	50
Elem Algebra (HS)	1	0	5	59
Intermed Algebra (HS)	0	0	3	22
Business Mathematics	0	14	0	1
Statistics & Probability	0	8	0	4
Technical Mathematics	8	0	0	1
Total	**11**	**23**	**17**	**137**

Note: 0 means less than 500 enrollments and this may cause column sums to seem inaccurate.

[1] A developmental studies department whose mathematics component is not supervised by the mathematics department would be an example.

TABLE TYE.17 Percentage of two-year colleges in which some of the precollege (remedial) mathematics course offerings are administered separately from, and not supervised by, the mathematics program, e.g. in a developmental studies department, with estimated percentages of enrollment outside of the mathematics program, by type of course, in fall 1990, 1995, 2000, and 2005.

Mathematics Outside of the Mathematics Program	1990 %	1995 %	2000 %	2005 %
Percentage of TYCs with some precollege mathematics courses outside of mathematics program control	na	29	29	31
Arithmetic/Pre-algebra	18	19	17	20
Elementary Algebra	13	12	12	15
Intermediate Algebra	9	4	4	7

Special Instructional Activities In Mathematics Programs

Teacher training

Enrollment data in Tables TYE.3 and TYE.5 give a partial perspective on the involvement of mathematics programs at two-year colleges in teacher education, especially in the preparation of future K–8 teachers. The expansion of two-year-college activity in this area has been rapid. Hence, the topic was one of the survey's Special Projects both in CBMS2000 and in CBMS2005. The reader should see Tables SP.2 and SP.4 in Chapter 2 for a comprehensive perspective on the mathematics education of future teachers at two-year and four-year institutions. For a more detailed discussion concerning two-year colleges, with an emphasis on the scope and organizational structure of teacher education in mathematics programs at two-year colleges, see the last section of Chapter 7.

Dual-enrollment courses

In fall 2000, so-called dual-enrollment courses were a growing phenomena that affected two-year college mathematics programs. Hence, in 2005 additional information was collected about these courses. A discussion of the 2005 survey results, including enrollment data and comparisons to what is happening in the same regard at four-year institutions, can be found with the Special Projects analysis in Chapter 2, Tables SP.16 and SP.17. Additional commentary on dual enrollment also can be found in Chapter 7 where it is discussed with emphasis on the credentials and the supervision of those who teach such courses.

These dual-enrollment courses earned credit both for high school graduation and at the two-year institution. In most cases, these courses were not "outside" the mathematics program in the sense of the CBMS survey. They had some level of supervision from the mathematics program, and most mathematics programs counted them among the courses offered by the program. However, these courses often were at the edge of mathematics program supervision since they often were taught by the regular high school mathematics faculty, who were hired and paid by the high school district.

Chapter 7

Faculty, Administration, and Special Topics in Mathematics Programs at Two-Year Colleges

This chapter continues the presentation of data and analysis about mathematics programs in public two-year colleges. It reports the number, teaching conditions, education, professional activities, age, gender, and ethnicity of the faculty in these mathematics programs in fall 2005. Also included is information on mobility into, within, and out of two-year college mathematics program teaching positions. Additional analysis of the items discussed in this chapter can be found in Chapters 1 and 2 where they are discussed from a comprehensive point of view in comparison to similar data for four-year colleges and universities. In particular, Chapter 2 discusses issues related to dual-enrollment courses and pre-service teacher training.

The data are compared with those from the 1975, 1980, 1985, 1990, 1995, and 2000 CBMS surveys. Unlike surveys prior to 1995, the mathematics faculty surveyed in 1995, 2000, and 2005 did not include faculty who taught in computer science programs that were separate from mathematics programs. Also, in contrast to previous surveys, the data is drawn from a survey of public two-year colleges only. A more detailed statement on these issues occurs at the beginning of Chapter 6. Information on the sampling procedure used in the 2005 survey can be found in Appendix II. A copy of the two-year college survey questionnaire for CBMS2005 can be found in Appendix V.

The term "permanent full-time" is used frequently below. Faculty members in this category at two-year colleges have an on-going stable relationship to the mathematics program similar to that of tenured and tenure-track faculty at four-year institutions. They occupy a recurring slot in the college's budget and are subject to the college's long-term evaluation and reappointment policies. They are the group of faculty primarily responsible for curriculum development, student advising, committee appointments, and other forms of college service. Full-time faculty who are not permanent are called "temporary full-time faculty."

The term "tenure" is not used because the majority of two-year colleges do not have traditional tenure systems, and the use of the word "tenure" in the survey questionnaire would have been confusing to respondents. At the majority of two-year colleges, faculty stability is embodied in a sequence of recurring contracts or appointments typically running from three to five years. Permanent full-time faculty members teach full course assignments, which distinguishes them from part-time or adjunct faculty. They also are distinguished from "temporary" full-time faculty who are meeting a short-term institutional need and do not participate in the college's on-going reappointment process.

The Table display code in this chapter is TYF, for "two-year faculty," since the chapter deals mostly with issues related to faculty.

Highlights of Chapter 7

- There were almost 8,800 permanent full-time faculty in public two-year college mathematics programs in the United States in fall 2005, a 26% increase from 2000 that strongly reversed the 8% decline that occurred between 1995 and 2000. Another 610 individuals were teaching as temporary full-time faculty, a 63% decrease from 2000 in those occupying temporary status and a sharp change from the 600% increase in temporary full-time faculty that occurred between 1995 and 2000. See Table TYF.1.
- Once again, in fall 2005 the number of part-time faculty in two-year college mathematics programs doubled the number of full-time faculty. Part-time faculty, if those paid by third parties such as school districts are included, made up 68% of the total faculty. When third party payees are omitted, part-time faculty made up 66% of the faculty. In 2000, this last percentage was 65%. About 44% of all sections were taught by part-time faculty members, a two-point drop from 2000. See Tables TYF.1 in this chapter and TYE.9 in Chapter 6.
- In light of the previous bullet, the data suggest that the large enrollment increase in mathematics and statistics that occurred in public two-year colleges from 2000 to 2005 was accompanied by a proportional growth in permanent full-time faculty and was not accommodated by employing a disproportional number of part-time faculty members. On enrollment, see Table TYE.2 in Chapter 6 and Table S.1 in Chapter 1.
- However, one should note that 53% of permanent full-time faculty in fall 2005 taught extra hours

for extra pay at their own college, little changed from the 52% reported in 2000. The average "extra" assignment for these faculty members was slightly more than one three-credit course, namely, 3.6 classroom contact hours weekly. This extra work accounted for about 4700 class sections, classified as being taught by full-time faculty, that otherwise would have required additional part-time staffing and would have raised the percentage of sections taught by part-time faculty to 50%. See Tables TYF.2 in this chapter and TYE.9 in Chapter 6.

- The average teaching assignment for permanent full-time faculty in classroom contact hours per week increased 3% in fall 2005 in comparison to fall 2000, from 14.8 hours to 15.3 hours. See Table TYF.2.
- In fall 2005, a masters degree was the terminal degree for 82% of permanent full-time mathematics faculty members at two-year colleges, up one point from 2000. An additional 16% held doctorates. In fall 2000, in a large and troubling increase, 19% of newly-hired permanent full-time faculty members were reported as holding only bachelors degrees. In 2005, this percentage for newly-hired faculty fell back sharply to 5%, but was still higher than the 1% reported in 1995. See Tables TYF.4, TYF.5, and TYF.19.
- Among part-time faculty in fall 2005, 22% had a bachelors degree as their highest degree, a status generally allowed by accrediting agencies for those who teach only precollege (remedial) courses. Among all degree types, 21% of part-time faculty had majors outside of mathematics, mathematics education, or statistics. See Tables TYF.6 and TYF.7.
- For the first time in a CBMS survey, the proportion of men and women among the permanent full-time faculty was exactly equal at 50%. Women made up 47% of the part-time faculty. See Tables TYF.8 and TYF.9.
- About 14% of permanent full-time faculty members in mathematics programs in fall 2005 were ethnic minorities, up slightly from the 13% reported in 2000. Ethnic minorities made up a higher proportion (23%) of the under-age-40 faculty than they did of the faculty as a whole. The percentage split between White (non-Hispanic) faculty and ethnic minority faculty almost exactly reflected the corresponding split for masters degrees awarded in mathematics and statistics in the United States in 2003–2004. See Tables TYF.10, TYF.11, TYF.12, and TYF.13.
- Among newly-hired permanent full-time faculty in fall 2005, 20% were ethnic minorities and 53% were women. See Table TYF.20.
- Among part-time faculty, 16% were ethnic minorities in fall 2005. See Tables TYF.14 and TYF.15.
- Distribution of faculty by age in fall 2005 was essentially identical to that in 2000, with 28% of the permanent full-time faculty over age 55 and 46% over age 50. The average age was 47.8. See Tables TYF.16 and TYF.17 in this chapter and Table S.18 in Chapter 1.
- There was a notable change in fall 2005 in the selection pattern for the 605 newly-hired permanent full-time faculty members. The percentage hired from graduate school jumped from 8% in 2000 (when the base was 572) to 23%, almost one-quarter of the new permanent full-time faculty hires. Additionally, 18% of these new full-time faculty arrived from teaching jobs at four-year institutions, up from 8%. Those hired from high school dropped to 13%, a decline of nine points. See Tables TYF.18 and TYF.19.
- Of the new hires in fall 2005, 22% were under age 30, 42% were under age 35, and 59% were under age 40. See Table TYF.21.
- Ready availability of computers or terminals continued to be a difficulty in fall 2005 for part-time faculty, with only 63% of institutions reporting these tools were in part-time faculty offices. In fall 2000, the CBMS survey reported essentially 100% availability in full-time faculty offices. Desk sharing remained common among part-time faculty, with sharing among three or more individuals reported in 65% of cases. See Tables TYF.23 and TYF.24.
- Unexpectedly, in fall 2005 the percentage of two-year colleges requiring periodic teaching evaluations for all full-time faculty members dropped from 98% to 89%. However, there was a jump in the percentage of colleges that used classroom visitation by an administrator as a part of the evaluation of full-time faculty members. See Tables TYF.25 and TYF.26.
- The percentage of two-year colleges requiring annual continuing education or professional development for permanent full-time faculty rose to 55%, up from 38% in 2000 and 20% in 1995.
- The three items reported by the highest percentage of mathematics program heads as being a major problem were (i) too many students needing remediation (63%), (ii) students not understanding the demands of college work (55%), and (iii) low student motivation (50%). When the "somewhat of a problem" category is included, the percentages for these items (in the same order) were 91%, 90%, and 81% of colleges. Too many students needing remediation and low student motivation also were at the top of the problems list in 2000. See Tables TYF.28 and TYF.29.

- In fall 2005, a traditional mathematics department was found in fewer than half (41%) of the two-year colleges. Only 2% of these were multicampus departmental arrangements. A combined mathematics/science department or division was the management structure at 36% of institutions. See Table TYF.30.
- Reflecting an expanded role for two-year colleges in teacher preparation, especially at the elementary school level, 38% of institutions assigned a mathematics faculty member to coordinate K–8 teacher education in mathematics, up from 22% in 2000. In what appears to be a new development, pre-service teachers could complete their entire mathematics course requirement at the two-year college in 30% of institutions. See Special Topics in Chapter 2, Tables SP.2 and SP.4.
- As reported in Chapter 6, about 42,000 students were dually enrolled in fall 2005 in a two-year college mathematics course that gave credit at both the high school and at the college. Such courses were taught on a high school campus by a high school faculty member. The academic control of such courses ranged from 89% of two-year college mathematics programs reporting they always approved the syllabus to 74% that they always chose the textbook. But only 52% said they controlled the choice of instructor, and only 37% reported control over the design of the final exam. In only 64% of cases was the usual department faculty teaching evaluation required in the dual-enrollment course. See Table SP.16 in Chapter 2.
- As noted in Chapter 6, with respect to the organization of mathematics instruction within two-year colleges, 31% of two-year colleges in fall 2005 reported some of their precollege (remedial) mathematics courses were administered separately from the mathematics program. This percentage was two points higher than the 29% reported in 2000. See Table TYE.17 in Chapter 6.

The Number and Teaching Assignments of Full-time and Part-time Mathematics Program Faculty

Number of permanent full-time faculty and part-time faculty

In fall 2005, the number of permanent full-time mathematics faculty at two-year colleges resumed the growth trend that had characterized every year from 1980 to 1995. There was a one-time 8% decline in permanent full-time faculty between 1995 and 2000. The growth from 2000 to 2005 was an eye-catching 26%, making the size of the permanent full-time faculty a record 8,793.

Another 610 individuals were reported as temporary full-time faculty, a 63% decrease in a category that had taken a worrisome 600% rise from 1995 to 2000. The strong movement to permanent full-time faculty that appeared in fall 2005 paralleled the large enrollment growth that occurred from 2000 to 2005. See Chapter 6 for two-year college enrollment data and the overall enrollment data summary in Chapter 1.

Part-time faculty members fell into two categories. Most were paid by the college. Some were paid by a third party. These latter most often were high school teachers in a school with which the college had a dual-enrollment agreement. (Dual enrollment is discussed later in this chapter and comprehensively in Chapter 2.) When both categories are included, part-time faculty numbered 20,142 or 68% of the total two-year college teaching staff. When third party payees are excluded, part-time faculty members were about 66% of total faculty, a percentage almost identical to the 65% reported in 2000.

Teaching assignment of permanent full-time and part-time faculty

The average required teaching assignment in weekly classroom contact hours for a permanent full-time mathematics faculty member at a public two-year college rose slightly in fall 2005 to 15.3 weekly

TABLE TYF.1 Number of full-time permanent and full-time temporary faculty, and number of part-time faculty paid by two-year colleges (TYC) and by a third party (e.g., dual-enrollment instructors), in mathematics programs at two-year colleges in fall 1990, 1995, 2000, and 2005. (Data for 2005 include only public two-year colleges.)

Two-Year Colleges	1990	1995	2000	2005
Full-time permanent faculty	7222	7578	6960	8793
Full-time temporary faculty	na	164	961	610
Part-time faculty paid by TYC	13680	14266	14887	18227
Part-time, paid by third party	na	na	776	1915

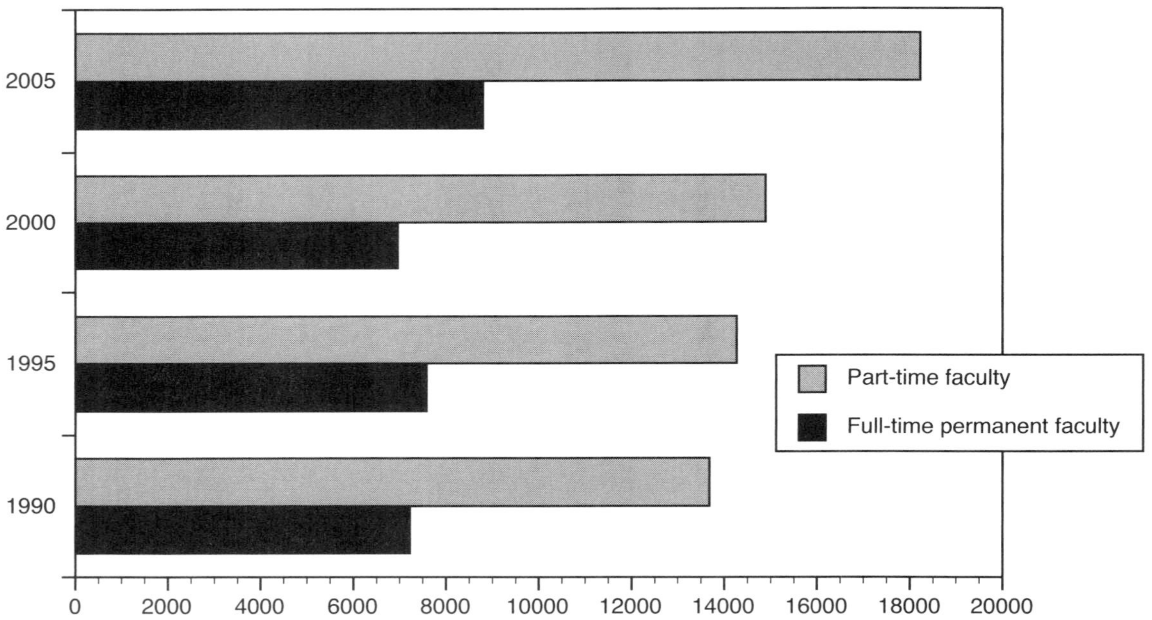

FIGURE TYF.1.1 Number of full-time permanent faculty and part-time faculty in mathematics programs in two-year colleges in fall 1990, 1995, 2000, and 2005. (Data for 2005 include public two-year colleges only.)

contact hours. This continued a twenty-year period of oscillation. In 2000 the average weekly contact hour assignment had been 14.8, but in 1995 it was reported as 15.8. In 1990, the number was 14.7 hours, but in 1985 it had been 16.1 hours.

About 80% of colleges had a teaching requirement for full-time faculty between 13 and 15 weekly contact hours. About 15% had higher weekly contact hour teaching assignments. Only 5% had teaching assignments below 13 weekly contact hours.

See Table TYF.2 for the following fall 2005 data. About 57% of part-time faculty members in two-year college mathematics programs taught six credit hours or more. This was up three percentage points from 2000. Office hours were required of part-time faculty in 37% of two-year colleges, exactly the same percentage as in 2000. The fall 2005 CBMS survey showed 54% of part-time faculty members were paid on the same pay scale as that for the extra-hours teaching of full-time faculty members. This percentage was noticeably lower than the 71% reported for fall 2000 and closer to the 60% reported in 1995. In fall 2005, 5% of colleges paid part-timers more, and 42% paid less, than full-time faculty were paid for extra courses. In fall 2000, these percentages were 2% and 27% respectively.

TABLE TYF.2 Teaching assignment for full-time permanent faculty, and teaching and other duties of part-time faculty, in mathematics programs at two-year colleges in fall 2005 with 2000 data in parentheses. (Data for 2005 include only public two-year colleges.)

Teaching assignment in contact hours	<10	10 to 12	13 to 15	16 to 18	19 to 21	>21
Percentage of two-year colleges	0 (0)	6 (12)	79 (72)	8 (13)	4 (3)	3 (0)
Average contact hours for full-time permanent faculty: 15.3 (14.8)						
Percentage of the full-time permanent mathematics faculty who teach extra hours for extra pay at their own two-year college: 53% (52%)						
Average number of extra hours for extra pay: 3.6 (3.6)						
Percentage of full-time permanent mathematics faculty who teach additional hours at another school: 7.6% (6%)						
Percentage of part-time faculty who teach 6 or more hours weekly: 57%						
Percentage of two-year colleges requiring part-time faculty to hold office hours: 37%						

	Pay scale for full-time faculty teaching extra hours for extra pay		
	Same	Part-time paid more	Part-time paid less
Pay scale for part-time faculty	54%	5%	42%

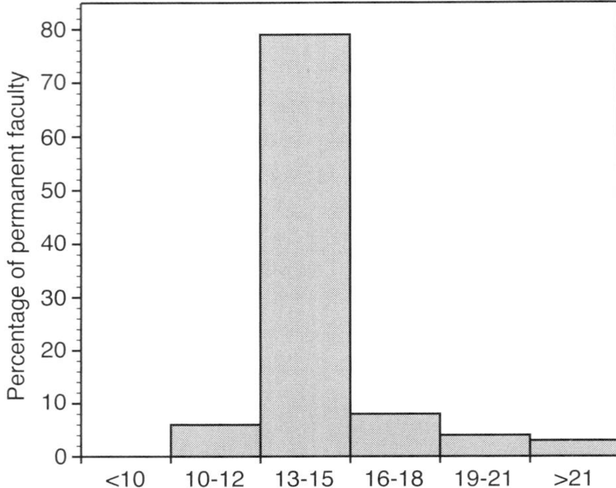

FIGURE TYF.2.1 Percentage of full-time permanent faculty having various teaching assignments in mathematics programs at public two-year colleges in fall 2005.

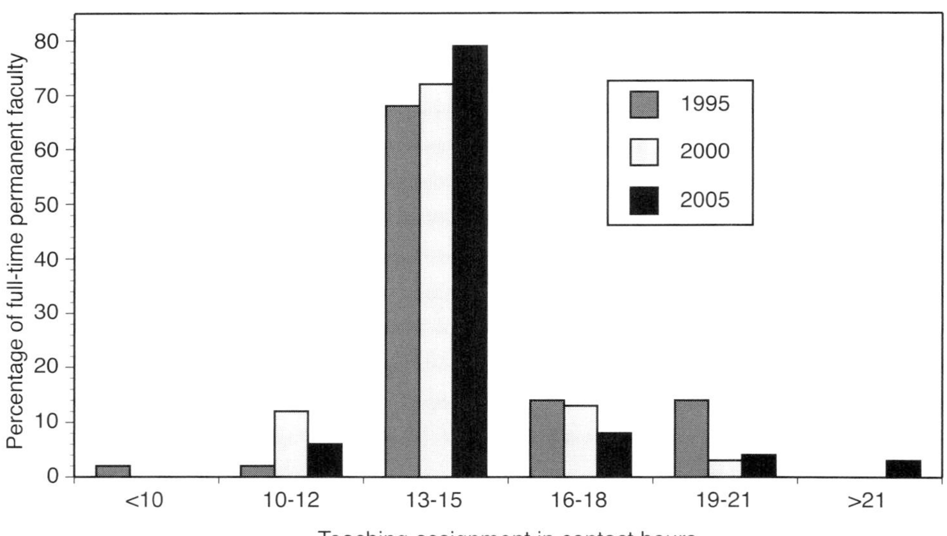

FIGURE TYF.2.2 Percentage of full-time permanent faculty with various teaching assignments in mathematics programs at two-year colleges in fall 1995, 2000, and 2005. (Data for fall 2005 include only public two-year colleges.)

Extra teaching by full-time faculty

Table TYF.2 shows that 53% of permanent full-time mathematics faculty members at two-year colleges taught extra hours for extra pay at their own colleges. This figure is essentially identical to the percentage in 2000, up only one percentage point. Almost 8% of permanent full-time faculty taught at other colleges, up two points from 2000. The average number of extra hours for extra pay taught by these full-time faculty members at their own colleges was 3.6, identical to the corresponding number in both 2000 and 1995.

As a fifteen-year trend, the percentage of permanent full-time mathematics faculty teaching extra courses for extra pay at their own colleges is up. From a 48% base in 1995, this percentage rose four points to 52% in 2000 and another point in 2005 to 53%.

The extra teaching for extra pay by permanent full-time faculty in fall 2005 accounted for about 4700 mathematics program class sections. These sections were classified as being taught by full-time faculty. Had it been necessary to find part-time faculty to teach these sections, the percentage of sections taught by part-time faculty in fall 2005 would have risen from about 44% to about 50%.

Other occupations of part-time faculty

In fall 2005, about 49% of part-time mathematics faculty members at two-year colleges were not employed full-time elsewhere and were not graduate students, up from 41% in 2000. In 1995, the percentage was 35%, and in 1990 and 1985 these percentages, respectively, were 27% and 21%. There is a clear trend in two-year college mathematics programs toward part-time faculty whose only employment is this teaching.

The percentage of part-time faculty who were employed full-time in a high school remained constant at 25%, after a steady decline from 37% in 1985, 30% in 1990, 28% in 1995, and finally to 25% in 2000 and 2005. This pattern reflects one of the most interesting historical trends in two-year college mathematics instruction. In the formative years of two-year colleges in the late 1960s, both full-time and part-time mathematics faculty were drawn in large numbers from secondary schools, in part because many secondary school faculty had earned the required masters degree in National Defense Education Act (NDEA) summer programs in the 1960s. This phenomenon (a decline in secondary schools as a source for two-year college mathematics faculty) also is reflected in Table TYF.18, which shows sources of newly appointed permanent full-time faculty in fall 2005.

TABLE TYF.3 Percentage of part-time faculty in mathematics programs at two-year colleges having various other occupations in fall 2000 and 2005. (Data for 2005 include only public two-year colleges.)

	Percentage of part-time faculty	
Other occupations of part-time faculty	2000	2005
Employed full-time in:		
a high school	25	25
another two-year college	2	2
another department at the same college	7	5
a four-year college	2	2
industry or other	20	14
Graduate student	3	3
No full-time employment and not a graduate student	41	49
	100%	**100%**
Number of part-time faculty	**14887**	**18227**

Educational Credentials of Faculty in Mathematics Programs

Highest degree of permanent full-time faculty

Table TYF.4 records that a masters degree was the terminal degree for 82% of permanent full-time mathematics faculty at two-year colleges, a percentage that has been essentially unchanged for 15 or more years. The percentage of faculty with a doctorate remained constant at 16%. The percentage of these faculty whose terminal degree was a bachelors dropped from 3% to 2%, most likely as a result of credential enforcement by accrediting agencies and of very different patterns in hiring new faculty than were present in 2000. As for the degrees of new hires in fall 2005, see Table TYF.19 and the additional discussion there.

Table TYF.5 gives the academic major of the highest degree of permanent full-time two-year college mathematics faculty. Table TYR.21 in the CBMS2000 report gives analogous data for fall 2000. Overall, the proportion of the faculty with a masters or doctorate whose major field was mathematics rose eight points to 70%. The percentage of the faculty whose most advanced degree included a major in mathematics education dropped six points to 18%, with four points of the drop at the masters level. The percentage of degrees with majors in statistics or other fields remained essentially constant.

TABLE TYF.4 Percentage of full-time permanent faculty in mathematics programs at two-year colleges by highest degree in fall 1990, 1995, 2000, and 2005. (Data for 2005 include only public two-year colleges.)

	Percentage of full-time permanent faculty			
Highest degree	1990	1995	2000	2005
Doctorate	17	17	16	16
Masters	79	82	81	82
Bachelors	4	1	3	2
Number of full-time permanent faculty	**100%** **7222**	**100%** **7578**	**100%** **6960**	**100%** **8793**

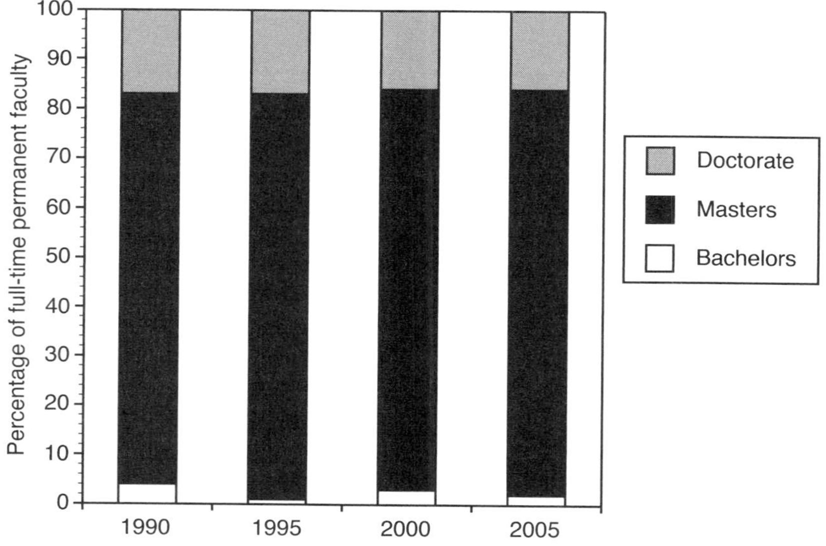

FIGURE TYF.4.1 Percentage of full-time permanent faculty in mathematics programs at two-year colleges by highest degree in fall 1990, 1995, 2000, and 2005. (Data for 2005 include only public two-year colleges.)

TABLE TYF.5 Percentage of full-time permanent faculty in mathematics programs at public two-year colleges by field and highest degree, in fall 2005.

	Percentage having as highest degree			
Field	Doctorate	Masters	Bachelors	**Total**
Mathematics	8	61	1	**70%**
Statistics	0	2	0	**2%**
Mathematics Education	4	14	0	**18%**
Other fields	3	5	1	**9%**
Total	16	82	2	**100%**

Note: 0 means less than half of 1% and round-off may make column sums seem inaccurate

Highest degree of part-time faculty

Tables TYF.6 and TYF.7 summarize data on the highest degrees held by part-time faculty members and on their fields of specialization. In fall 2005, a doctoral degree was the highest degree held by 6% of part-time faculty, the same percentage as fall 2000. A masters degree was highest for 72%, up two percentage points from 2000. A bachelors was the highest degree for 22%, down two percentage points from fall 2000. The percentage of part-time faculty with only bachelors degrees was 27% in 1990, but fell to 18% in 1995 and then rose to 24% in 2000. The turn in fall 2005 again is downward, if only slightly. Generally, accrediting agencies permit faculty who teach only precollege (remedial) courses to hold a bachelors as the highest degree.

In fall 2005, the percentage of part-time faculty whose most advanced degree included mathematics or mathematics education as the major field of study rose a combined five percentage points, from 71% in 2000 to 76% in 2005. All but one point of this gain was at the expense of "other" fields (excluding statistics). See Table TYF.7.

In 2000, the CBMS survey reported that there had been a ten percentage point decline from 1995 in the percentage of masters-level mathematics program faculty holding degrees in mathematics, and a five percentage point increase in bachelors-level faculty who held their degrees outside of the mathematical sciences. It was suggested in 2000 that these trends deserved monitoring. Happily, in 2005, the proportion of masters degrees in mathematics is up three points and the proportion of bachelors degrees outside of mathematical sciences is down four points.

In 1995, 58% of all part-time faculty members in two-year college mathematics programs held their highest degree (Ph.D., MA, or BA) in mathematics. In 2000, the percentage had dropped to 45%. Again, as part of an increase in overall faculty preparedness, in 2005 that figure is back up to 49%.

TABLE TYF.6 Percentage of part-time faculty in mathematics programs at two-year colleges (including those paid by a third party, as in dual enrollment courses) by highest degree, in fall 1990, 1995, 2000, and 2005. (Data for 2005 include only public two-year colleges.)

	Percentage of part-time faculty			
Highest degree	1990	1995	2000	2005
Doctorate	8	7	6	6
Masters	65	76	70	72
Bachelors	27	18	24	22
Number of part-time faculty	**100%** **13680**	**100%** **14266**	**100%** **14887**	**100%** **20142**

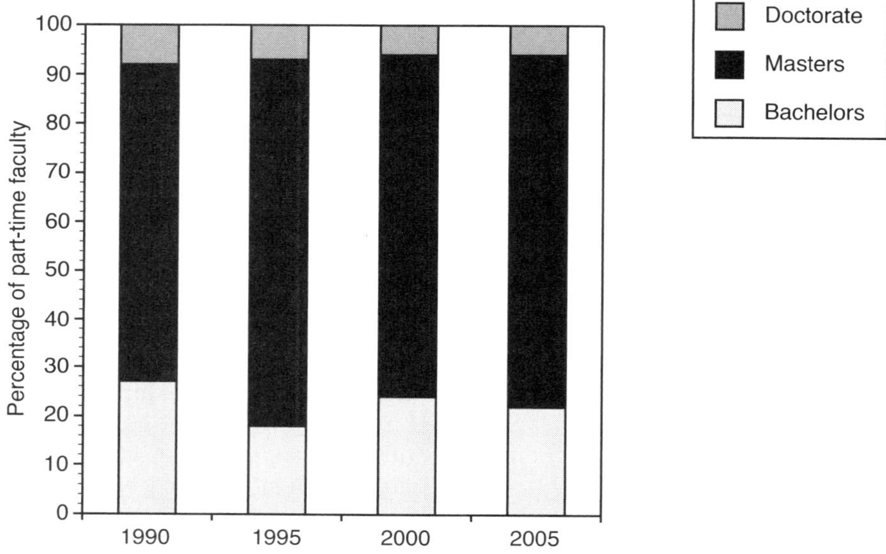

FIGURE TYF.6.1 Percentage of part-time faculty in mathematics programs at two-year colleges (including those paid by a third party, as in dual enrollment courses) by highest degree in fall 1990, 1995, 2000, and 2005. (Data for 2005 include only public two-year colleges.)

TABLE TYF.7 Percentage of part-time faculty in mathematics programs at two-year colleges (including those paid by a third party, as in dual enrollments) by field and highest degree, in fall 2005, with 2000 data in parentheses. (Data for 2005 include only public two-year colleges.)

	Percentage having as highest degree			
Field	Doctorate	Masters	Bachelors	**Total**
Mathematics	2	36	11	**49%**
Mathematics Education	1	20	7	**27%**
Statistics	0	2	0	**3%**
Other fields	3	14	4	**21%**
Total	6 (6)	72 (70)	22 (24)	**100%**

Note: 0 means less than half of 1% and round-off may make row totals seem inaccurate.

Gender, Ethnic Composition, and Age of Permanent Full-time Mathematics Program Faculty

Gender of permanent full-time faculty and part-time faculty

An increase in the percentage of women among permanent full-time mathematics faculty at two-year colleges has been reported in every CBMS study since 1975. In fall 2000, the percentage of women faculty reached 49%. In fall 2005, 50% of permanent full-time mathematics faculty members at the nation's public two-year colleges were women. This proportion of women among permanent full-time faculty was noticeably higher than the percentage of women (44%) among U.S. citizen/resident alien mathematics masters degree recipients in 2003–2004, the last year for which firm data were available. See Table TYF.9.

Table TYF.9 also reports that in fall 2005, the percentage of women among part-time faculty was 47%. This was up from 43% in fall 2000.

CBMS2000 had pointed out that it might be difficult over the long term to maintain the equal split of men and women among the two-year college permanent full-time mathematics faculty since in that year the proportion of women in the under-40 age group only was 45%, less than their representation in the entire permanent full-time faculty. Alleviating this concern, in fall 2005, the proportion of women in the under-40 age group rose to 49%. See the data in Table S.17 in Chapter 1, where the reader can find a comprehensive review of mathematics faculty gender patterns at institutions of all levels, two-year and four-year. As regards two-year colleges, also see Table TYF.17 in this chapter.

In fall 2000, the percentage of women among newly-hired permanent full-time mathematics faculty was 42%, another factor that seemed to threaten the long-term trend toward gender equality. But by fall 2005, the percentage of women among new hires had risen to 53%. See Table TYF.20.

Here is some information from an historical perspective about the participation of women in mathematics at the masters degree level that further emphasizes their high faculty level at two-year colleges. In each CBMS report from 1970 to 1985, the percentage of women among mathematics masters degree recipients in the United States was reported as 35% or less. In 1995, based on NCES data for 1992–1993, CBMS reported the percentage of women mathematics masters degree recipients as 41%. That was the same figure NCES reported for 1997–1998 and also reported in CBMS2000. The percentage of U.S. masters degrees among women in fall 2000 was 44%. Yet in fall 2005, women made up 50% of the permanent full-time mathematics faculty at two-year colleges.

TABLE TYF.8 Number and percentage of full-time permanent faculty in mathematics programs at two-year colleges by gender, in fall 1990, 1995, 2000, and 2005. (Data for 2005 include only public two-year colleges.)

	1990	1995	2000	2005
Men	4767 (66%)	4579 (60%)	3537 (51%)	4420 (50%)
Women	2455 (34%)	2999 (40%)	3423 (49%)	4373 (50%)
Total	**7222** **(100%)**	**7578** **(100%)**	**6960** **(100%)**	**8793** **(100%)**

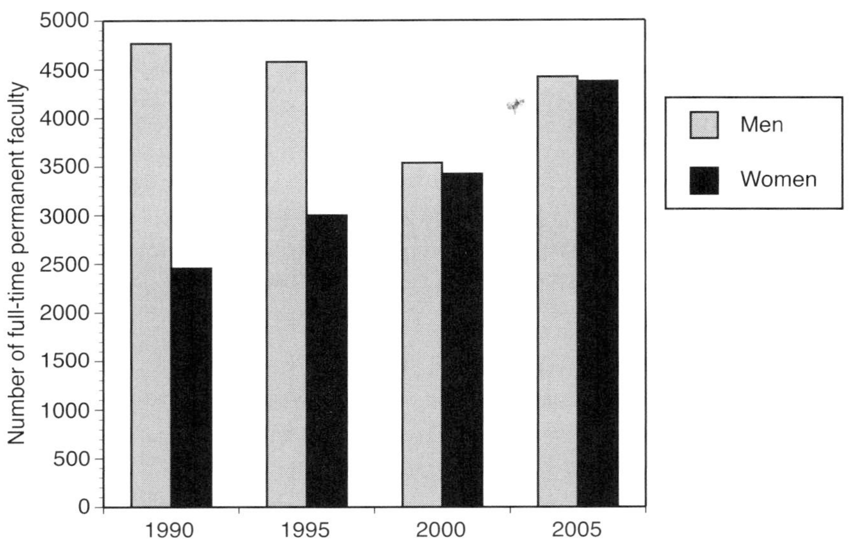

FIGURE TYF.8.1 Number of full-time permanent faculty in mathematics programs at two-year colleges by gender in fall 1990, 1995, 2000, and 2005. (Data for 2005 include only public two-year colleges.)

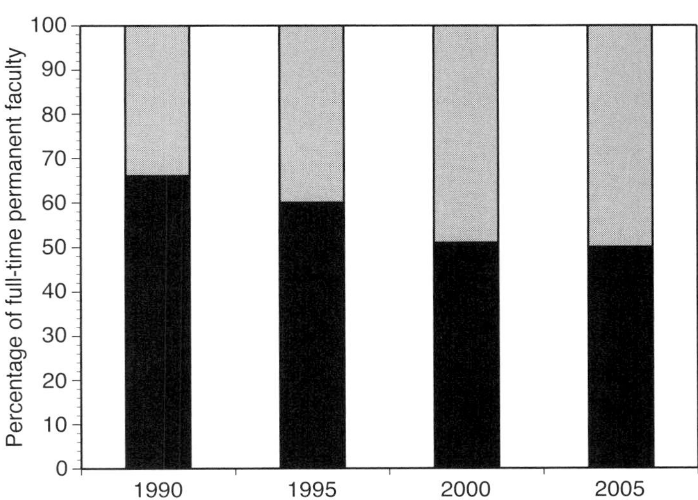

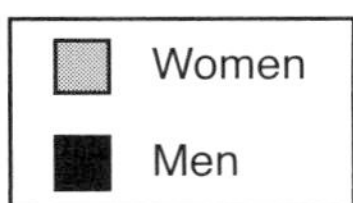

FIGURE TYF.8.2 Percentage of full-time permanent faculty in mathematics programs at two-year colleges by gender in fall 1990, 1995, 2000, and 2005. (Data for 2005 include only public two-year colleges.)

TABLE TYF.9 Percentage of full-time permanent faculty and part-time faculty in mathematics programs at public two-year colleges by gender, in fall 2005. Also masters degrees in mathematics and statistics granted in the U.S. to citizens and resident aliens, by gender, in 2003-04. Part-time faculty paid by a third party are not included.

	Percentage of		
	Full-time permanent faculty	Part-time faculty	Masters degrees in mathematics & statistics granted in the U.S. in 2003–04 to citizens and resident aliens[1]
Men	50	53	56%
Women	50	47	44%
Total	**100%**	**100%**	**100%**
Number	**8793**	**18227**	**2475**

[1] Table 265, Digest of Education Statistics, 2005, National Center for Education Statistics IPEDS Annual Completion Survey. (These figures include resident aliens but do not include a total of 1716 nonresident aliens who received masters degrees.)

Ethnicity among permanent full-time and part-time faculty

Demographic data about ethnic minority faculty among permanent full-time mathematics faculty members at two-year colleges are given in Tables TYF.10, TYF.11, TYF.12, and TYF.13. The minority groups referenced in the survey are listed in TYF.11. Tables TYF.10 and TYF.11 provide an historical perspective, while Tables TYF.12 and TYF.13 present more detailed information on the ethnic profile of the permanent full-time mathematics faculty in fall 2005, including information about both age and gender.

From 1995 to 2000, the overall number of permanent full-time mathematics faculty in two-year colleges decreased by about 8%. Although the total number of ethnic minority faculty also declined, the percentage of ethnic minorities among the permanent full-time mathematics faculty remained at about 13%. Similarly, the dramatic increase in the overall size of the permanent full-time mathematics faculty from 2000 to 2005 was matched by a proportional growth in the size of the ethnic minority faculty. In fall 2005, ethnic minority faculty constituted 14% of the permanent full-time faculty. This percentage was still two points below the ethnic minority faculty proportion in 1990.

The relative sizes of most ethnic groups within the permanent full-time faculty changed little between 2000 and 2005, but the percentage of Black (non-Hispanic) faculty (constant at 5%) was surpassed by the percentage of Asian/Pacific Islanders (6%, up two points), who were the largest ethnic minority group in fall 2005.

Table TYF.12 gives the percentage of women within ethnic groups of the permanent full-time faculty. CBMS2000 had reported a significant drop in the percentage of female Black (non-Hispanic) permanent full-time faculty, from 42% in fall 1995 to 28% in fall 2000. That figure was back up to 47% in fall 2005. The percentage of Asian/Pacific Islander faculty who are women rose 16 points to 52%, the highest percentage of women in any of the ethnic groups, slightly larger proportionally than women within White (non-Hispanic) faculty. Native Americans (American Indians/Eskimo/Aleut) had the largest loss in percentage share of faculty and of women among ethnic faculty, dropping to less than 0.5% in both categories. Finally, a word of caution is in order. Compared to CBMS1995, both CBMS2005 and CBMS2000 reported large increases in the percentages of women whose ethnicity was unknown.

Between 1995 and 2000, the percentage of ethnic minority permanent full-time mathematics faculty under the age of 40 did not change, remaining at 20%. However, Table TYF.13 shows that in fall 2005 this number rose to 23%, noticeably higher than the percentage of ethnic faculty (14%) among all permanent full-time faculty members. Data on ethnicity of newly-hired faculty in fall 2005 are given in Table TYF.20.

TABLE TYF.10 Percentage and number of ethnic minority full-time permanent faculty in mathematics programs at two-year colleges, in fall 1990, 1995, 2000, and 2005. (Data for 2005 include only public two-year colleges.)

	1990	1995	2000	2005
Percentage of ethnic minorities among full-time permanent faculty	16	13	13	14
Number of full-time permanent ethnic minority faculty	1155	948	909	1198
Number of full-time permanent faculty	7222	7578	6960	8793

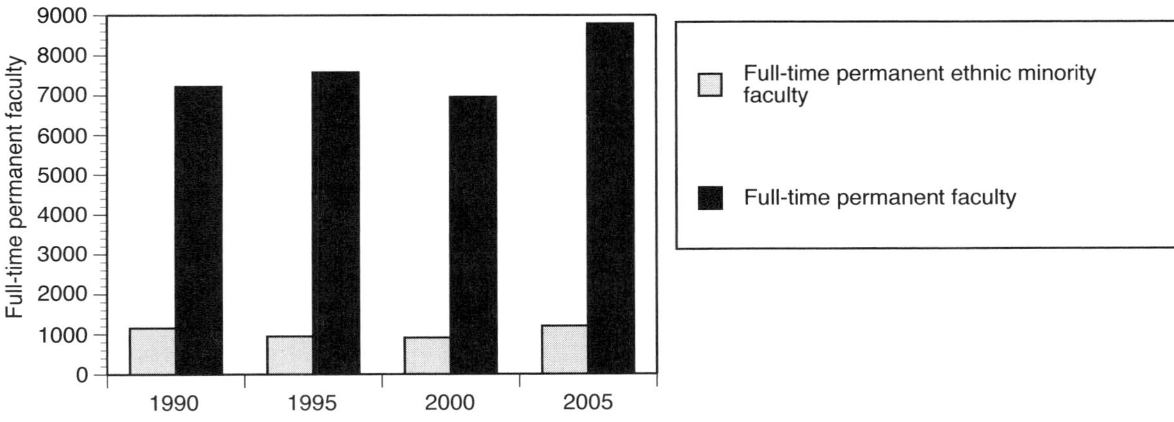

FIGURE TYF.10.1 Number of ethnic minority full-time permanent faculty and number of all full-time permanent faculty in mathematics programs at two-year colleges in fall 1990, 1995, 2000, and 2005. (Data for 2005 include only public two-year colleges.)

TABLE TYF.11 Percentage of full-time permanent faculty in mathematics programs at two-year colleges by ethnicity, in fall 1990, 1995, 2000, and 2005. (Data for 2005 include only public two-year colleges.)

	Percentage of full-time permanent faculty			
Ethnic Group	1990	1995	2000	2005
American Indian/Eskimo/Aleut	1	0	1	0
Asian/Pacific Islander	4	4	4	6
Black (non-Hispanic)	4	5	5	5
Mexican American/Puerto Rican/ other Hispanic	7	3	3	3
White (non-Hispanic)	84	87	85	84
Status unknown	na	1	2	2
Number of full-time permanent faculty	**100%** **7222**	**100%** **7578**	**100%** **6960**	**100%** **8793**

Note: 0 means less than half of 1%.

TABLE TYF.12 Number and percentage of full-time permanent faculty in mathematics programs at public two-year colleges by ethnic group and percentage of women within each ethnic group, in fall 2005.

Ethnic group	Number of full-time permanent faculty	Percentage of ethnic group in full-time permanent faculty	Percentage of women in ethnic group
American Indian/Eskimo/Aleut	27	0	0
Asian/Pacific Islander	538	6	52
Black (non-Hispanic)	413	5	47
Mexican American/Puerto Rican/ other Hispanic	280	3	43
White (non-Hispanic)	7353	84	51
Status not known	182	2	34
Total	**8793**	**100%**	**50%**

Note: 0 means less than one-half of one percent.

TABLE TYF.13 Percentage of full-time permanent faculty and of full-time permanent faculty under age 40 in mathematics programs at public two-year colleges by ethnic group, in fall 2005. Also U.S. masters degrees in mathematics and statistics granted in the U.S. to citizens and resident aliens by ethnic group in 2003–2004.

Ethnic Group	Percentage among all full-time permanent faculty	Percentage among full-time permanent faculty under age 40	Masters degrees in mathematics and statistics granted in the U.S. in 2003–04 to citizens and resident aliens[1]
Ethnic minorities	14	23	22
White (non-Hispanic)	84	76	78
Unknown	2	1	0
Total	**100%**	**100%**	**100%**
Number	**8793**	**2209**	**2475**

[1] Table 265, Digest of Education Statistics, 2005, National Center for Education Statistics IPEDS Annual Completion Survey. (These figures include resident aliens but do not include a total of 1716 nonresident aliens who received masters degrees.)

In fall 2005, about 16% of part-time faculty members were ethnic minorities, which was up three percentage points from 2000. The comparable figure in 1995 was 13%, the same as in 2000. Among the permanent full-time faculty, Asian/Pacific Islanders and Blacks (non-Hispanic) were the two largest groups.

TABLE TYF.14 Percentage of ethnic minority part-time faculty in mathematics programs at public two-year colleges, in fall 2005.

Percentage of ethnic minorities among part-time faculty	16
Number of part-time faculty	18227

TABLE TYF.15 Number and percentage of part-time faculty in mathematics programs at public two-year colleges by ethnic group and percentage of women within ethnic groups, in fall 2005.

Ethnic group	Number of part-time faculty	Percentage of ethnic group among all part-time faculty	Percentage of women within ethnic group
American Indian/Eskimo/Aleut	106	1	18
Asian/Pacific Islander	1045	6	46
Black (non-Hispanic)	1181	6	47
Mexican American/Puerto Rican/ other Hispanic	521	3	45
White (non-Hispanic)	14833	81	48
Status not known	541	3	45
Total	**18227**	**100%**	**47%**

Age distribution of permanent full-time faculty

In fall 1990, CBMS reported that the average age of the permanent full-time mathematics faculty at two-year colleges was 45.4 years. In five-year steps, corresponding to CBMS reports in 1995 and 2000, this average age rose successively to 47.2 and 47.6 years. In fall 2005 the average faculty age was 47.8, again slightly up. (See Table S.18 in Chapter 1.) During this fifteen-year period (1990 to 2005), the two-year college mathematics faculty, as a cohort, has been getting older, but the rate of this aging has slowed from the rate for 1990 to 1995. For comparison, Chapter 4 gives age and other demographic data about mathematics faculty in four-year institutions.

The percentage of permanent full-time faculty under age 40 slid gradually from 47% in 1975 to 21% in 1995. It rose to almost 26% in 2000 and in 2005 maintained its level at just over 25%. Among ethnic minority faculty, 23% were under age 40 in fall 2005, as reported in Table TYF.13. At the other end of the age range, the percentage of permanent full-time faculty over age 54 had grown from 12% in 1975 to 18% in 1995, reached 27% in 2000, and was at 29% in fall 2005.

While the size of the permanent full-time faculty grew about 26% from 2000 to 2005, this growth was by no means equally distributed among the age categories. As would be expected, there was a 64% growth in faculty under 30, double the 32% growth in the faculty age 55 and over.

Women were a majority in the 45–54 age group, just as they were in 2000. They made up only 43% of the over-54 age group. Otherwise, in terms of age, as reported in TYF.17, their distribution in the faculty matched that of men.

TABLE TYF.16 Percentage and number of full-time permanent faculty in mathematics programs at two-year colleges by age, in fall 1990, 1995, 2000, and 2005. (Data for 2005 include only public two-year colleges.)

	Percentage of full-time permanent faculty				Number of full-time permanent faculty			
Age	1990	1995	2000	2005	1990	1995	2000	2005
<30	5	5	4	5	361	358	290	478
30–34	8	8	9	8	578	580	615	716
35–39	10	8	13	12	722	633	890	1037
40–44	21	14	11	13	1517	1044	763	1163
45–49	22	22	15	15	1589	1672	1075	1298
50–54	21	26	20	18	1517	1933	1418	1574
55–59	8	13	16	17	578	966	1146	1528
>59	5	5	11	11	360	391	763	999
Total	**100%**	**100%**	**100%**	**100%**	**7222**	**7577**	**6960**	**8793**

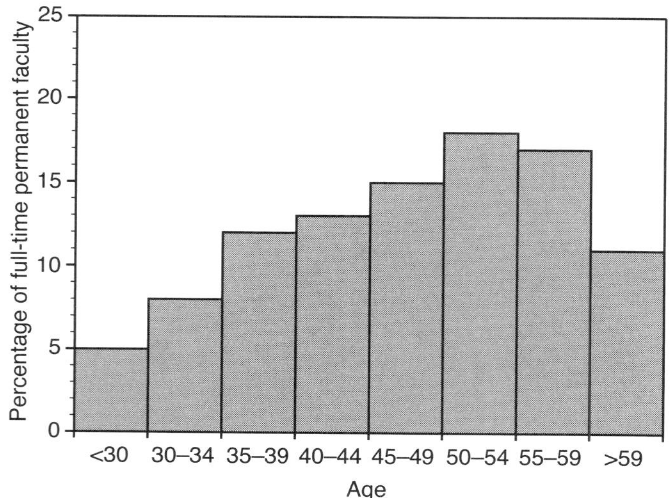

FIGURE TYF.16.1 Percentage distribution of full-time permanent faculty in mathematics programs at public two-year colleges by age in fall 2005.

TABLE TYF.17 Percentage of full-time permanent faculty in mathematics programs at public two-year colleges by age and by gender and percentage of women by age, in fall 2005.

Age	Percentage of full-time permanent faculty		Percentage of women in age group
	Women	Men	
<35	7	7	49
35–44	13	12	50
45–54	18	15	55
>54	12	16	43
Total	**50%**	**50%**	

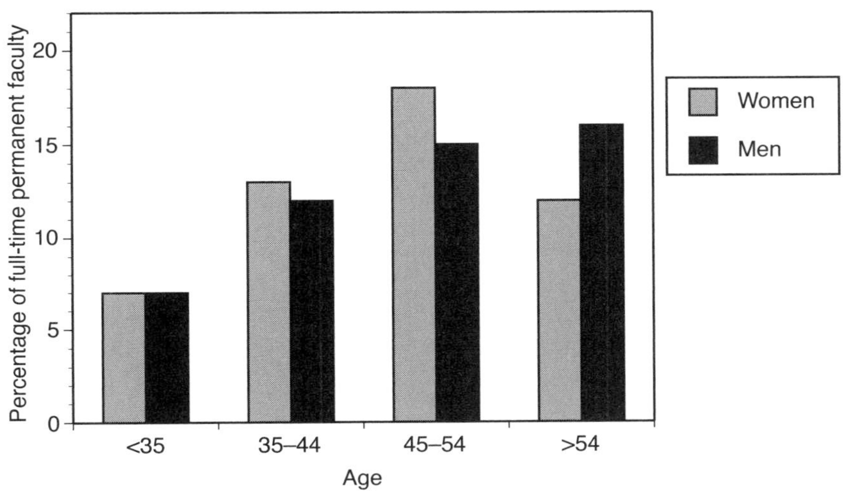

FIGURE TYF.17.1 Percentage of full-time permanent faculty in mathematics programs at public two-year colleges by gender and age in fall 2005.

Demographics of Permanent Full-time Faculty Newly Hired by Mathematics Programs for Fall 2005

Number and source of new permanent full-time faculty

Two-year college mathematics programs hired about 600 new permanent full-time faculty members for fall 2005. This was about the same size as the new faculty cohort in fall 2000 and was a second strong increase (as recorded by CBMS surveys) over the 305 new hires reported for fall 1995. In fact, the dramatic total increase in faculty size (by 1,833 permanent full-time positions) as well as the on-going replacement of exiting faculty suggest permanent faculty positions in the range of 500 persons per year were being filled throughout the period 2000 to 2005.

For fall 2005, hiring patterns moved back toward those of 1995. In 1995, 30% of new faculty members were hired directly out of graduate school, about the same percentage as in 1990. In 2000, this fell to 8%. In 2005, graduate school as a faculty source rose to 23%. Similarly, the percentage of new hires previously teaching at a four-year institution dropped eight percentage points to 10% in 2000. In 2005, this percentage was back up to 18%. Hiring from among part-time faculty at the same institution almost doubled, to 34%, in 2000. It remained high at 29% in 2005 but had moved back toward the 19% level of 1995.

In 2000, the percentage of secondary school teachers among newly-hired faculty rose from 4% to 22%, an anomaly in the long-term pattern that was more characteristic of the earliest years of two-year college hiring. This percentage for new hires fell back to 13% in 2005. In 1979, about 60% of all two-year college mathematics faculty had come from secondary schools [MALL].

TABLE TYF.18 Percentage of newly appointed full-time permanent faculty in mathematics programs at two-year colleges coming from various sources, in fall 2000 and 2005. (Data for 2005 include only public two-year colleges.)

	Percentage of new faculty from	
Source	2000	2005
Graduate school	8	23
Teaching in a four-year college or university	10	18
Teaching in another two-year college	19	11
Teaching in a secondary school	22	13
Part-time or full-time temporary employment at the same college	34	29
Nonacademic employment	6	5
Unemployed	0	0
Unknown	1	1
Total number hired	**100%** **572**	**100%** **605**

Educational credentials of newly-hired permanent full-time faculty

The masters degree was held by 84% of newly-hired permanent full-time faculty in fall 2005. This percentage was 18 points higher than in 2000. Combined with a 14-point drop from 2000 (to 5% in 2005) in the number of newly-hired permanent full-time faculty whose highest degree was a bachelors degree, this 84% suggests a strong return to the masters degree as the standard entry-level credential for two-year college permanent full-time mathematics faculty.

In 2000, the CBMS report voiced concern at the high level of permanent full-time faculty being hired with no degree beyond the bachelors, a change from historical practice being implemented at a time when large numbers of retiring faculty were being replaced with new hires. If continued over time, the 2000 report expressed concern that there could be a rapid drop in the percentage of masters degrees among permanent full-time mathematics faculty within two-year college mathematics programs. This could lead to a two-tiered faculty structure within the programs, to an overall change in program philosophy and cohesiveness, and to conflicts with four-year colleges and universities on course comparability and transferability. Fortunately, the 2005 data indicate a return to traditional practice. For example, 80% of new hires in fall 1995 held a masters degree, compared to 84% in 2005.

It is important to note again the likely influence of accrediting agencies in the return to "masters-degree-minimum" hiring. Anecdotal evidence indicates that these agencies were very active during the period 2000 to 2005 regarding verification of faculty credentials. Most accrediting agencies require that two-year college faculty who teach courses that transfer for baccalaureate degree credit hold a masters degree with an 18 semester-hour graduate credit concentration in the academic field in which they are teaching. Accrediting agencies usually allow faculty who teach precollege (remedial) or developmental courses to hold only a bachelors degree, provided the major is in the subject that they are teaching.

In fall 2005, about 12% of the newly-hired permanent full-time mathematics faculty held a doctorate, a one-point drop from fall 2000 but seven percentage points below 1995. The 13% doctorate level for new hires in 2000 had reversed the trend reported in the 1995 CBMS survey of two-year colleges hiring more new permanent full-time faculty members with doctorates than they had previously. Prior to 1995, CBMS surveys found that two-year colleges hired very few permanent full-time faculty members with doctorates and that faculty earned their doctorates while on the job. The 1990 survey found, for example, that 2% of new hires had doctorates, rising to 19% in 1995. During the decade from 1995 to 2005, this number seemed to stabilize in the neighborhood of 12%.

TABLE TYF.19 Percentage of full-time permanent faculty newly hired for mathematics programs at two-year colleges by highest degree, in fall 2000 and 2005. (Data for 2005 include only public two-year colleges.)

	Percentage of new hires	
Highest degree	2000–2001	2005–2006
Doctorate	13	12
Masters	66	84
Bachelors	19	5
Unknown	2	0
Total	**100%**	**100%**

Note: 0 means less than one-half of one percent and round-off may make column totals seem inaccurate.

Gender, ethnicity, and age of newly-hired permanent full-time faculty

For 2005, about 53% of new mathematics faculty hires were women, up 11 percentage points from 2000. As noted earlier in this chapter, this bodes well for maintaining a 50-50 split between women and men in the permanent full-time faculty. Table TYF.20 shows White (non-Hispanic) faculty comprised 80% of new hires for 2005, down 6 points from 2000. Overall, 19% of new hires in 2005 were ethnic minorities, up six points from 2000 but a four-percentage-point drop from 1995.

Table TYF.21 gives the percentage of new hires whose ages fall in five-year intervals beginning at age 30. As would be expected, almost 60% of new hires were under age 40, but this was ten percentage points lower than in 2000, when 70% of new hires were under age 40. In 2005, 30% of new hires were between age 40 and 50, a sharp rise from the 11% in 2000. This may reflect the already noted 18% of new hires who came to two-year colleges from four-year institutions, up eight points from 2000. The reduced percentage of new hires between 30 and 39 years old is interesting. This number dropped to 32% from 58% in 2000, but the percentage of new hires under age 35, rising from 31% in 2000 to 42% in 2005, is consistent with other CBMS2005 data (Table TYF.18) showing that graduate school is the largest source of new hires other than a college's own current part-time faculty.

Information about gender, ethnicity, and age of new hires was not collected in CBMS surveys prior to 1995.

TABLE TYF.20 Percentage of full-time permanent faculty newly hired for mathematics programs at two-year colleges by ethnic group, in fall 2000 and 2005. Also percentage of women within each ethnic group in fall 2005. (Data for 2005 include only public two-year colleges.)

	Percentage of new hires		Percentage of women in ethnic group for 2005–2006 new hires
Ethnic group	2000–2001	2005–2006	
Asian/Pacific Islander	7	7	49
Black (non-Hispanic)	1	1	100
Mexican American/Puerto Rican/other Hispanic	5	11	62
White (non-Hispanic)	86	80	52
Unknown	1	1	31
Percentage of women among all new hires	**42%**	**53%**	--

TABLE TYF.21 Percentage of full-time permanent faculty newly hired for mathematics programs at two-year colleges by age, in fall 2000 and 2005. (Data for 2005 includes only public two-year colleges.)

	Percentage of new hires	
Age	2000	2005
<30	11	22
30–34	21	20
35–39	37	17
40–44	5	15
45–49	6	15
50–54	12	5
55–59	6	0
>59	3	6
Total	**100%**	**100%**

Outflow of Permanent Full-time Mathematics Faculty

During academic year 2004–2005, 439 people left their permanent full-time mathematics faculty positions at two-year colleges. This was 9% more than the 401 who left during 1999–2000. Using 8,793 as the estimate of permanent full-time faculty in fall 2005, 439 was almost 5% of the faculty, down from about 5.7% in 1999–2000. However, one should note that the percentage for 2004–2005 is strongly affected by an increased denominator in the percentage calculation, from 6,960 in 2000 to 8,793 in 2005. For the long-term historical pattern, the outflow in academic year 1994–1995 was 402 people or about 5.3% of the fall 1995 permanent full-time faculty. In 1989–1990, the outflow was 317 (4.4%), and in 1984–1985 it was 449 (7.1%).

In 2004–2005, about 67% of those who left a permanent faculty position were accounted for by death or retirement. This was a sharp rise from 1999–2000 when about 41% of the outflow left for these reasons but comparable to the 68% in 1994–1995. No information was available for about 24% of the departures.

TABLE TYF.22 Outflow of full-time permanent faculty from mathematics programs at public two-year colleges, in 2004–2005.

Status	Number
Died or retired	292
Teaching in a four-year college or university	9
Teaching in another two-year college	14
Teaching in a secondary school	2
Left for a nonacademic position	5
Returned to graduate school	3
Other	107
Unknown	7
Total	**439**

Resources Available to Mathematics Program Faculty

Computer and office facilities for part-time faculty

To gauge the extent to which two-year colleges were making computer technology available to faculty members, in 1995 the CBMS survey first collected information on the availability of office computers and other computer facilities to full-time faculty members. By 2000, office computers for permanent full-time faculty were nearly universal. So, in 2005, the CBMS survey asked about office computers only for part-time faculty. About two-thirds of colleges reported computers available in part-time offices with the remaining one-third reporting shared computer access near the office. Only 2% reported no convenient access to computers or terminals for part-time faculty.

Between 1995 and 2000, there was an eight-percentage-point jump in the number of part-time faculty who shared a desk with two or more people. In 2005, this figure jumped another 14 points to 65% with a seven-point drop to 5% of part-time faculty who had their own desk. In 1995, 18% of part-time faculty had their own desk.

TABLE TYF.23 Percentage of part-time faculty in mathematics programs at two-year colleges by desk availability, in fall 2000 and 2005. (Data for 2005 include only public two-year colleges.)

	Percentage of part-time faculty	
Desk availability	2000	2005
Have their own desk	12	5
Share a desk with one other person	5	7
Share a desk with two or more other people	51	65
Have no desk, or unknown	31	23

TABLE TYF.24 Percentage of part-time faculty in mathematics programs at public two-year colleges by access to computer facilities in fall 2005.

Computer facilities for part-time faculty	Percentage of part-time faculty
Computer or terminal in office	63
No computer or terminal in office, but shared computers or terminals nearby	35
No convenient access or no access at all to computers or terminals	2

Teaching evaluation

In fall 2005 there was an unexpected nine-percentage-point drop, to 89%, in the percentage of two-year colleges that periodically evaluated the teaching of permanent full-time mathematics faculty members. In fall 2000, this figure was 98%, and in fall 1995, it was 100%. In 2005, periodic teaching evaluation was required for part-time faculty at 89% of colleges, a proportion almost identical to the 88% reported in 2000. Data on evaluation of part-time faculty were not collected in the 1995 survey.

In 2005, there was a strong jump in the percentage of colleges that used classroom visitation by a division or department chair or other administrator as a component of full-time faculty evaluation. In 2005, the percentage rose to 61% from 52% in 2000. Simultaneously, the percentage of colleges using classroom observation by other faculty (not administrators) dropped 12 points to 52%. Together, these facts suggest a move in fall 2005 towards a somewhat less collegial evaluation system for full-time faculty.

The most common method of evaluating teaching remained the use of evaluation instruments completed by students. For full-time faculty, this was up to 96%, from 90% in 2000. It had been 97% in 1995. To evaluate part-time faculty, a student questionnaire was used by 94% of colleges (up from 87% in 2000). Self-evaluation portfolios were used as a component of the evaluation of full-time faculty by 46% of colleges, both in 2005 and in 2000. For full-time faculty, evaluation of written materials—such as syllabi or course examinations—rose from 48% to 55%. The use of such written materials for part-time faculty evaluation rose nine points from 2000 to 49% in 2005. For part-time faculty, observation of classes by an administrator remained very low, 33% in 2005 (up from 28% in 2000). However, observation of classes taught by part-time faculty by non-administrative faculty rose from 60% of colleges in 2000 to 64% in 2005. It is common for full-time faculty at two-year colleges to have a major involvement in orienting, assisting, supervising, and evaluating part-time faculty.

TABLE TYF.25 Percentage of two-year colleges that require periodic teaching evaluations for all full-time or part-time faculty, in fall 2000 and 2005. (Data for 2005 include only public two-year colleges.)

Teaching evaluation	Percentage of two-year colleges in fall 2000	Percentage of two-year colleges in fall 2005
that require teaching evaluations for all full-time faculty	98	89
that require teaching evaluations for all part-time faculty	88	89

TABLE TYF.26 Percentage of mathematics programs at public two-year colleges using various methods of evaluating teaching of part-time and full-time faculty, in fall 2005.

	Percentage of programs using evaluation method for	
Method of evaluating teaching	Part-time faculty	Full-time faculty
Observation of classes by other faculty	64	52
Observation of classes by division head (if different from chair) or other administrator	33	61
Evaluation forms completed by students	94	96
Evaluation of written course material such as lesson plans, syllabus, or exams	49	55
Self-evaluation such as teaching portfolios	19	46
Other methods	0	5

Note: 0 means less than one-half of one percent.

Professional development obligations and activities of permanent full-time faculty

In fall 2005, as reported in Table TYF.27, some form of continuing education or professional development was required of permanent full-time faculty members at 55% of two-year colleges. This percentage had been 38% in 2000. The fall 2005 percentage was almost triple the 1995 percentage of 20%. This decade-long increase in required professional development for permanent full-time faculty parallels the increased faculty use of various professional development opportunities, also reported in Table TYF.27. Slightly more than half of the permanent full-time faculty met part of their professional development obligation through activities provided by their own colleges. This figure was 36% in 2000. About 38% (perhaps overlapping with the previous category) participated in activities provided by professional societies, up from 31% in 2000.

Direct comparison of CBMS2005 and CBMS2000 data to the professional development data from CBMS1995 is not possible due to changes in the format of the two-year college questionnaire for 2005 and 2000. The 1995 survey asked about participation in a wide variety of specific professional development activities, while the CBMS2005 and CBMS2000 questionnaires asked about broad categories of activities. Even so, one important observation is possible concerning involvement in professional societies by full-time mathematics faculty at two-year colleges. The 1995 CBMS survey found that over 70% of permanent full-time mathematics faculty participated in professional meetings, while CBMS2005 reported only 38% (31% in 2000) used this resource to fulfill professional development responsibilities. This likely reflects a concern expressed by 44% of program heads (TYF.29) about the level of travel funding for faculty. Nonetheless, attendance at the annual conference sponsored by the American Mathematical Association of Two-Year Colleges (AMATYC) has remained strong throughout the period 2000 to 2005, numbering about 1,200 each year, though generally not increasing to the same extent that full-time faculty size increased.

TABLE TYF.27 Percentage of two-year colleges that require some form of continuing education or professional development for full-time permanent faculty, and percentage of faculty using various methods to fulfill those requirements, in mathematics programs at two-year colleges in fall 2000 and 2005. (Data for 2005 include only public two-year colleges.)

Faculty Development	Fall 2000	Fall 2005
Percentage of institutions requiring continuing education or professional development for full-time permanent faculty	38%	55%
How Faculty Meet Professional Development Requirements	Percentage of permanent faculty in fall 2000	Percentage of permanent faculty in fall 2005
Activities provided by employer	36	53
Activities provided by professional associations	31	38
Publishing books or research or expository papers	3	6
Continuing graduate education	8	7

Problems in Mathematics Programs

In every CBMS survey since 1985, 60% or more of mathematics program heads classified the need for too much student remediation as a major problem for their programs. In fall 2005, this figure was 63%. The fall 2000 figure was 62%. A new category was introduced in 2005, namely, students' lack of understanding of the demands of college work. This showed up as second in the ranking of major problems, reported by 55% of mathematics program heads. Low student motivation ranked third, as reported by 50% of mathematics program heads. This had been the second category in both 2000 (47%) and 1995 (51%). Rounding out the top five in 2005 were lack of student progress from developmental to advanced courses (34%), need to use too many part-time faculty (30%), and a fifth-place tie between low faculty salaries and inadequate travel funds (22% each). These were the same topics that ranked in the top five in 2000. All other major problems listed showed a much lower percentage of mathematics programs than these five. See Tables TYF.28 and TYF.29 both for the historical perspective on these issues and the fall 2005 ratings. These tables also include data on the extent to which program heads thought these matters were somewhat of a problem, though not a major one.

Administration of Mathematics Programs

Between 1995 and 2000, two-year colleges (like four-year institutions) made a major shift to the semester system. In fall 2000, 93% of two-year colleges operated under the semester structure, up from 73% in 1995. The use of the semester system had become so widespread after 2000 that CBMS2005 elected to omit this question from the survey in 2005.

In fall 2000, as in 1995, about 43% of two-year college mathematics programs were administered as departments, with 10% of these being multi-campus departmental systems. In 2005, 41% reported a departmental structure, with only 2% of these being part of a multi-campus organization. A division structure, where mathematics is combined with science or other disciplines, was found in 53% of two-year colleges, down slightly from the 55% reported in 2000.

Historically, mathematics courses at two-year colleges have been taught in many different administrative units other than in mathematics programs. This practice continued in fall 2005, as shown in Table TYE.17 at the end of Chapter 6. The location of precollege (remedial) mathematics courses within a college's academic structure always has been of special interest. In fall 2005, about 31% of colleges reported that some precollege mathematics courses were taught outside of the mathematics program,

TABLE TYF.28 Percentage of program heads classifying various problems as "major" in mathematics programs at two-year colleges, in fall 1990, 1995, 2000, and 2005. (Data for 2005 include only public two-year colleges,)

	Percentage of program heads classifying problem as major			
Problem	1990	1995	2000	2005
Maintaining vitality of faculty	22	11	9	2
Dual-enrollment courses	na	na	8	5
Staffing statistics courses	na	4	2	3
Students don't understand demands of college work	na	na	na	55
Need to use part-time faculty for too many courses	na	30	39	30
Faculty salaries too low	na	31	36	22
Class sizes too large	10	11	10	5
Low student motivation	38	51	47	50
Too many students needing remediation	65	63	62	63
Lack of student progress from developmental to advanced courses	na	na	na	34
Low success rate in transfer-level courses	na	15	8	7
Too few students who intend to transfer actually do	na	7	2	4
Inadequate travel funds for faculty	26	21	15	22
Inadequate classroom facilities for use of technology	na	na	na	12
Inadequate computer facilities for part-time faculty use	na	na	na	9
Inadequate computer facilities for student services	na	23	3	1
Commercial outsourcing of instruction	na	na	1	0
Heavy classroom duties prevent personal & teaching enrichment by faculty	na	na	na	14
Coordinating mathematics courses with high schools	9	8	6	7
Lack of curricular flexibility because of transfer rules	10	6	1	7
Use of distance education	na	na	10	6

Note: 0 means less than one-half of one percent.

TABLE TYF.29 Percentage of program heads of mathematics programs at public two-year colleges classifying various problems by severity in fall 2005.

	Percentage of program heads classifying problems as		
Problem	minor or no problem	somewhat of a problem	major problem
Maintaining vitality of faculty	77	21	2
Dual-enrollment courses	74	21	5
Staffing statistics courses	88	9	3
Students don't understand demands of college work	10	35	55
Need to use part-time faculty for too many courses	38	32	30
Faculty salaries too low	32	46	22
Class sizes too large	72	23	5
Low student motivation	20	31	50
Too many students needing remediation	8	28	63
Lack of student progress from developmental to advanced	29	37	34
Low success rate in transfer-level courses	58	35	7
Too few students who intend to transfer actually do	73	23	4
Inadequate travel funds for faculty	56	22	22
Inadequate classroom facilities for use of technology	74	14	12
Inadequate computer facilities for part-time faculty use	72	18	9
Inadequate computer facilities for student services	89	10	1
Commercial outsourcing of instruction	98	2	0
Heavy classroom duties prevent personal & teaching enrichment by faculty	47	39	14
Coordinating mathematics courses with high schools	77	17	7
Lack of curricular flexibility because of transfer rules	77	17	7
Use of distance education	83	11	6

Note: 0 means less than one-half of 1% and round-off may make row sums seem inaccurate.

most likely in a developmental studies unit or in a laboratory setting. This was very similar to the 29% reported in 2000 and the 30% found in 1995.

Topics of Special Interest for Mathematics Programs

In each CBMS survey cycle, certain topics of special interest are chosen for data collection and comprehensive analysis across both two-year and four-year colleges. In fall 2005, six such topics were chosen. They are discussed in Chapter 2 of this report. Two of them, pre-service education of K–8 teachers and faculty who teach dual-enrollment courses, are relevant to the current chapter. The special interest topic that deals with resources available to undergraduates (such as placement testing and tutoring labs) was covered in Chapter 6.

Scope and organization of pre-service mathematics education for K–8 teachers

CBMS2005 expanded an inquiry begun in 2000 about the level of involvement of two-year college mathematics programs in the mathematical education of future mathematics teachers. These data are

TABLE TYF.30 Percentage of mathematics programs at public two-year colleges by type of administrative structure, in fall 2005.

	Percentage of Mathematics Programs	
Administrative structure	On their own campus	As part of a multicampus organization
Mathematics department	39	2
Mathematics and science department or division	35	1
Other department or division structure	15	2
None of the above or unknown	6	

reported primarily among the special topics in Chapter 2, especially in Tables SP.2 and SP.4.

Anecdotal evidence has suggested a growing involvement in teacher education at two-year colleges as more students turned to them, especially in summer sessions, to take required mathematics courses. Regarding the Mathematics for Elementary Teachers course, fall 2005 survey data confirm this involvement, reporting 29,000 students enrolled. This number was an attention-getting 61% increase from the 18,000 reported in 2000. See Table TYE.3 in Chapter 6.

CBMS2005 determined that 66% of two-year colleges offered the course Mathematics for Elementary School Teachers either in academic year 2004–2005 or in academic year 2005–2006. CBMS2000 showed this availability percentage was 49% for the combination of years 1999–2000 and 2000–2001. See Table TYE.5 in Chapter 6. The growth in fall term offerings for this course at two-year colleges, beginning in 1990 for five-year CBMS intervals, is reported in TYE.6 as successively 32%, 43%, 49%, and 59%.

Table SP.2 reports on organized programs at two-year colleges in which students can obtain their entire mathematics course requirement for teacher licensure. These data confirm that two-year colleges are involved in teacher education primarily at the K–8 level, though it is also creditable to assert that future secondary school teachers often take their lower-division mathematics courses at two-year colleges. The single largest component, reported by 30% of two-year colleges, is the program for pre-service elementary school teachers. Pre-service middle school licensure-oriented programs were reported at 19% of colleges. The flexible nature of two-year colleges makes them an attractive venue for in-service teacher education and for retraining by career switchers moving into teaching. Between 15% and 20% of two-year colleges reported programs at the elementary or middle school levels for these populations.

Table SP.4 reports on other involvements two-year college mathematics programs have with K–8 teacher education. Almost 40% report that a faculty member is assigned to coordinate mathematics education for future K–8 teachers. About 11% designate special sections of courses other than Mathematics for Elementary School Teachers for attendance by future teachers. Among mathematics departments, 9% offer mathematics pedagogy courses for future K–8 teachers, and 10% of colleges offer such pedagogy courses outside of the mathematics department.

The conclusion in Chapter 2 is that, given the large number of two-year colleges in the United States, even when the percentage of colleges involved in the education of future K–8 teachers is small, the cumulative impact of two-year colleges on the next generation of K–8 teachers can be significant. As a harbinger of this potential impact, in January 2007 the two principal higher education governing boards in Florida agreed the state's two-year colleges could offer certain bachelors degrees, education being one.

Credentials and supervision of dual-enrollment faculty

Dual enrollment is a credit structure that allows high school students to receive simultaneous high school and college credit for courses that were taught at a high school by a high school teacher. Data in Chapter 2 (Tables SP.16 and SP.17) show how large the dual-enrollment system had become by fall 2005 when (for example) just over 19% of all two-year college enrollments in the Precalculus course were dually enrolled and 18% of all Calculus I students were dually enrolled.

A faculty member teaching a dual-enrollment course usually was classified as a part-time faculty member at the two-year college that awarded college credit for the course, even though the salary was paid completely by a third party, e.g., the local school district. CBMS2000, the last available survey with relevant data, reported that nine out of ten of these "third-party" faculty members met the same academic credential requirements as regular part-time faculty. Given the enhanced monitoring of academic credentials by accrediting agencies mentioned above, just after Table TYF.3, it is unlikely the degree requirements for these "third party" faculty members have fallen off since 2000.

In fall 2005, 42,000 dual-enrolled students were taught by "third party" part-time faculty. Only 12% of colleges assigned their own direct-pay full-time or part-time faculty to teach dual-credit classes on a high school campus. These direct-pay faculty members taught about 2000 additional such students. See Tables SP.16 and SP.17 in Chapter 2.

In the 2000 survey, CBMS first investigated the extent to which two-year college mathematics programs retained control of various aspects of these dual-enrollment courses. This exploration was expanded in the 2005 survey. Overall, the conclusion in Chapter 2 is that the supervisory record for dual-enrollment courses will not be entirely reassuring to those who expect colleges to control the content and depth of the courses for which they are granting credit. See Table SP.16 in Chapter 2.

As presented in SP.16, only 52% of two-year college mathematics programs reported they always had full control over the selection of instructors for dual-enrollment courses, down almost ten points from the 2000 report (61%). In 74% of cases, the textbook used by a dual-enrollment instructor always was controlled by the college mathematics program, down five points from 2000. Only 37% of two-year college mathematics programs reported controlling the final examinations in their dual-enrollment courses, a very large decline of 20 percentage points from 2000. However, 89% of colleges reported they always had syllabus design or syllabus approval for dual-enrollment courses, up from 82% in 2000. In only 64% of cases was the college's usual teaching evaluation for part-time faculty required in dual-enrollment courses. This was down from 67% in 2000.

In spite of some of the issues raised in the preceding paragraph, as reported in Tables TYF.28 and TYF.29, among all survey respondents (who, it should be noted, include respondents from colleges that do not have dual-enrollment arrangements), only 5% of mathematics program heads in two-year colleges saw dual-enrollment courses as a major problem, down three points from 2000. Another 8% found dual-enrollment arrangements somewhat of a problem, down 13 points from 2000. In CBMS2000, the latest available satisfaction data from the subset of colleges that reported they actually had functioning dual-enrollment programs, only about 13% said dual enrollment was a major problem, and only an additional 14% said it was a moderate problem. In this group of actual users of dual enrollment in fall 2000, about 72% said dual enrollment was only a minor problem or no problem.

Bibliography for CBMS2005

[A] Ashburn, E., Two-year College Students Rarely Use Advisers, *Chronicle for Higher Education*, December 1, 2006.

[BI] Bryant, R. and Irwin, M., 1999–2000 Taulbee Survey: Current and Future Ph.D. Output Will Not Satisfy Demand for Faculty, *Computing Research News*, March, 2001, 5–11.

[CBMS1995] Loftsgaarden, D., Rung, D., and Watkins, A., *Statistical Abstract of Undergraduate Programs in the Mathematical Sciences in the United States, Fall 1995 CBMS Survey*, MAA Reports, Number 2, Mathematical Association of America, Washington, D.C., 1997.

[CBMS2000] Lutzer, D., Maxwell, J., and Rodi, S., *Statistical Abstract of Undergraduate Programs in the Mathematical Sciences in the United States, Fall 2000 CBMS Survey*, American Mathematical Society, Providence, R.I., 2002.

[CCSSE] Community College Survey of Student Engagement, http://www.ccsse.org/publications/CCSSENationalReport2006.pdf.

[CUPM] Committee for the Undergraduate Program in Mathematics, Assessment of Student Learning for Improving the Undergraduate Major in Mathematics, *Focus: The Newsletter of the Mathematical Association of America*, 15 (3), June 1995, pp. 24–28.

[GKM] *Assessment Practices in Undergraduate Mathematics*, ed. by B. Gold, S. Keith, and W. Marion, MAA Notes #49, Mathematical Association of America, Washington, D.C., 1999.

[JDC] Annual Reports of the Joint Data Committee, *Notices of the American Mathematical Society*, published annually, available at http://www.ams.org/employment/surveyreports.html.

[LM] Lutzer, D. and Maxwell, J., Staffing Shifts in Mathematical Sciences Departments, 1990–2000, *Notices of the American Mathematical Society*, 50 (2003), 683–686.

[M] Madison, B., Assessment of Undergraduate Mathematics, in L. A. Steen, ed., *Heeding the Call for Change: Suggestions for Curricular Action*, Mathematical Association of America, Washington, D.C., 1992, pp. 137–149.

[MAAGuidelines] *Guidelines for Programs and Departments in the Undergraduate Mathematical Sciences*, Revised Edition, February 2003, Mathematical Association of America, Washington, D.C.; http://www.maa.org/guidelines/guidelines.html

[MALL] McKelvey, R., Albers, D., Liebeskind, S., and Loftsgaarden, D., An inquiry into the graduate training needs of two-year college teachers of mathematics, Rocky Mountain Mathematics Consortium, 1979; ERIC document ED168629.

[MET] *The Mathematical Education of Teachers*, Volume 2 in *CBMS Issues in Mathematical Education* series, American Mathematical Society, Providence, R.I., 2001.

[NCES1] *Projections of Educational Statistics to 2015*, National Center for Educational Statistics, U.S. Department of Education, available at http://nces.ed.gov/programs/projections/tables/asp.

[NCES2] *Background Characteristics, Work Activities, and Compensation of Instructional Faculty and Staff: Fall 2003*, National Center for Educational Statistics, U.S. Department of Education, available at http://nces.ed.gov/pubs2006/2006176.pdf.

[NCES3] 2005 Digest of Educational Statistics, National Center for Educational Statistics, U.S. Department of Education, available at http://nces.ed.gov/programs/digest/d05/tables/dt05 252.asp.

[NCES4] 2005 Digest of Educational Statistics IPEDS Annual Completion Survey, National Center for Educational Statistics, U.S. Department of Education, available at http://nces.ed.gov/programs/digest/d05/lt3.asp and http://nces.ed.gov/programs/digest/d05/tables/dt05 265.asp.

[SMO] Schaeffer, R., Mendenhall, W., and Ott, L., *Elementary Survey Sampling*, Third Edition (1986), PWS-KENT Publishing Co., Boston, MA.

[V] Vegso, J., Drop in CS Bachelor's Degree Production, *Computing Research News*, Vol. 18/No. 2, March 2006.

[Z] Zweben, S., 2004–2005 Taulbee Survey, *Computing Research News*, Vol. 18/No. 3, May 2006.

Appendix I

Enrollments in Department Courses in Four-Year Colleges and Universities: 1995, 2000, 2005

TABLE A.1 Enrollment (in 1000s) in mathematics courses: in fall 1995, 2000, and 2005, [with SE for 2005 totals]. Roundoff may cause marginal totals to appear incorrect.

				Fall 2005 Enrollment (in 1000s)						
				Mathematics Departments				Statistics Departments		
Courses	1995	2000	2005	Univ (PhD)	Univ (MA)	Coll (BA)	Subtotal Math Depts	Univ (PhD)	Univ (MA)	Subtotal Stat Depts
Precollege										
1 Arithmetic	7	10	14 [4.7]	4	1	10	14 [4.7]			
2 Genl Math (Basic Skills)	13	13	16 [4.6]	1	3	11	16 [4.6]			
3 High School Elem Algebra	56	70	59 [9.8]	10	23	26	59 [9.8]			
4 High School Intermed Alg	131	117	105 [11.6]	38	29	38	105 [11.6]			
5 Other precollege level	15	8	7 [2.4]	1	4	2	7 [2.4]			
Subtotal Precollege Lvl	**222**	**218**	**201 [18.8]**	**55 [7.1]**	**60 [10.2]**	**87 [14.0]**	**201 [18.8]**			
Introductory (incl. pre-Calc)										
6 Coll Algebra	195	211	201 [17.2]	75	64	63	201 [17.2]			
7 Trigonometry	42	33	30 [3.5]	17	6	7	30 [3.5]			
8 Coll Alg & Trig combined	45	37	34 [6.8]	18	7	9	34 [6.8]			
9 Elem Fnctns [1]	86	105	93 [8.9]	47	20	25	93 [8.9]			
10 Intro Math Modeling	(na)	13	8 [3.1]	1	4	3	8 [3.1]			
11 Math Lib Arts	74	86	123 [11.7]	31	37	55	123 [11.7]			
12 Finite Math	59	82	94 [16.1]	43	18	33	94 [16.1]			
13 Business Math	40	53	38 [5.8]	16	12	10	38 [5.8]			
14 Math Elem Sch Tchrs	59	68	72 [6.5]	15	20	37	72 [6.5]			
15 Other Intro level math	14	36	12 [2.5]	6	1	5	12 [2.5]			
Subtotal Intro Level	**614**	**723**	**706 [29.0]**	**269 [17.2]**	**190 [10.9]**	**248 [20.6]**	**706 [29.0]**			

[1] Elementary Functions, Precalculus, and Analytic Geometry.

TABLE A.1, Cont. Fall term mathematics course enrollment (in 1000s) [with SE for 2005 totals].

				Fall 2005 Enrollments (in 1000s)						
				Mathematics Departments				Statistics Deptartments		
Courses	1995	2000	2005	Univ (PhD)	Univ (MA)	Coll (BA)	Subtotal Math Depts	Univ (PhD)	Univ (MA)	Subtotal Stat Depts
Calculus Level										
16 Mainstream Calc I	192	192	201 [9.6]	105	30	65	201 [9.6]			
17 Mainstream Calc II	83	87	85 [4.9]	54	12	19	85 [4.9]			
18 Mainstream Calc III,IV	62	73	74 [4.0]	51	9	14	74 [4.0]			
19 Non-mainstrm Calc I	98	105	108 [8.6]	61	21	26	108 [8.6]			
20 Non-mainstrm Calc II	14	10	11 [2.0]	10	0	0	11 [2.0]			
21a Diff Eq & Lin Alg (comb)	na	na	9 [2.2]	6	1	2	9 [2.2]			
21b Differential Equations	33	34	36 [2.8]	26	4	5	36 [2.8]			
22 Discrete Math	16	20	17 [1.9]	6	3	8	17 [1.9]			
23 Linear/Matrix Algebra	33	41	37 [2.6]	22	6	10	37 [2.6]			
24 Other calculus level	9	7	9 [2.7]	4	0	5	9 [2.7]			
Subtotal calculus level	**539**	**570**	**586 [23.6]**	**345 [17.4]**	**88 [7.5]**	**154 [14.0]**	**586 [23.6]**			
Advanced Level										
25 Intro to Proofs	7	10	12 [1.3]	6	3	4	12 [1.3]			
26 Mod Alg I & II	13	11	11 [1.1]	4	2	5	11 [1.1]			
27 Nmbr Theory	2	4	3 [0.5]	1	1	1	3 [0.5]			
28 Combinatorics	2	3	3 [0.5]	2	0	1	3 [0.5]			

Note: 0 means less than 500 enrollments.

TABLE A.1, Cont. Fall term mathematics course enrollment (in 1000s) [with SE for 2005 totals].

				Fall 2005 Enrollments (1000s)						
				Mathematics Departments				Statistics Departments		
Courses	1995	2000	2005	Univ (PhD)	Univ (MA)	Coll (BA)	Subtotal Math Depts	Univ (PhD)	Univ (MA)	Subtotal Stat Depts
29 Actuarial Mathematics	1	1	2 [0.5]	1	0	1	2 [0.5]			
30 Logic/ Foundations	3	2	1 [0.4]	1	0	0	1 [0.4]			
31 Discrete Structures	3	5	3 [0.7]	1	1	1	3 [0.7]			
32 Hist of Mathematics	3	2	6 [1.0]	1	2	3	6 [1.0]			
33 Geometry	6	6	8 [1.0]	3	2	4	8 [1.0]			
34 Math for HS Teachers	5	7	8 [2.2]	2	4	2	8 [2.2]			
35 Adv Calc I, & II, Real Analysis I&II	11	10	15 [1.2]	7	2	6	15 [1.2]			
36 Adv Math for Engr & Physics	8	5	6 [1.1]	4	1	0	6 [1.1]			
37 Adv Linear Algebra	4	3	4 [0.7]	3	1	0	4 [0.7]			
38 Vector Analysis	3	2	2 [0.8]	1	0	1	2 [0.8]			
39 Adv Diff Eqns	3	2	1 [0.2]	1	0	0	1 [0.2]			
40 Partial Diff Eqns	1	2	3 [0.5]	2	0	1	3 [0.5]			
41 Numerical Analysis	6	5	5 [0.5]	3	1	0	5 [0.5]			
42 Appl Math (Math Modeling)	4	2	2 [0.3]	1	1	0	2 [0.3]			
43 Complex Variables	2	3	3 [0.5]	2	0	1	3 [0.5]			
44 Topology	1	2	1 [0.3]	1	0	1	1 [0.3]			
45 Math of Finance	na	na	1 [0.4]	1	0	0	1 [0.4]			

Note: 0 means less than 500 enrollments.

TABLE A.1, Cont. Fall term mathematics course enrollment (in 1000s) [with SE for 2005 totals].

				Fall 2005 Enrollment (in 1000s)						
				Mathematics Departments				Statistics Departments		
Courses	1995	2000	2005	Univ (PhD)	Univ (MA)	Coll (BA)	Subtotal Math Depts	Univ (PhD)	Univ (MA)	Subtotal Stat Depts
46 Cryptology	na	na	0 [0.2]	0	0	0	0 [0.2]			
47 Biomathematics	na	na	1 [0.2]	1	0	0	1 [0.2]			
48 Senior Sem/Ind Study in Math	3	3	3 [0.5]	1	1	2	3 [0.5]			
46 Other Adv Level Courses	5	10	5 [0.7]	2	1	2	5 [0.7]			
Operations Research										
58 Intro Oper Res	1	1	1 [0.2]	0	0	0	1 [0.2]			
59 Int to LinearProgramming	1	1	1 [0.4]	1	0	0	1 [0.4]			
60 Other Oper Research	0	0	0 [0.2]	0	0	0	0 [0.2]			
Subtotal Advanced Math	**96**	**102**	**112 [6.2]**	**52**	**24**	**36**	**112 [6.2]**			
Mathematics Total	**1471**	**1614**	**1606 [45.3]**	**719 [25.8]**	**362 [18.1]**	**525 [32.5]**	**1606 [45.3]**			

Note: 0 means less than 500 enrollments.

TABLE A.2 Enrollment (in 1000s) in statistics courses in fall 1995, 2000, and 2005 in mathematics and statistics departments [with SE for totals]. Roundoff may cause marginal totals to appear incorrect.

				Fall 2005 Enrollment (in 1000s)						
				Mathematics Departments				Statistics Departments		
Statistics Courses	1995	2000	Total 2005	Univ (PhD)	Univ (MA)	Coll (BA)	Subtotal Math Depts	Univ (PhD)	Univ (MA)	Subtotal Stat Depts
Lower Level Statistics										
1 Elem Statistics. (no Calc prereq)	132	155	167 [14.3]	23	25	76	124 [13.8]	31	11	43 [3.7]
2 Prob.&Statistics (no Calc. prereq)	26	17	21 [5.5]	4	7	7	19 [5.5]	2	1	3 [0.6]
3 Other elem. level statistics	6	17	13 [2.5]	2	0	2	5 [1.5]	8	1	9 [2.0]
Subtotal, Elem Level Statistics	**164**	**190**	**202 [14.9]**	**30**	**32**	**86**	**148 [14.2]**	**42**	**13**	**54 [4.3]**
Upper Level Statistics										
4.Math Statistics (Calc Prereq)	16	18	12 [2.1]	2	4	3	9 [2.0]	3	0	3 [0.3]
5 Probability (Calc Prereq)	10	17	10 [1.0]	4	1	2	7 [0.9]	2	0	3 [0.4]
Prob & Statistics Combined	na	na	16 [2.0]	5	2	3	10 [1.9]	5	0	6 [0.7]
6 Stochastic Processes	0	1	1 [0.2]	0	0	0	0 [0.1]	0	0	1 [0.2]
7 Applied Statistical Analysis	9	6	7 [1.2]	1	1	0	3 [0.8]	3	1	4 [1.0]
8 Design & Anal of Experiments	1	2	1 [0.2]	0	0	0	0 [0.2]	1	0	1 [0.2]
9 Regressn & Correlation	1	2	3 [0.5]	0	0	0	1 [0.3]	2	0	2 [0.4]
10 Biostatistics	(na)	2	2 [0.6]	0	0	0	1 [0.5]	1	0	1 [0.4]
11 Nonparametric Statistics	(na)	1	0 [0.1]	0	0	0	0 [0.1]	0	0	0 [0.04]

Note: 0 means less than 500 enrollments.

TABLE A.2, Cont. Fall term statistics course enrollment (in 1000s) [with SE for 2005 totals].

				Fall 2005 Enrollment (in 1000s)						
				Mathematics Departments				Statistics Departments		
Statistics Courses	1995	2000	Total 2005	Univ (PhD)	Univ (MA)	Coll (BA)	Subtotal Math Depts	Univ (PhD)	Univ (MA)	Subtotal Stat Depts
12 Categorical Data Analysis	(na)	0	0 [0.1]	0	0	0	0 [0.1]	0	0	0 [0.1]
13 Survey Design & Analysis	(na)	0	1 [0.2]	0	0	0	0 [0.2]	0	0	0 [0.06]
14 Stat Software & Computing	(na)	1	1 [0.2]	0	0	0	0 [0.1]	0	0	1 [0.1]
15 Data Management	(na)	0	0 [0.0]	0	0	0	0 [0.0]	0	0	0 [0.0]
16 Senior Sem/ Indep Stdy in Statistics	0	0	0 [0.1]	0	0	0	0 [0.02]	0	0	0 [0.04]
17 Other Upper Level Statistics	7	5	3 [0.5]	1	0	0	1 [0.3]	2	0	2 [0.5]
Subtotal Upper Level Statistics	**44**	**45**	**57 [3.7]**	**15 [1.7]**	**9 [2.0]**	**10 [1.7]**	**34 [3.1]**	**20 [2.0]**	**3 [0.5]**	**23 [2.0]**
Statistics Total	**208**	**235**	**259 [15.4]**	**44 [4.4]**	**42 [6.7]**	**96 [12.2]**	**182 [14.6]**	**62 [4.2]**	**16 [2.8]**	**78 [5.0]**

Note: 0 means less than 500 enrollments.

TABLE A.3 Enrollment (in 1000s) in computer science courses in fall 1995, 2000, and 2005 [with SE for 2005 totals]. Roundoff may cause marginal totals to appear incorrect.

				Fall 2005 Enrollments (in 1000s)				
				Mathemtics Departments				
CS Courses	1995	2000	2005 Total	Univ (PhD)	Univ (MA)	Coll (BA)	Subtotal Math Depts	Subtotal Stat Depts
General Education CS Courses								
Computers & Society	14	4	5 [1.8]	0	2	2	4 [1.6]	1 [0.9]
Intro. to Software Pkgs	18	25	12 [4.1]	0	7	5	12 [4.1]	0 [0.1]
Other CS general ed courses	6	6	11 [4.8]	0	0	11	11 [4.8]	0 [0.0]
Subtotal general education courses	**38**	**35**	**28 [6.2]**	**1**	**8**	**17**	**26 [6.2]**	**1 [0.9]**
Lower-level CS Courses								
Computer Programming I *	17	23	10 [1.8]	2	1	7	10 [1.8]	--
Computer Programming II *	5	6	2 [0.6]	0	0	2	2 [0.6]	--
Discrete Structures for CS	2	4	1 [0.5]	0	0	1	2 [0.5]	--
Other Lower-level CS courses	13	22	4 [1.1]	0	1	2	4 [1.1]	--
Subtotal lower-level CS	**37**	**55**	**18 [2.9]**	**2**	**3**	**12**	**17 [2.9]**	**0 [0.1]**
All intermediate-level courses	**13**	**18**	**8 [1.4]**	**1**	**1**	**6**	**8 [1.4]**	**0 [0.2]**
All upper-level CS courses	**12**	**17**	**5 [1.3]**	**1 [0.5]**	**1 [0.3]**	**3 [1.1]**	**5 [1.3]**	**0 [0.0]**
Total Computer Science	**100**	**123**	**59 [9.9]**	**5 [2.0]**	**13 [4.2]**	**39 [8.7]**	**57 [9.8]**	**2 [1.1]**

* For 1995 and 2000, this course category was described in the 1991 ACM/IEEE CS curriculum report. For 2005, these courses were described in the 2001 ACM/IEEE report "Model Curricula for Computing".

Appendix II, Part I
Sampling and Estimation Procedures

Leela M. Aertker and **Robert P. Agans**
The Survey Research Unit
The University of North Carolina, Chapel Hill, North Carolina

Overview

A stratified, simple random sample was employed in the CBMS 2005 survey, and strata were based on three variables: curriculum, highest degree level offered, and total institutional enrollment. A paper-and-pencil data collection method was implemented between the months of September 2005 and May 2006, and all resulting estimates were generated in an SAS-Callable version of SUDAAN using a stratified-sampling-without-replacement design. This report is divided into the following two sections: Sampling Approach and Survey Design.

Sampling Approach

A stratified, simple random sample of 600 two-year and four-year colleges and universities was employed in CBMS 2005. A compromise mix of statistically optimum Neyman allocations based on two key outcome variables was used to determine targeted sample sizes for the 24 sampling strata.

Target Population and Sampling Frames

The target population of the CBMS 2005 survey consisted of undergraduate mathematics and statistics programs at two-year and four-year colleges and universities in the United States. In most cases, these programs were established academic departments whereas others were fledgling departments or other types of curriculum concentrations. A total of 2,459 programs were identified as eligible for participation in the survey. Sample selection was made from a merged program frame of 1,417 mathematics programs at four-year colleges and universities, 67 statistics programs at four-year colleges and universities, and 975 mathematics programs at two-year colleges.

Selection of Stratification Variables

Prior to selecting the sample for the CBMS 2005 and CBMS 2000 surveys, the stratification variables used in the CBMS 1995 survey were examined to determine their significance in predicting specific key outcome variables in each of the programs surveyed and thus, their utility for stratification in future CBMS surveys. This was done because the utility of a variable for stratification in generating estimates from a stratified sample depends on its statistical correlation with important measurements made on the sample.

Stratification in the CBMS 1995 survey was accomplished as follows: universities and colleges were separately divided into 20 strata based on curriculum (four-year mathematics programs, four-year statistics programs, or two-year mathematics programs), control (publicly or privately funded), level (the highest degree offered—BA, MA, or PhD), and enrollment (total institutional enrollment for Fall 1995). Our analysis of the CBMS 1995 data showed that curriculum, level, and enrollment would be the best stratification variables for producing estimates for future CBMS target populations. It was, therefore, decided not to stratify by each program's public or private classification as only minimal strength in predicting key outcome variables was gained by using this stratification variable.

The final stratum designations for the CBMS 2005 survey follow the exact stratum designations for the CBMS 2000 survey and very closely follow the stratum designations for the CBMS 1995 survey with the exception of control as a stratification variable. The four-year mathematics programs were divided into 12 strata, the four-year statistics programs were divided into five strata, and the two-year programs were divided into seven strata. Table A2.1 displays the overall stratum breakdown (24 strata total).

Allocation Process

For purposes of consistency in design development strategy, the same approach as used in CBMS 2000 was followed to determine the allocation of the CBMS 2005 sample. For CBMS 2005, stratum designations were assigned, key outcome variables were selected, and a multi-variable Neyman allocation was implemented in two iterations so that comparable precision was produced for each frame with the same number of schools expected to respond as in CBMS 2000.

Three program frames were sent to us by the study directors. Each frame included colleges and universities who were thought to offer undergraduate programs in four-year mathematics, four-year statistics, and two-year mathematics programs. The goal of sample selection was to select a representative sample of programs from each of the three frames. The sample was stratified by curriculum (four-year mathematics programs, four-year statistics programs, or two-year mathematics programs), level (the highest degree offered—BA, MA, or PhD), and enrollment (total institutional enrollment for Fall 2005).

The same key outcome variables from CBMS 2000 were once again proposed by the study directors in CBMS 2005; namely, total fall enrollment and number of full-time faculty. An additional outcome variable, number of baccalaureate degrees awarded, was also proposed, but this information was only collected for strata involving four-year institutions (i.e., strata 1–17). The variances of the two key outcome variables that were considered for purposes of allocation decisions, total fall enrollment and total full-time faculty, were estimated for each stratum using CBMS 2000 respondent data.

A multi-variable Neyman allocation was implemented to determine the optimum sample sizes for the strata within each frame, which would produce the most cost-effective allocation of the sample. This type of allocation samples more intensely from strata with more diversity or variability. The sample allocation intended to produce estimates of comparable precision for each of the three frames (four-year mathematics programs, four-year statistics programs, or two-year mathematics programs). This was done so that estimates aimed at the three frames would have approximately equal precision.

For CBMS 2005, it was determined that the same number of schools would be selected as in CBMS 2000 (i.e., n = 600). Due to refusals and unforeseen ineligibles, not all institutions selected would consequently respond. Thus, we intended to select a sample for CBMS 2005 that was *expected* to produce the same number of participating institutions as in CBMS 2000 (i.e., m = 392). The simple variance of each key outcome variable in each frame was calculated by using CBMS 2000 respondent data. The expected number of participating programs in each frame (m_g) was determined by the constraint that the variances of each frame were equivalent ($V_1 = V_2 = V_3$). A weighted average of the subgroup allocations was computed; however, this compromise mix of subgroup allocations called for sampling more four-year statistics programs than were on the frame. Therefore, the expected number to respond in the four-year statistics programs was set to the maximum expected number to respond (m_2 = 47) based on a realistic response rate for the particular subgroup.

The number expected to respond in the four-year and two-year mathematics program frames was then determined by the constraint that the variances of the four-year and two-year mathematics programs were equivalent ($V_1 = V_3$). A compromise mix of the expected number of programs to respond in the subgroup allocations was determined by giving the subgroup allocation based on total fall enrollment a relative weight of 0.75 and the subgroup allocation based on the number of full-time faculty a relative weight of 0.25. A larger relative weight was given to the subgroup allocation based on total fall enrollment since this variable, according to the study directors, was more salient to the study. The resulting subgroup allocation was as follows: expected number to respond for the four-year mathematics programs (m_1) = 202, expected number to respond for the two-year mathematics programs (m_3) = 143, and expected number to respond for the four-year statistics programs (m_2) = 47.

Separate Neyman allocations were then conducted for the four-year and two-year mathematics programs. The first Neyman allocation iteration produced two different sets of allocations among the strata—one based on total fall enrollment and the other based on full-time faculty. A minimum expected number of seven responding programs in each stratum was set unless seven exceeded the total stratum size times the CBMS 2000 response rate. In the latter case, the minimum expected number was the maximum number of expected respondents. By applying this rule, we set the minimum expected number of responding programs and computed a second iteration of the Neyman allocation for the 15 strata whose first iteration allocations exceeded the minimum standard.

The final sample allocation was anchored to the allocation produced by the key outcome variable, total fall enrollment, since this outcome variable was more salient to the study, according to the study directors. Modifications to the allocation based on total fall enrollment were made in consideration of sample size needs vis-à-vis the allocation based on total full-time faculty. Accordingly, a weighted average of the two second iteration allocations was computed based on total fall enrollment (given a relative weight of 0.75) and total full-time faculty (given a relative weight of 0.25) to produce the compromise mix of allocations in the four-year and two-year mathematics categories. Once the optimum allocation was determined, the number of selected programs in each stratum was calculated based on CBMS 2000 response rates. To obtain comparable precision for estimates aimed at the three frames, more participating four-year statistics programs were called for than were on the frame. Thus, for the four-year statistics frame, we simply took the maximum number of programs expected to respond and selected all programs in the frame. Table A2.1 lists the final agreed allocation and the sampling rate of the 600 selected programs for the CBMS 2005 survey.

Table A2.1 Stratum Designations and Final Agreed Allocation for the CBMS 2005 Study

Stratum	Curriculum	Level	Enrollment	Final Agreed Allocation	Sampling Rate
1	Four-Year Math	PhD	0 – 14,999	37	0.3627
2			15,000 – 24,999	54	0.8438
3			25,000 – 34,999	15	0.7500
4			35,000 +	6	1.0000
5		MA	0 – 6,999	17	0.2208
6			7,000 – 14,999	21	0.2414
7			15,000 +	12	0.4800
8		BA	0 – 999	16	0.0874
9			1,000 – 1,499	17	0.0846
10			1,500 – 2,499	30	0.1024
11			2,500 – 4,999	26	0.1130
12			5,000 +	41	0.3178
13	Four-Year Statistics	PhD	0 – 14,999	20	1.0000
14			15,000 – 24,999	23	1.0000
15			25,000 – 34,999	9	1.0000
16			35,000 +	3	1.0000
17		MA/BA	All	12	1.0000
18	Two-Year Schools	N/A	0 – 999	12	0.1519
19			1,000 – 1,999	16	0.1096
20			2,000 – 3,999	35	0.1378
21			4,000 – 7,999	64	0.2540
22			8,000 – 14,999	51	0.3312
23			15,000 – 19,999	26	0.6500
24			20,000+	37	0.7400
				600 programs	

Sample Selection

The SurveySelect procedure in SAS Version 8.2 was used to select the allocation from the merged program frame. We employed a stratified simple random sample design with three stratification variables (i.e., curriculum, level, and enrollment). The N= option specified the sample sizes for each of the 24 strata.

Survey Design

This section describes data collection, analysis procedures, and final weight construction.

Survey Implementation

Data collection occurred over a nine-month period. An advance letter was sent out to all respondents informing them that they were selected to participate and that they would receive the CBMS 2005 questionnaire within the next couple of weeks. All questionnaires were mailed out August 29, 2005 and a postcard was sent out at the end of October to either remind participants to respond or to thank them for their participation. A second batch of questionnaires was mailed out to all nonrespondents in the beginning of November. Questionnaires were accepted until an extended deadline of May 15, 2006.

Data Analysis

SUDAAN is a statistical package of choice when analyzing data from complex sample surveys. This software is advantageous since it allows the user to compute not only estimates such as totals and ratios, but also the standard errors of those estimates in accordance with the sample design. Many statistical packages are capable of computing population estimates, but the standard errors are based on simple random sampling; thus, they produce standard errors that are inappropriate for more complex designs. SUDAAN uses first-order Taylor series approximation procedures in generating the standard errors, which tend to be more accurate than estimates from other statistical packages. The sample design used in this study and incorporated in SUDAAN was stratified sampling without replacement (STRWOR).

For quality control purposes, all questionnaires were doubly entered by data entry personnel at the Survey Research Unit (SRU) at the University of North Carolina at Chapel Hill, and most discrepancies between the two files were settled by review of the original document. In a few cases, however, the respondents had to be contacted to clarify discrepancies. The bulk of data cleaning occurred between the months of May and July 2006. Data analysis took place between the months of May and August 2006.

Sample weights

For any respondent in the hth stratum, the nonresponse adjusted sample weight was computed as follows:

- Raw Weight = N_h / n_h
- Response Rate (RR) = m_h / (n_h – i_h)
- Adjusted weight = Raw Weight * (1/RR)

where,

N_h = the total number of programs in the h_{th} stratum

n_h = the number of selected programs in the h_{th} stratum

m_h = the number of (eligible) respondents in the h_{th} stratum

i_h = the number of study ineligibles in the sample for the h_{th} stratum

See Tables A2.2, A2.3, and A2.4 for the weights used in the four-year mathematics, four-year statistics, and two-year mathematics categories, respectively.

Table A2.2 Nonresponse Adjusted Sample Weights Used in the Four-Year Mathematics Questionnaire

Stratum	Total (N_h)	Number Selected (n_h)	Number of completes (m_h)	Number of ineligibles (i_h)	Response rate (RR)	Program level raw weight	Program level adjusted weight
1	102	37	30	0	0.811	2.757	3.400
2	64	54	34	1	0.642	1.185	1.847
3	20	15	10	0	0.667	1.333	2.000
4	6	6	3	1	0.600	1.000	1.667
5	77	17	14	0	0.824	4.529	5.500
6	87	21	14	0	0.667	4.143	6.214
7	25	12	6	0	0.500	2.083	4.167
8	183	16	8	0	0.500	11.438	22.875
9	201	17	8	0	0.471	11.824	25.125
10	293	30	14	0	0.467	9.767	20.929
11	230	26	13	1	0.520	8.846	17.012
12	129	41	22	0	0.537	3.146	5.864
Total	**1417**	**292**	**176**	**3**	**0.609**	-	-

Table A2.3 Nonresponse Adjusted Sample Weights Used in the Statistics Questionnaire

Stratum	Total (N_h)	Number Selected (n_h)	Number of completes (m_h)	Number of ineligibles (i_h)	Response rate (RR)	Program level raw weight	Program level adjusted weight
13	20	20	12	0	0.600	1.000	1.667
14	23	23	12	2	0.571	1.000	1.750
15	9	9	7	0	0.778	1.000	1.286
16	3	3	2	0	0.667	1.000	1.500
17	12	12	6	0	0.500	1.000	2.000
Total	**67**	**67**	**39**	**2**	**0.600**	-	-

Table A2.4 Nonresponse Adjusted Sample Weights Used in the Two-Year Mathematics Questionnaire

Stratum	Total (N_h)	Number Selected (n_h)	Number of completes (m_h)	Number of ineligibles (i_h)	Response rate (RR)	Program level raw weight	Program level adjusted weight
18	79	12	6	0	0.500	6.583	13.167
19	146	16	9	0	0.563	9.125	16.222
20	254	35	18	0	0.514	7.257	14.111
21	252	64	30	0	0.469	3.938	8.400
22	154	51	29	0	0.569	3.020	5.310
23	40	26	15	1	0.600	1.538	2.564
24	50	37	23	0	0.622	1.351	2.174
Total	**975**	**241**	**130**	**1**	**0.542**	-	-

Analysis Plan

To expedite analysis, protocols were developed in advance. Each protocol identified the variables involved, any mathematical transformations, the type of parameter being estimated, the procedure used to estimate the parameter, the units in which the estimate was to be reported, and any domain variables used to compartmentalize the variables. All protocols were subject to review by the CBMS director and approved before any estimates were generated. Table A2.5 is an example of the protocol used to construct a portion of the table FY.1 on page 114. All variables and resulting calculations were defined in an attempt to eliminate ambiguity.

TABLE A2.5 Example of Analysis Protocol: Portion of Table FY.1 (page 114).

Key[1]	Variable	Description	Analysis[2] Numerator	Denominator	Parameter Type	SUDAAN Procedure[3]	Unit	Domain Variable[4]
4M	C64	No. of Mathematics for Liberal Arts sections	C64					
4M	C65	No. of sections taught by tenured/tenure eligible faculty	C65					
4M	PC65	Percentage of sections taught by tenured/tenure eligible faculty	C65	C64	Percentage	Ratio	Sections	HDO_Math
4M	C66_67	No. of sections taught by other full-time (total) faculty	SUM(C66,C67);					
4M	PC66_67	Percentage of sections taught by other full-time (total) faculty	C66_67	C64	Percentage	Ratio	Sections	HDO_Math
4M	C66	No. of sections taught by other full-time (doctoral) faculty	C66					
4M	PC66	Percentage of sections taught by other full-time (doctoral) faculty	C66	C64	Percentage	Ratio	Sections	HDO_Math
4M	C68	No. of sections taught by part-time faculty	C68					
4M	PC68	Percentage of sections taught by part-time faculty	C68	C64	Percentage	Ratio	Sections	HDO_Math
4M	C69	No. of sections taught by graduate teaching assistants	C69					
4M	PC69	Percentage of sections taught by graduate teaching assistants	C69	C64	Percentage	Ratio	Sections	HDO_Math
4M	C6UNK	No. of sections taught by unknown faculty	C64 - (SUM (C65, C66,C67,C68,C69));					
4M	PC6UNK	Percentage of sections taught by Unknown faculty	C6UNK	C64	Percentage	Ratio	Sections	HDO_Math
4M	C63	Total On-campus enrollment in Mathematics for Liberal Arts	C63					
4M	AC64	Average size of Mathematics for Liberal Arts sections	C63	C64	Average	Ratio	Sections	HDO_Math

[1] Key: 4M=4-Yr Math, 4S=4-Yr Stat, 2M=2-Yr Math
[2] Blank boxes in the questionnaire tables were interpreted zeros
[3] Estimates weighted by final adjusted weight to produce national estimates
[4] HDO_Math (Highest Degree Offered for Four Year Math Departments)

Manipulation Checks

Because of the complex nature of the questionnaire, several manipulation checks were performed on the data before analyses proceeded. If a discrepancy could not be settled by reviewing the questionnaire, the respondent was called or emailed to settle it. No imputations were made for missing data. In fact, blank boxes in questionnaire tables were interpreted as zeros since many respondents refused to fill in all of the boxes. Hence, it was impossible to tell the difference between missing values and zeros in the questionnaire tables.

Generation of Information Products

All analyses were generated using a SAS-Callable version of SUDAAN (Version 9.01). To ease interpretation, the SUDAAN output was exported to Excel spreadsheets and sent to the CBMS director, which were transferred into production table shells. See Table A.2.6 for an example of the SUDAAN output that refers to the percentage of sections of one particular course taught by faculty with various appointments and the average section size in four-year mathematics departments by school type (or highest degree offered—HDO). All estimates were produced in a similar manner.

TABLE A2.6 Example of SUDAAN Output: Portion of Table FY.1 (page 114).

Estimate Description	Highest Degree Offered PhD	MA	BA	OVERALL
Four-Year Mathematics	Percentage (SE)	Percentage (SE)	Percentage (SE)	Percentage (SE)
Mathematics for Liberal Arts				
Percentage of sections taught by tenured/tenure eligible faculty	18.04% (3.09%)	35.55% (5.66%)	42.93% (5.70%)	36.71% (3.53%)
Percentage of sections taught by other full-time (total) faculty	18.93% (3.99%)	13.12% (4.09%)	15.65% (3.71%)	15.50% (2.45%)
Percentage of sections taught by other full-time (doctoral) faculty	5.29% (1.55%)	3.81% (1.74%)	4.11% (1.82%)	4.22% (1.15%)
Percentage of sections taught by part-time faculty	27.62% (3.55%)	37.90% (6.15%)	31.93% (7.03%)	32.86% (4.29%)
Percentage of sections taught by graduate teaching assistants	24.86% (4.72%)	3.47% (2.40%)	0% (0%)	5.13% (1.36%)
Percentage of sections taught by Unknown faculty	10.54% (4.55%)	9.96% (4.69%)	9.49% (4.03%)	9.80% (2.70%)
	Average (SE)	Average (SE)	Average (SE)	Average (SE)
Average size of Mathematics for Liberal Arts sections	45.95 (3.10)	33.87 (2.29)	24.75 (1.18)	30.83 (1.07)

Appendix II, Part II

Sampling and Estimation Procedures
Four-Year Mathematics and Statistics Faculty Profile

James W. Maxwell
American Mathematical Society

Overview

In all previous CBMS surveys, data on the faculty were collected on the CBMS form. For CBMS 2005, the information on the faculty at four-year colleges and universities provided in this report is derived from a separate survey conducted by the American Mathematical Society under the auspices of the AMS-ASA-IMS-MAA-SIAM Data Committee. The "Departmental Profile – Fall 2005" is one of a series of surveys of mathematical sciences departments at four-year institutions conducted annually as part of the *Annual Survey of the Mathematical Sciences*. In 2005 this survey was expanded to gather data on the age and the race/ethnicity of the faculty, in addition to the usual data collected annually on rank, tenure status and gender. The information on the four-year mathematics and statistics faculty derived from this data is presented in Chapters 1 and 4 of this report.

Using the faculty data collected in the 2005 Annual Survey reduced the size of the 2005 CBMS survey form. Furthermore, it eliminated the collection of the same faculty data on both surveys. Coordination between the administrators of the Annual Survey and the CBMS survey allowed for minimizing the number of departments that were asked to complete both surveys.

Target Populations and Survey Approach

The procedures used to conduct the 2005 Departmental Profile survey are very similar to those used in CBMS 2005, described in detail in the preceding pages of this appendix. The primary characteristic used to group the departments for survey and reporting purposes is the highest mathematical sciences degree offered by the department: doctoral, masters, or bachelors, the same groupings used by CBMS 2005. There are some notable differences. The Departmental Profile survey uses a census of the doctoral mathematics and statistics departments, and it surveys only the doctoral statistics departments. There were twelve departments in the CBMS 2005 sample frame of statistics departments that offered at most a bachelors or masters degree. These departments are not represented in the description of the faculty at the doctoral statistics departments.

Comparison of the Annual Survey Sample Frame with the CBMS Sample Frame

Table AS.1 demonstrates that the sample frames of four-year mathematics departments used in the two surveys are in extremely close alignment. As a consequence of this alignment, the distinction between the terms "Bachelors", "Masters" and "Doctoral" mathematics departments as defined in the two surveys is immaterial. Furthermore, the estimates produced from each of the surveys may be applied interchangebly to these groupings of departments.

Table AS.1 Comparability of 2005 Annual Survey Sample Frame and the 2005 CBMS Sample Frame for Four-Year Mathematics Departments

Dept. Grouping	Annual Survey Count	CBMS Count	Overlap Count
Bachelors Depts.	1036	1036	1030
Masters Depts.	190	189	188
Doctoral Depts.	196	192	188
Total	1422	1417	1406

Sampling Masters and Bachelors Departments at Four-Year Institutions

While the Annual Survey employs a census of the doctoral mathematics and statistics departments, it uses a stratified, random sample of the masters and bachelors departments. The masters and bachelors departments are stratified by control (public or private) and by total institutional undergraduate enrollment. Table AS.2 summarizes the stratifications used for the Departmental Profile and the allocation of the sample to the strata for the masters and bachelors departments.

Table AS.2 Stratum Designations and Allocations for the 2005 Departmental Profile Survey

Stratum	Curriculum	Level	Institutional Enrollment	Sample Allocation	Sampling Rate
1	Four-Year Math	PhD	All	196	1.0000
2	(Public)	MA	0 – 5,999	12	0.4444
3			6,000 – 8,999	21	0.5526
4			9,000 – 11,999	21	0.5833
5			12,000 – 17,999	22	0.5641
6			18,000 +	10	0.5559
7	(Private)	MA	0 – 3,999	6	0.5000
8			4,000 – 7,999	5	0.4545
9			8,000 +	3	0.3333
10	(Public)	BA	0 – 1,999	22	0.3548
11			2,000 – 3,999	31	0.3605
12			4,000 – 6,999	40	0.5063
13			7,000 – 11,999	21	0.7778
14			12,000 +	10	0.6667
15			Military academies	2	0.6667
16	(Private)		0 – 999	48	0.2667
17			1,000 – 1,499	49	0.3161
18			1,500 – 1,999	65	0.4815
19			2,000 – 3,999	70	0.3483
20			4,000 – 6,999	24	0.3582
21			7,000 – 8,999	10	0.6250
22			9,000 +	4	0.4000
23	Four-Year Statistics	PhD	All	56	1.0000
				748 departments	

Survey Implementation

Departmental Profile forms were mailed in late September 2005 with a due date of October 30th to all doctoral-granting mathematics and statistics departments and to a sampling of the masters- and bachelors-granting departments of mathematical sciences at four-year colleges and universities in the U.S. A second mailing of forms was sent to non-responders in early November with a due date of December 6th. A third mailing was sent via email at the end of January 2006 providing a link to an interactive PDF version of the form with a due date in early February. The final effort to obtain responses took place during February through March in the form of phone calls to non-responding departments. The final efforts were concentrated on the stata with the lowest response rates.

Data Analysis

The data analysis used with the 2005 Departmental Profile survey parallels that used by CBMS 2005. The only notable variation is that if a non-responding department had completed a Departmental Profile survey within the previous three years, data from that survey was used to replace as much of the missing data as feasible. This previously reported data consisted of the department's counts of faculty by rank, tenure-status and gender. This technique was not possible for data on faculty age and race/ethnicity since this information is not a part of previous Departmental Profile surveys.

The use of a department's prior-year faculty data to replace missing data for fall 2005 is supported by a review of annual faculty data from departments responding to the Departmental Profile in multiple years. Analysis of these data series demonstrates that the year-to-year variations in a given department's faculty data are highly likely to be smaller than the department's variation from the mean data for that department's stratum. Since the technique used to estimate a total for a stratum is equivalent to replacing the missing data with the average for the responding departments in that stratum, using prior responses to the same question is likely to produce a more accurate estimate of the total.

Table AS.3 lists the program-level adjusted sample weights used to produce the estimates within each stratum of counts of faculty by rank, type-of-appointment and gender. The column "Number of Completes" displays the total of the forms returned plus the responses from prior years when available. (Compare with Table A2.2 in Appendix II.) The adjusted weights used to produce estimates of age distribution and race/ethnicity distributions are slightly higher since responses to those items were not available for prior years.

The standard errors reported for the faculty data are computed using the formulas described on pages 83–84 and 97–98 of [SMO]. For the doctoral mathematics departments, use of prior-year responses produced a 100% response rate for certain items, hence the contribution of the doctoral mathematics departments to the standard errors for those items was zero.

Table AS.3 Nonresponse Adjusted Sample Weights Used with the 2005 Departmental Profile Questionnaire.

Stratum	Total	Number Selected	Number of completes	Number of Prior-year Resp. used	(Final) Response rate	Program level raw weight	Program level adjusted weight
1	196	196	163	33	1.000	1.000	1.000
2	27	12	5	3	0.667	2.250	3.375
3	38	21	13	3	0.762	1.810	2.375
4	36	21	13	2	0.714	1.714	2.400
5	39	22	12	5	0.773	1.773	2.294
6	18	10	7	3	1.000	1.800	1.800
7	12	6	2	1	0.500	2.00	4.000
8	11	5	4	0	0.800	2.200	2.750
9	9	3	3	0	1.000	3.000	3.000
10	62	22	2	2	0.182	2.818	15.500
11	86	31	13	1	0.452	2.774	6.143
12	79	40	23	6	0.725	1.975	2.724
13	27	21	14	2	0.762	1.286	1.688
14	15	10	5	1	0.600	1.500	2.500
15	3	2	2	0	1.000	1.500	1.500
16	180	48	15	1	0.333	3.750	11.250
17	155	49	16	3	0.388	3.163	8.158
18	135	65	26	6	0.492	2.077	4.219
19	201	70	34	5	0.557	2.871	5.154
20	67	24	13	5	0.750	2.792	3.722
21	16	10	4	1	0.500	1.600	3.200
22	10	4	2	0	0.500	2.500	5.000
23	56	56	39	16	0.982	1.000	1.018

Appendix III

List of Responders to the Survey

Two-Year Respondents

American River College
Mathematics

Arkansas State University - Mountain Home
Mathematics

Butler County Community College
Mathematics

Cerritos College
Mathematics

Chabot College
Science & Mathematics

City College Of San Francisco
Mathematics

City Colleges Of Chicago - Olive-Harvey College
Mathematics

Cochise College
Mathematics & Science

College Of Southern Idaho
Mathematics

College Of The Sequoias
Mathematics

Columbus State Community College
Mathematics

Community College Of Allegheny County
Mathematics

Community College Of Denver
Center For Arts & Science

Community College Of Philadelphia
Mathematics

Corning Community College
Mathematics

Cosumnes River College
Science, Mathematics & Engineering

Crafton Hills College
Mathematics

CUNY Queensborough Community College
Mathematics & Computer Science

Cuyahoga Community College District
Institutional Planning & Evaluation

Cypress College
Science, Engineering & Mathematics

Darton College
Science & Mathematics

Delta College
Mathematics & Computer Science

Diablo Valley College
Mathematics & Computer Science

Dodge City Community College
Mathematics

Eastern New Mexico University - Roswell Campus
Mathematics

Eastfield College
Academic Support & Mathematics

El Paso Community College
Mathematics

Elgin Community College
Mathematics

Evergreen Valley College
Mathematics & Science

Florida Community College at Jacksonville
Mathematics

Foothill College
Physical Sciences, Mathematics & Engineering

Fort Peck Community College
Mathematics

Fox Valley Technical College
Mathematics

Fresno City College
Mathematics

Gavilan College
Natural Science

Genesee Community College
Mathematics & Science

Georgia Perimeter College
Mathematics & Science

Glendale Community College
Mathematics

Green River Community College
Mathematics

Greenfield Community College
Mathematics

Greenville Technical College
Mathematics

Grossmont College
Mathematics

Harrisburg Area Community College - Harrisburg
Mathematics

Hill College
Mathematics & Science

Hocking College
Arts & Sciences

Illinois Eastern Community Colleges - Olney Central
Mathematics

Iowa Lakes Community College
Mathematics

Itasca Community College
Mathematics & Science

J. Sargeant Reynolds Community College
Mathematics & Science

Johnson County Community College
Mathematics

Joliet Junior College
Mathematics

Kankakee Community College
Mathematics, Science & Engineering

Lake Land College
Mathematics & Physical Science

Lake Tahoe Community College
Mathematics

Lansing Community College
Mathematics & Computer Science

Laramie County Community College
Mathematics

Lewis & Clark Community College
Mathematics

Lord Fairfax Community College
Mathematics

Macomb Community College
Mathematics

Manatee Community College
Mathematics

Martin Community College
College Transfer

McLennan Community College
Mathematics

Mesa Community College
Mathematics

Metropolitan Community College Area
Mathematics, Science & Health Centers

Miami University - Hamilton
Mathematics

Mid Plains Community College Area
Mathematics

Middle Georgia College
Mathematics & Engineering

Middlesex County College
Mathematics

Midland College
Mathematics & Science

Mid-South Community College
Learning Assessment & Support

Monroe Community College
Mathematics

Montgomery College
Mathematics

Moraine Valley Community College
Mathematics & Computer Science

Motlow State Community College
Mathematics

Mt. Hood Community College
Mathematics

Murray State College
Mathematics & Science

New Mexico State University - Alamogordo
Mathematics, Statistics & Developmental Mathematics

North Florida Community College
Mathematics

North Harris Montgomery Community College District
Mathematics

North Lake College
Mathematics, Science & Sports Science

Northampton County Area Community College
Mathematics

Northcentral Technical College
General Education

Northern Essex Community College
Mathematics

Oakland Community College
Mathematics

Ocean County College
Mathematics

Ohlone College
Mathematics

Orange Coast College
Mathematics

Palomar College
Mathematics

Pellissippi State Technical Community College
Mathematics

Piedmont Community College
General Education & Business Technology

Piedmont Virginia Community College
Mathematics, Science & Human Services

Pima Community College
Mathematics & Engineering

Polk Community College
Mathematics, Science & Health

Portland Community College
Mathematics

Ranger College
Mathematics

Raritan Valley Community College
Mathematics

Renton Technical College
Mathematics

Rio Hondo College
Mathematics & Science

Rio Salado Community College
Mathematics

Sacramento City College
Mathematics

Saint Louis Community College - Florissant Valley
Mathematics

San Diego Mesa College
Mathematics

San Jacinto College - North Campus
Mathematics

San Joaquin Delta College
Mathematics

Santa Monica College
Mathematics

Schoolcraft College
Mathematics

Seminole Community College
Mathematics

Seward County Community College
Natural Science & Mathematics

Sierra College
Mathematics

Skyline College
Mathematics

Somerset Technical College
Mathematics & Natural Science

Southeastern Illinois College
Mathematics & Science

Southwestern Indian Polytechnic Institute
Mathematics & Science

Spokane Falls Community College
Mathematics

Suffolk County Community College
Mathematics

SUNY Ulster County Community College
Mathematics

Thomas Nelson Community College
Mathematics

Tri-County Technical College
Mathematics

Trident Technical College
Mathematics

Tulsa Community College
Science & Mathematics

Tunxis Community College
Mathematics

Tyler Junior College
Mathematics

University of Montana - Helena College Of Technology
General Education

University of South Carolina at Lancaster
Mathematics, Science & Nursing

University of Wisconsin Colleges
Mathematics

Virginia Highlands Community College
Science & Engineering Technology

Volunteer State Community College
Mathematics

Waubonsee Community College
Technology, Mathematics & Physical Science

Whatcom Community College
Mathematics

Yavapai College
Mathematics

Four-Year Mathematics Respondents

Ashland University
Mathematics & Computer Science

Assumption College
Mathematics & Computer Science

Auburn University
Mathematics & Statistics

Augsburg College
Mathematics

Baker College
General Education

Bellarmine University
Mathematics

Bethany University
School of Arts & Sciences

Bowling Green State University
Mathematics & Statistics

Brigham Young University
Mathematics

California State University, San Bernardino
Mathematics

California State University, San Marcos
Mathematics

Calvin College
Mathematics & Statistics

Carnegie Mellon University
Mathematical Sciences

Centenary College of Louisiana
Mathematics

Central Michigan University
Mathematics

Central Washington University
Mathematics

Chestnut Hill College
Mathematical Sciences

College of Charleston
Mathematics

College of New Jersey
Mathematics & Statistics

College of William & Mary
Mathematics

Colorado School of Mines
Mathematical & Computer Science

Colorado State University - Pueblo
Mathematics & Physics

Columbia College Chicago
Science & Mathematics

Cornell College
Mathematics

Dartmouth College
Mathematics

Davidson College
Mathematics

East Carolina University
Mathematics

Eastern Kentucky University
Mathematics & Statistics

Eastern Mennonite University
Mathematical Sciences

Eastern Michigan University
Mathematics

Eastern New Mexico University
Mathematical Sciences

Edinboro University of Pennsylvania
Mathematics & Computer Science

Evangel University
Science & Technology

Fairmont State University
Computer Science, Mathematics & Physics

Florida State University
Mathematics

Fontbonne University
Mathematics & Computer Science

Friends University
Mathematics

George Mason University
Mathematical Sciences

Georgetown College
Mathematics, Physics & Computer Science

Georgia Institute of Technology
School of Mathematics

Goucher College
Mathematics & Computer Science

Grand Valley State University
Mathematics

Guilford College
Mathematics

Hope College
Mathematics

Humboldt State University
Mathematics

Huston-Tillotson University
Mathematics

Illinois State University
Mathematics

Indiana University - Purdue University Indianapolis
Mathematical Sciences

Indiana Wesleyan University
Mathematics

James Madison University
Mathematics & Statistics

Lake Forest College
Mathematics & Computer Science

Lamar University
Mathematics

Le Moyne College
Mathematics & Computer Science

Lehigh University
Mathematics

Linfield College
Mathematics

Long Island University, C. W. Post Campus
Mathematics

Loyola Marymount University
Mathematics

Lynchburg College
Mathematics

Manchester College
Mathematics & Computer Science

Marquette University
Mathematics, Statistics & Computer Science

Mercy College
Mathematics & Computer Information Science

Miami University, Oxford
Mathematics

Michigan State University
Mathematics

Midwestern State University
Mathematics

Millersville University of Pennsylvania
Mathematics

Minnesota State University, Mankato
Mathematics & Statistics

Missouri State University
Mathematics

Morgan State University
Mathematics

Mount Union College
Mathematics

Muskingum College
Mathematics & Computer Science

New Jersey Institute of Technology
Mathematical Sciences

New Mexico Institute of Mining & Technology
Mathematics

New York Institute of Technology, Old Westbury Campus
Mathematics

Nicholls State University
Mathematics & Computer Science

North Carolina State University
Mathematics

North Dakota State University
Mathematics

Northeastern University
Mathematics

Northern Illinois University
Mathematical Sciences

Oakland University
Mathematics & Statistics

Ohio State University, Columbus
Mathematics

Oklahoma Panhandle State University
Mathematics & Physics

Oklahoma State University
Mathematics

Pacific University
Mathematics & Computer Science

Penn State University
Mathematics

Plymouth State University
Mathematics

Queens College
Mathematics

Rensselaer Polytechnic Institute
Mathematical Sciences

Rowan University
Mathematics

Rutgers University - New Brunswick
Mathematics

Saint Josephs University
Mathematics & Computer Science

San Francisco State University
Mathematics

Simons Rock College of Bard
Mathematics

Southern Illinois University - Carbondale
Mathematics

Southern New Hampshire University
Mathematics & Science

Southern Utah University
Mathematics

Southwest Baptist University
Mathematics

Southwestern Oklahoma State University
Mathematics

Stephen F. Austin State University
Mathematics & Statistics

SUNY at Oswego
Mathematics

SUNY College at Cortland
Mathematics

SUNY Fredonia
Mathematical Sciences

Temple University
Mathematics

Texas A&M University, College Station
Mathematics

Texas Christian University
Mathematics

The Citadel
Mathematics & Computer Science

Trinity University (Texas)
Mathematics

Troy University, Dothan Campus
Mathematics

University at Buffalo, SUNY
Mathematics

University of Akron
Theoretical & Applied Mathematics

University of Alabama
Mathematics

University of Alabama at Birmingham
Mathematics

University of Alaska - Anchorage
Mathematical Sciences

University of Alaska - Fairbanks
Mathematics & Statistics

University of Arkansas
Mathematical Sciences

University of California, Los Angeles
Mathematics

University of California, Riverside
Mathematics

University of California, Santa Barbara
Mathematics

University of Cincinnati
Mathematical Sciences

University of Colorado at Boulder
Mathematics

University of Connecticut
Mathematics

University of Dayton
Mathematics

University of Delaware
Mathematical Sciences

University of Florida
Mathematics

University of Georgia
Mathematics

University of Illinois at Chicago
Mathematics, Statistics, & Computer Science

University of Illinois at Urbana-Champaign
Mathematics

University of Louisiana at Lafayette
Mathematics

University of Maine at Augusta
Mathematics

University of Mary Washington
Mathematics

University of Maryland, Baltimore County
Mathematics & Statistics

University of Massachusetts - Amherst
Mathematics & Statistics

University of Michigan
Mathematics

University of Minnesota
School of Mathematics

University of Minnesota - Crookston
Mathematics

University of Missouri - Rolla
Mathematics & Statistics

University of Missouri - St. Louis
Mathematics & Computer Science

University of Montana
Mathematical Sciences

University of Nebraska - Kearney
Mathematics & Statistics

University of Nebraska - Lincoln
Mathematics

University of New Hampshire
Mathematics & Statistics

University of New Mexico
Mathematics & Statistics

University of Northern Colorado
School of Mathematical Sciences

University of North Carolina Chapel Hill
Mathematics

University of Oklahoma
Mathematics

University of Rhode Island
Mathematics

University of South Florida
Mathematics

University of Southern Mississippi
Mathematics

University of St. Thomas (St. Paul)
Mathematics

University of Tennessee
Mathematics

University of Tennessee at Martin
Mathematics & Statistics

University of Texas at Arlington
Mathematics

University of Toledo
Mathematics

University of Utah
Mathematics

University of Virginia
Mathematics

University of Washington
Mathematics

University of Wisconsin - Oshkosh
Mathematics

University of Wisconsin - River Falls
Mathematics

University of Wyoming
Mathematics

University of Iowa
Mathematics

Vanderbilt University
Mathematics

Virginia Intermont College
Arts & Sciences

Virginia Polytechnic Institute & State University
Mathematics

Walla Walla College
Mathematics

Washington State University
Mathematics

Washington University (St. Louis)
Mathematics

Wayne State College
Physical Sciences & Mathematics

West Virginia University
Mathematics

Western Washington University
Mathematics

Westminster College
Mathematical Sciences

Wichita State University
Mathematics & Statistics

Wilkes University
Mathematics & Computer Science

William Carey College
Mathematics & Physics

William Woods University
Arts & Sciences

Xavier University
Mathematics & Computer Science

York College of Pennsylvania
Physical Science

Youngstown State University
Mathematics & Statistics

Four-Year Statistics Respondents

Brigham Young University
Statistics

California Polytechnic State University - San Luis Obispo
Statistics

California State University, East Bay
Statistics

Carnegie Mellon University
Statistics

Case Western Reserve University
Statistics

Colorado State University
Statistics

Columbia University
Statistics

Duke University
Institute of Statistics & Decision Sciences

Florida State University
Statistics

George Washington University
Statistics

Iowa State University
Statistics

Kansas State University
Statistics

Louisiana State University, Baton Rouge
Experimental Statistics

Ohio State University, Columbus
Statistics

Oregon State University
Statistics

Penn State University, University Park
Statistics

Purdue University, West Lafayette
Statistics

Rutgers University - New Brunswick
Statistics

Southern Methodist University
Statistical Science

St. Cloud State University
Statistics & Computer Networking

Stanford University
Statistics

Temple University
Statistics

Texas A&M University, College Station
Statistics

University of California, Davis
Statistics

University of California, Los Angeles
Statistics

University of California, Santa Barbara
Statistics & Applied Probability

University of Chicago
Statistics

University of Connecticut, Storrs
Statistics

University of Denver
Statistics & Operations Technology

University of Illinois at Urbana-Champaign
Statistics

University of Iowa
Statistics & Actuarial Science

University of Minnesota - Twin Cities
School of Statistics

University of Pennsylvania
Statistics

University of Pittsburgh, Pittsburgh
Statistics

University of South Carolina, Columbia
Statistics

University of Wisconsin, Madison
Statistics

Virginia Commonwealth University
Statistical Sciences & Operations Research

Virginia Polytechnic Institute & State University
Statistics

Yale University
Statistics

Appendix IV

Four-Year Mathematics Questionnaire

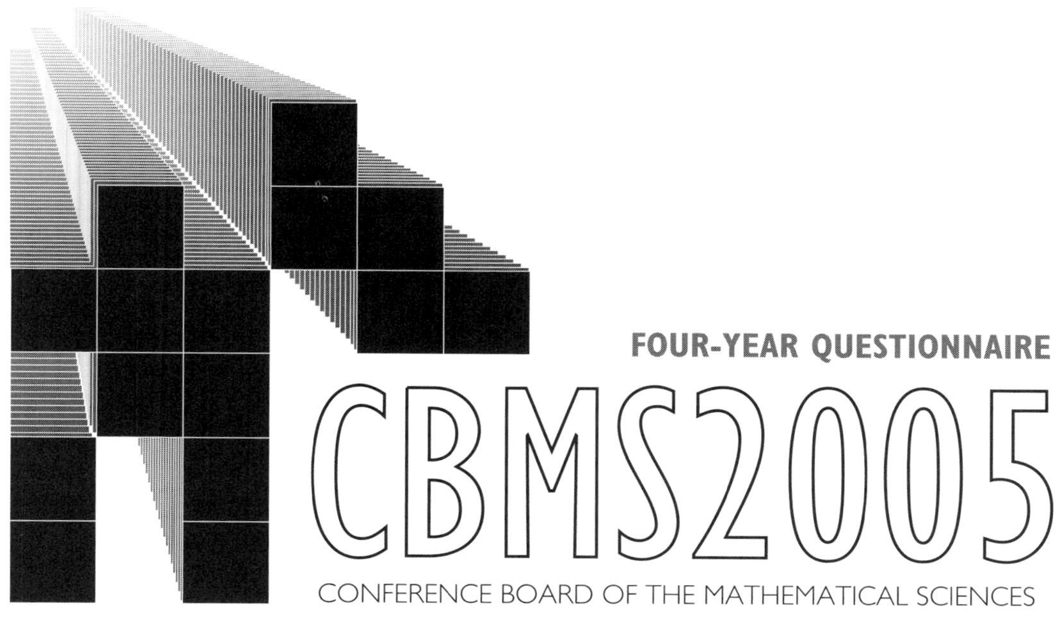

SURVEY OF UNDERGRADUATE PROGRAMS IN THE MATHEMATICAL SCIENCES

General Information

Mathematics Questionnaire

As part of a random sample, your department has been chosen to participate in the NSF-funded CBMS2005 National Survey of Undergraduate Mathematical Sciences. Even though it is a very complicated survey, the presidents of all U.S. mathematical sciences organizations have endorsed it and ask for your cooperation.

We assure you that no individual departmental data, except the names of responding departments, will be released.

This survey provides data about the nation's undergraduate mathematical and statistical effort that is available from no other source. You can see the results of a similar survey five years ago by going to www.ams.org/cbms where the CBMS 2000 report is available on-line.

This survey studies the undergraduate programs in universities and colleges that offer at least a bachelors degree. Many of the departments in our random sample also offer higher degrees in mathematical sciences.

We have classified your department as belonging to a university or four-year college. If this is not correct, please contact David Lutzer, Survey Director, at 757-221-4006 or at Lutzer@math.wm.edu.

If you have any questions while filling out this survey form, please call the Survey Director, David Lutzer, at 757-221-4006 or contact him by e-mail at Lutzer@math.wm.edu.

Please report on undergraduate programs in the broadly defined mathematical sciences including applied mathematics, statistics, operations research, and computer science that are under the direction of your department. Do not include data for other departments or for branches or campuses of your institution that are budgetarily separate from your own.

Please return your completed questionnaire by October 15, 2005 in the enclosed envelope to:

CBMS Survey
UNC-CH Survey Research Unit
730 Martin Luther King, Jr. Blvd
Suite 103, CB#2400, UNC-CH
Chapel Hill, NC 27599-2400

Please retain a copy of your responses to this questionnaire in case questions arise.

A. General Information

Mathematics Questionnaire

PLEASE PRINT CLEARLY

A1. Name of your institution: ______________________________

A2. Name of your department: ______________________________

A3. We have classified your department as being part of a university or four-year college. Do you agree?

Yes.......................... ☐ (1) ⟶ If "Yes", go to A4 below.

No.......................... ☐ (2) ⟶ If "No", please call David Lutzer, Survey Director, at 757-221-4006 before proceeding any further.

A4. Your institution is ....public ☐ (1) ; ....private ☐ (2)

A5. Which programs leading to the following degrees does your department offer? Please check at least one box in each row.

Program	None (1)	Baccalaureate Degree (2)	Masters Degree (3)	Doctoral Degree (4)
a) Mathematics (including applied)				
b) Statistics				
c) Mathematics Education				
d) Computer Science				
e) Other (please specify below)				

If you offer bachelors, masters, or doctoral degrees in a mathematical science other than those in A5-a, b, c, and d, please enter the name(s) of the fields here: ______________________________

A6. Responses to this question will be used to project total enrollment in the current (2005-2006) academic year based on the pattern of your departmental enrollments in 2004-2005. Do NOT include any numbers from dual-enrollment courses[1] in answering question A6.

a) Previous fall (2004) total student enrollment in your department's undergraduate courses (remember: do not include dual-enrollment courses[1]): .... ☐ (1)

b) Previous academic year (2004-2005) total enrollment in your department's undergraduate courses, <u>excluding</u> dual enrollments[1] and <u>excluding</u> enrollments in summer school 2005: ☐ (2)

c) Total enrollment in your department's undergraduate courses in summer school 2005: .... ☐ (3)

d) Total enrollment in Calculus II in Winter/Spring term of 2005: .... ☐ (4)

e) Total number of sections in Calculus II in Winter/Spring term of 2005: .... ☐ (5)

[1] *In this question, the term "dual-enrollment courses" is used to mean courses taught on a high school campus, by high school teachers, for which high school students may obtain high school credit and simultaneously college credit through your institution.*

A. General Information cont.

A7. Which of the following best describes your institution's academic calendar? Check only one box.

a) Semester	
b) Trimester	
c) Quarter	
d) Other (please specify below)	

Academic calendar description if not a), b), or c): ______________________________

A8. If your college or university does not recognize tenure, check the following box [ ] and follow the special instructions in subsequent sections for counting departmental faculty of various types.

A9. Contact person in your department: [ ]

A10. Contact person's e-mail address: [ ]

A11. Contact person's phone number including area code: [ ]

A12. Contact person's mailing address: [ ]

B. Dual Enrollment Courses

In this questionnaire the term dual enrollment courses refers to courses conducted on a high school campus and taught by high school teachers, for which high school students may obtain high school credit and simultaneously college credit through your institution.

B1. Does your department participate in any dual enrollment programs of the type defined above?

Yes............................ ☐ (1) ⟶ If "Yes", go to B2.

No.............................. ☐ (2) ⟶ If "No", go to B6.

B2. Please complete the following table concerning your dual enrollment program (as defined above) for the previous term (spring 2005) and the current fall term of 2005.

Course	Total Dual Enrollments Last Term =Spring 2005 (1)	Number of Dual-Enrollment Sections Last Term =Spring 2005 (2)	Total Dual Enrollments This Term =Fall 2005 (3)	Number of Dual-Enrollment Sections This Term =Fall 2005 (4)
a) College Algebra				
b) Pre-calculus				
c) Calculus I				
d) Statistics				
e) Other				

B3. For the dual enrollment courses in B2, to what extent are the following the responsibility of your department?

	Never Our Responsibility (1)	Sometimes Our Responsibility (2)	Always Our Responsibility (3)
a) Choice of textbook			
b) Design/approval of syllabus			
c) Design of final exam			
d) Choice of instructor			

B4. Does your department have a teaching evaluation program in which your part-time department faculty are required to participate?

Yes............................ ☐ (1) ⟶ If "Yes", go to B5.

No.............................. ☐ (2) ⟶ If "No", go to B6.

B5. Are instructors in the dual-enrollment courses reported in B2 required to participate in the teaching evaluation program for part-time departmental faculty described in B4?

Yes............................ ☐ (1)

No.............................. ☐ (2)

B. Dual Enrollment Courses cont.

Mathematics Questionnaire

B6. Does your department assign any of its own full-time or part-time faculty to teach courses conducted on a high school campus for which high school students may receive both high school and college credit (through your institution)?

Yes........................... ☐ (1) ⟶ If "Yes", go to B7.

No............................ ☐ (2) ⟶ If "No", go to Section C.

B7. How many students are enrolled in the courses conducted on a high school campus and taught by your full-time or part-time faculty and through which high school students may receive both high school and college credit (through your institution)? .... ☐

*In subsequent sections we ask about course enrollments in your department and we ask that you **not** include any of the enrollments reported in this section B.*

C. Mathematics Courses (Fall 2005)

The following instructions apply throughout sections C, D, E, and F (pages 6-20).

- If your departmental course titles do not match exactly with the ones that we suggest, please use your best judgment to match them.
- Report distance-learning enrollments separately from other enrollments. A *distance-learning* section is one in which a majority of students receive the majority of their instruction by Internet, TV, correspondence courses, or other methods where the instructor is NOT physically present.
- Do NOT include any dual-enrollment sections or enrollments in these tables. (In this questionnaire, a *dual-enrollment* section is one that is conducted on a high-school campus, taught by a high-school teacher, and which allows students to receive high-school credit and simultaneously college credit from your institution for the course. These courses were reported in Section B.)
- For some courses (e.g., C-16, below) we ask you to list those lecture sections with several recitation/problem/laboratory sessions separately from other sections of the course that do not have such recitation/problem/laboratory sessions.
- Except in C16-2, C17-2, C18-2, C19-2, and D1-2, please count any lecture course along with its associated recitation/problem/laboratory sessions as one section of the course. Special instructions for C16-2, C17-2, C18-2, C19-2, and D1-2 are given in footnotes.
- Report a section of a course as being taught by a *graduate teaching assistant (GTA)* if and only if that section is taught *independently* by the GTA, i.e., when it is the GTA's own course and the GTA is the instructor of record.
- If your institution does not recognize tenure, report sections taught by your permanent full-time faculty in column (5) and sections taught by other full-time faculty in columns (6) or (7) as appropriate.
- Full-time faculty teaching in your department and holding joint appointments with other departments should be counted in column (5) if they are tenured, tenure-eligible, or permanent in your department. Faculty who are not tenured, tenure-eligible, or permanent in your department should be counted in column (8) if their fall 2005 teaching in your department is less than or equal to 50% of their total fall teaching assignment, and they should be reported in column (6) or (7) otherwise. (Example: If a tenured physics professor with a joint appointment in your department teaches a total of two courses in fall 2005, with exactly one being in your department, then that person would be counted as part-time in your department.)
- Do not fill in any shaded rectangles.
- Any unshaded rectangle that is left blank will be interpreted as reporting a count of zero.
- Except where specifically stated to the contrary, the tables in Sections C, D, E, and F deal with enrollments in fall term 2005.

C. Mathematics Courses (Fall 2005) cont.

♦Cells left blank will be interpreted as zeros

Name of Course (or equivalent) (1)	Total distance-education enrollment[a] (2)	Total enrollment NOT in Col (2) and NOT dual enrollments[b] (3)	Number of sections corresponding to Column (3) (4)	Of the number in Column 4, how many sections are taught by: Tenured or Tenure-eligible Faculty (5)	Other Full-time Faculty with Ph.D. (6)	Other Full-time Faculty without Ph.D. (7)	Part-time Faculty (8)	Graduate Teaching Assist.[c] (9)
MATHEMATICS								
PRECOLLEGE LEVEL								
C1. Arithmetic/Basic Math								
C2. Pre-algebra								
C3. Elementary Algebra (high school level)								
C4. Intermediate Algebra (high school level)								
C5. Other precollege level courses								
INTRODUCTORY LEVEL, INCLUDING PRE-CALCULUS								
C6. Mathematics for Liberal Arts								
C7. Finite Mathematics								
C8. Business Mathematics (non-Calculus)								

[a] A majority of students receive the majority of their instruction via Internet, TV, correspondence courses, or other method where the instructor is NOT physically present.

[b] Do not include any dual-enrollments courses, i.e., courses taught on a high school campus by a high school instructor, for which high school students may obtain both high school credit and simultaneously college credit through your institution.

[c] Sections taught independently by GTAs .

C. Mathematics Courses (Fall 2005) cont.

◆**Cells left blank will be interpreted as zeros**

Name of Course (or equivalent) (1)	Total distance-education enrollment[a] (2)	Total enrollment NOT in Col (2) and NOT dual enrollments[b] (3)	Number of sections corres-ponding to Column (3) (4)	Of the number in Column 4, how many sections are taught by: Tenured or Tenure-eligible Faculty (5)	Other Full-time Faculty with Ph.D. (6)	Other Full-time Faculty without Ph.D. (7)	Part-time Faculty (8)	Graduate Teaching Assist.[c] (9)	Of the number in Column 4, how many sections: Use graphing calculators (10)	Include writing components such as reports or projects (11)	Require computer assign-ments (12)	Use on-line homework generating and grading packages (13)	Assign group projects (14)
MATHEMATICS													
INTRODUCTORY LEVEL, INCLUDING PRE-CALCULUS, CONT.													
C9. Mathematics for Elementary School Teachers I, II													
C10. College Algebra (beyond C4)													
C11. Trigonometry													
C12. College Algebra & Trigonometry (combined)													
C13. Elementary Functions, Pre-calculus, Analytic Geometry													
C14. Introduction to Mathematical Modeling													
C15. All other introductory level pre-calculus courses													

[a] A majority of students receive the majority of their instruction via Internet, TV, correspondence courses, or other method where the instructor is NOT physically present.

[b] Do not include any dual-enrollments courses, i.e., courses taught on a high school campus by a high school instructor, for which high school students may obtain both high school credit and simultaneously college credit through your institution.

[c] Sections taught independently by GTAs.

C. Mathematics Courses (Fall 2005) cont.

◆Cells left blank will be interpreted as zeros

Name of Course (or equivalent)	Total distance-education enrollment[a]	Total enrollment NOT in Col (2) and NOT dual enrollments[b]	Number of sections corres-ponding to Column (3)	Of the number in Column 4, how many sections are taught by: Tenured or Tenure-eligible Faculty	Other Full-time Faculty with Ph.D.	Other Full-time Faculty without Ph.D.	Part-time Faculty	Graduate Teaching Assist.[c]	Of the number in Column 4, how many sections: Use graphing calculators	Include writing components such as reports or projects	Require computer assign-ments	Use on-line homework generating and grading packages	Assign group projects
(1)	(2)	(3)	(4)	(5)	(6)	(7)	(8)	(9)	(10)	(11)	(12)	(13)	(14)
MATHEMATICS													
MAINSTREAM [d] CALCULUS I													
C16-1. Lecture with separately scheduled recitation/problem/laboratory sessions[e]													
C16-2. Number of recitation/problem/laboratory sessions associated with courses reported in C16-1. See example[f] below.													
C16-3. Other sections with enrollment of 30 or less													
C16-4. Other sections with enrollment above 30													
MAINSTREAM [d] CALCULUS II													
C17-1. Lecture with separately scheduled recitation/problem/laboratory sessions[e]													
C17-2. Number of recitation/problem/laboratory sessions associated with courses reported in C17-1. See example[f] below.													
C17-3. Other sections with enrollment of 30 or less													
C17-4. Other sections with enrollment above 30													

[a] A majority of students receive the majority of their instruction via Internet, TV, correspondence courses, or other method where the instructor is NOT physically present.

[b] Do not include any dual-enrollments courses, i.e., courses taught on a high school campus by a high school instructor, for which high school students may obtain both high school credit and simultaneously college credit through your institution.

[c] Sections taught independently by GTAs.

[d] A calculus course is mainstream if it leads to the usual upper division mathematical sciences courses.

[e] Report a calculus class along with its recitation/problem/laboratory sessions as one section in C16-1, C17-1, C18-1, and C19-1.

[f] Example: suppose your department offers four 100-student sections of a course and that each is divided into five 20-student discussion sessions that meet separately from the lectures. Report 4•5=20 recitation/problem/laboratory sessions associated with the course, even if each discussion meets several times per week.

C. Mathematics Courses (Fall 2005) cont.

Mathematics Questionnaire

◆**Cells left blank will be interpreted as zeros**

Name of Course (or equivalent) (1)	Total distance-education enrollment[a] (2)	Total enrollment NOT in Col (2) and NOT dual enrollments[b] (3)	Number of sections corres-ponding to Column (3) (4)	Of the number in Column 4, how many sections are taught by: Tenured or Tenure-eligible Faculty (5)	Other Full-time Faculty with Ph.D. (6)	Other Full-time Faculty without Ph.D. (7)	Part-time Faculty (8)	Graduate Teaching Assist.[c] (9)	Of the number in Column 4, how many sections: Use graphing calculators (10)	Include writing components such as reports or projects (11)	Require computer assign-ments (12)	Use on-line homework generating and grading packages (13)	Assign group projects (14)
MATHEMATICS													
MAINSTREAM [d] CALCULUS III (and IV, etc)													
C18-1. Lecture with separately scheduled recitation/problem/laboratory sessions[e]													
C18-2. Number of recitation/problem/laboratory sessions associated with courses reported in C18-1. See example[f] below.													
C18-3. Other sections with enrollment of 30 or less													
C18-4. Other sections with enrollment above 30													
NON-MAINSTREAM [d] CALCULUS I													
C19-1. Lecture with separately scheduled recitation/problem/laboratory sessions[e]													
C19-2. Number of recitation/problem/laboratory sessions associated with courses reported in C19-1. See exampe[f] below.													
C19-3. Other sections with enrollment of 30 or less													
C19-4. Other sections with enrollment above 30													

[a] A majority of students receive the majority of their instruction via Internet, TV, correspondence courses, or other method where the instructor is NOT physically present.

[b] Do not include any dual-enrollments courses, i.e., courses taught on a high school campus by a high school instructor, for which high school students may obtain both high school credit and simultaneously college credit through your institution.

[c] Sections taught independently by GTAs.

[d] A calculus course is mainstream if it leads to the usual upper division mathematical sciences courses.

[e] Report a calculus class along with its recitation/problem/laboratory sessions as one section in C16-1, C17-1, C18-1, and C19-1.

[f] Example: suppose your department offers four 100-student sections of a course and that each is divided into five 20-student discussion sessions that meet separately from the lectures. Report 4∗5=20 recitation/problem/laboratory sessions associated with the course, even if each discussion meets several times per week.

C. Mathematics Courses (Fall 2005) cont.

◆**Cells left blank will be interpreted as zeros**

Name of Course (or equivalent) (1)	Total distance-education enrollment[a] (2)	Total enrollment NOT in Col (2) and NOT dual enrollments[b] (3)	Number of sections corresponding to Column (3) (4)	Of the number in Column 4, how many sections are taught by: Tenured or Tenure-eligible Faculty (5)	Other Full-time Faculty with Ph.D. (6)	Other Full-time Faculty without Ph.D. (7)	Part-time Faculty (8)	Graduate Teaching Assist.[c] (9)
MATHEMATICS								
CALCULUS LEVEL, CONT.								
C20. Non-Mainstream[d] Calculus, II, III, etc.								
C21. Differential Equations and Linear Algebra (combined)								
C22. Differential Equations								
C23. Linear Algebra or Matrix Theory								
C24. Discrete Mathematics								
C25. Other calculus-level courses								

[a] A majority of students receive the majority of their instruction via Internet, TV, correspondence courses, or other method where the instructor is NOT physically present.

[b] Do not include any dual-enrollments courses, i.e., courses taught on a high school campus by a high school instructor, for which high school students may obtain both high school credit and simultaneously college credit through your institution.

[c] Sections taught independently by GTAs.

[d] A calculus course is mainstream if it leads to the usual upper division mathematical sciences courses.

C. Mathematics Courses (Fall 2005) cont.

In reporting on advanced courses, please pay special attention to the following instructions:

- If an undergraduate course contains a mixture of graduate and undergraduate students, report them all in Column (2).
- If your institution does not recognize tenure, report sections taught by your permanent faculty in Column (4).
- Make sure that no course is reported in more than one row.

◆Cells left blank will be interpreted as zeros

Name of Course (or equivalent) (1)	Total enrollment Fall 2005 (2)	Number of sections corresponding to Column (2) (3)	Number of sections corresponding to Column (3) taught by Tenured or Tenure-eligible Faculty (4)	Was this course taught in ANY term of the previous academic year? Y(es) / N(o) (5)	Will this course be offered in the next term (Spring 2006)? Y(es) / N(o) (6)
MATHEMATICS					
ADVANCED UNDERGRADUATE LEVEL					
C26. Introduction to Proofs					
C27-1. Modern Algebra I					
C27-2. Modern Algebra II					
C28. Number Theory					
C29. Combinatorics					
C30. Actuarial Mathematics					
C31. Logic/Foundations (not C26)					
C32. Discrete Structures					
C33. History of Mathematics					
C34. Geometry					

C. Mathematics Courses (Fall 2005) cont.

◆**Cells left blank will be interpreted as zeros**

Name of Course (or equivalent) (1)	Total enrollment Fall 2005 (2)	Number of sections corresponding to Column (2) (3)	Number of sections corresponding to Column (3) taught by Tenured or Tenure-eligible Faculty (4)	Was this course taught in ANY term of the previous academic year? Y(es) / N(o) (5)	Will this course be offered in the next term (Spring 2006)? Y(es) / N(o) (6)
MATHEMATICS					
ADVANCED UNDERGRADUATE LEVEL, CONT.					
C35. Mathematics for Secondary School Teachers I and II (methods, special content, etc.)					
C36-1. Advanced Calculus and/or Real Analysis, I					
C36-2 Advanced Calculus and/or Real Analysis, II					
C37. Advanced Mathematics for Engineering and Physics, I and II					
C38. Advanced Linear Algebra (beyond C21, C23)					
C39. Vector Analysis					
C40. Advanced Differential Equations (beyond C22)					
C41. Partial Differential Equations					
C42. Numerical Analysis I and II					
C43. Applied Mathematics (Modeling)					

C. Mathematics Courses (Fall 2005) cont.

◆**Cells left blank will be interpreted as zeros**

Name of Course (or equivalent) (1)	Total enrollment Fall 2005 (2)	Number of sections corresponding to Column (2) (3)	Number of sections corresponding to Column (3) taught by Tenured or Tenure-eligible Faculty (4)	Was this course taught in ANY term of the previous academic year? Y(es) / N(o) (5)	Will this course be offered in the next term (Spring 2006)? Y(es) / N(o) (6)
MATHEMATICS					
ADVANCED UNDERGRADUATE LEVEL, CONT.					
C44. Complex Variables					
C45. Topology					
C46. Mathematics of Finance (not C30, C43)					
C47. Codes and Cryptology					
C48. Biomathematics					
C49. Senior Seminar/Independent Study in Mathematics					
C50. All other advanced level mathematics (excluding Probability, Statistics, or Operations Research courses)					

D. Probability & Statistics Courses (Fall 2005)

Please refer to the course reporting instructions at the beginning of Section C.

D. Does your department offer any Probability and/or Statistics Courses?

Yes............................ ☐ (1) ⟶ If "Yes", go to D1-1, below.

No............................. ☐ (2) ⟶ If "No", go to Section E.

◆Cells left blank will be interpreted as zeros

Name of Course (or equivalent) (1)	Total distance-education enrollment[a] (2)	Total enrollment NOT in Col (2) and NOT dual enrollments[b] (3)	Number of sections corres-ponding to Column (3) (4)	Of the number in Column 4, how many sections are taught by: Tenured or Tenure-eligible Faculty (5)	Other Full-time Faculty with Ph.D. (6)	Other Full-time Faculty without Ph.D. (7)	Part-time Faculty (8)	Graduate Teaching Assist.[c] (9)	Of the number in Column 4, how many sections: Use graphing calculators (10)	Include writing components such as reports or projects (11)	Require computer assign-ments (12)	Use on-line homework generating and grading packages (13)	Assign group projects (14)
PROBABILITY & STATISTICS													
ELEMENTARY LEVEL													
D1. Elementary Statistics (no calculus prerequisite):													
D1-1. Lecture with separately scheduled recitation/problem/laboratory sessions[d]													
D1-2. Number of recitation/problem/ laboratory sessions associated with courses reported in D1-1[e]													
D1-3. Other sections with enrollment of 30 or less													
D1-4. Other sections with enrollment above 30													
D2. Probability & Statistics (no calculus prerequisite)													
D3. Other elementary level Probability & Statistics courses													

[a] A majority of students receive the majority of their instructor via Internet, TV, correspondence courses, or other methods where the instructor is <u>NOT</u> physically present.

[b] Do not include any dual-enrollments courses, i.e., courses taught on a high school campus by a high school instructor, for which high school students may obtain both high school credit and simultaneously college credit through your institution.

[c] Sections taught independently by GTAs.

[d] A class along with its recitation/problem/laboratory sessions is to be counted as one section in D1-1.

[e] Example: suppose your department offers four 100-student sections of a course and that each is divided into five 20-student discussion sessions that meet separately from the lectures. Report 4•5=20 recitation/problem/laboratory sessions associated with the course, even if each discussion meets several times per week.

D. Probability & Statistics Courses (Fall 2005) cont.

◆**Cells left blank will be interpreted as zeros**

Name of Course (or equivalent) (1)	Total enrollment Fall 2005 (2)	Number of sections corresponding to Column (2) (3)	Number of sections corresponding to Column (3) taught by Tenured or Tenure-eligible Faculty (4)	Was this course taught in ANY term of the previous academic year? Y(es) / N(o) (5)	Will this course be offered in the next term (Spring 2006)? Y(es) / N(o) (6)
PROBABILITY & STATISTICS					
INTERMEDIATE AND ADVANCED LEVEL					
D4. Mathematical Statistics (calculus prerequisite)					
D5. Probability (calculus prerequisite)					
D6. Combined Probability & Statistics (calculus prerequisite)					
D7. Stochastic Processes					
D8. Applied Statistical Analysis					
D9. Design & Analysis of Experiments					
D10. Regression (and Correlation)					
D11. Biostatistics					
D12. Nonparametric Statistics					
D13. Categorical Data Analysis					
D14. Sample Survey Design & Analysis					
D15. Statistical Software & Computing					
D16. Data Management					
D17. Senior Seminar/ Independent Studies					
D18. All other upper level Probability & Statistics courses					

E. Operations Research Courses (Fall 2005)

Please refer to the course reporting instructions at the beginning of Section C.

E. Does your department offer any Operations Research courses?

Yes............................ ☐ (1) ⟶ If "Yes", go to E1, below.

No.............................. ☐ (2) ⟶ If "No", go to Section F.

◆Cells left blank will be interpreted as zeros

Name of Course (or equivalent) (1)	Total enrollment Fall 2005 (2)	Number of sections corresponding to Column (2) (3)	Number of sections corresponding to Column (3) taught by Tenured or Tenure-eligible Faculty (4)	Was this course taught in ANY term of the previous academic year? Y(es) / N(o) (5)	Will this course be offered in the next term (Spring 2006)? Y(es) / N(o) (6)
OPERATION RESEARCH					
E1. Intro. to Operations Research					
E2. Intro. to Linear Programming					
E3. All other O.R. courses					

F. Computer Science Courses (Fall 2005)

- Please refer to the course reporting instructions at the beginning of Section C.
- In December 2001, a joint IEEE Computer Society/ACM Task Force issued its recommendations on "Model Curricula for Computing." That report replaced the curricular recommendations published by ACM in 1991 and is available from http://www.computer.org/education/cc2001/. Course numbers and, to the degree possible, course names in the table below are taken from the detailed course outlines in the appendices of that CC2001 report.

F. Does your department offer any Computer Sciences courses?

Yes.......................... ☐ (1) ⟶ If "Yes", go to F1, below.

No........................... ☐ (2) ⟶ If "No", go to Section G

◆Cells left blank will be interpreted as zeros

				Of the number in Column 4, how many sections are taught by:				
Name of Course (or equivalent) (1)	Total distance-education enrollment[a] (2)	Total enrollment NOT in Col (2) and NOT dual enrollments[b] (3)	Number of sections corresponding to Column (3) (4)	Tenured or Tenure-eligible Faculty (5)	Other Full-time Faculty with Ph.D. (6)	Other Full-time Faculty without Ph.D. (7)	Part-time Faculty (8)	Graduate Teaching Assist.[c] (9)
COMPUTER SCIENCE								
GENERAL EDUCATION COURSES								
F1. Computers and Society, Issues in CS								
F2. Intro. to Software Packages								
F3. Other CS General Education Courses								

[a] A majority of students receive the majority of their instruction via Internet, TV, correspondence courses, or other method where the instructor is NOT physically present.

[b] Do not include any dual-enrollments (see Section B).

[c] Sections taught independently by GTAs.

F. Computer Science Courses (Fall 2005) cont.

◆**Cells left blank will be interpreted as zeros**

Name of Course (or equivalent) (1)	Total distance-education enrollment[a] (2)	Total enrollment NOT in Col (2) and NOT dual enrollments[b] (3)	Number of sections corresponding to Column (3) (4)	Of the number in Column 4, how many sections are taught by: Tenured or Tenure-eligible Faculty (5)	Other Full-time Faculty with Ph.D. (6)	Other Full-time Faculty without Ph.D. (7)	Part-time Faculty (8)	Graduate Teaching Assist.[c] (9)
COMPUTER SCIENCE								
INTRODUCTORY CS COURSES								
F4. Computer Programming I (CS101 or 111)[d]								
F5. Computer Programming II (CS102 or 112 and 113)[d]								
F6. Discrete Structures for CS (CS105, 106, or 115)[d], but not courses C24 or C32 in Section C above								
F7. All other introductory Level CS courses								
INTERMEDIATE LEVEL								
F8. Algorithm Design and Analysis (CS210)[d]								
F9. Computer Architecture (CS220, 221, or 222)[d]								
F10. Operating Systems (CS225, 226)[d]								

[a] A majority of students receive the majority of their instruction via Internet, TV, correspondence courses, or other method where the instructor is NOT physically present.

[b] Do not include any dual-enrollments (see Section B).

[c] Sections taught independently by GTAs.

[d] Course numbers from CC2001.

F. Computer Science Courses (Fall 2005) cont.

◆Cells left blank will be interpreted as zeros

Name of Course (or equivalent) (1)	Total distance-education enrollment[a] (2)	Total enrollment NOT in Col (2) and NOT dual enrollments[b] (3)	Number of sections corresponding to Column (3) (4)	Of the number in Column 4, how many sections are taught by: Tenured or Tenure-eligible Faculty (5)	Other Full-time Faculty with Ph.D. (6)	Other Full-time Faculty without Ph.D. (7)	Part-time Faculty (8)	Graduate Teaching Assist.[c] (9)
COMPUTER SCIENCE								
INTERMEDIATE LEVEL CONT.								
F11. Net-centric Computing (CS230)[d]								
F12. Programming Language Translation (CS240)[d]								
F13. Human-Computer Interaction (CS250)[d]								
F14. Artificial Intelligence (CS260, 261, 262)[d]								
F15. Databases (CS270, 271)[d]								
F16. Social and Professional Issues in Computing (CS280)[d]								
F17. Software Development (CS290, 291, 292)[d]								
F18. All other intermediate Level CS courses								
UPPER LEVEL								
F19. All upper level CS Courses (numbered 300 or above in CC2001)								

[a] A majority of students receive the majority of their instruction via Internet, TV, correspondence courses, or other method where the instructor is NOT physically present.

[b] Do not include any dual-enrollments (see Section B).

[c] Sections taught independently by GTAs.

[d] Course numbers from CC2001.

G. Faculty Profile (Fall 2005)

Mathematics Questionnaire

G1. Number of faculty in your department in fall 2005

NOTES for G1:

- In responding to questions in this section, use the same rules for distinguishing between full-time and part-time faculty that you used in sections C, D, E, and F. Often, one easy way to distinguish between full-time and part-time faculty is to ask whether a given faculty member participates in the same kind of insurance and retirement programs as does your department chair. Part-time faculty are often paid by the course and do not receive the same insurance and retirement benefits as does the department chair.

- If your institution does not recognize tenure, please report departmental faculty who are permanent on line G1-(a) and report all other faculty on lines G1-(c), (d), or (e) as appropriate.

(a) Number of full-time tenured faculty (not including visitors or those on leave) in fall 2005 .... ☐ (1)

(b) Number of full-time tenure-eligible-but-not-tenured faculty (not including visitors or those on leave) in fall 2005 .... ☐ (2)

(c) Number of tenured or tenure-eligible faculty on leave in fall 2005 .... ☐ (3)

(d) Number of post-docs in your department in fall 2005 (where a postdoctoral appointment is a temporary position primarily intended to provide an opportunity to extend graduate training or to further research) .... ☐ (4)

(e) Number of full-time faculty in your department in fall 2005 <u>not</u> included in (a), (b),(c), or (d) and who hold <u>visiting appointments</u> .... ☐ (5)

(f) Number of full-time faculty in your department in fall 2005 who are <u>not</u> in (a), (b), (c), (d), or (e) ☐ (6)

(g) Number of part-time faculty in your department in fall 2005 .... ☐ (7)

G2. What is the expected (or average) teaching assignment for the tenured and tenure-eligible faculty reported G1-(a), (b)? (If your institution does not recognize tenure, report on those faculty who are "permanent full-time.")

(a) Expected classroom contact hours per week for tenured and tenure-eligible faculty in fall 2005 .... ☐ (1)

(b) Expected classroom contact hours per week for tenured and tenure-eligible faculty last year in winter/spring 2005 .... ☐ (2)

H. Undergraduate Program (Fall 2005)

If you do not offer a major in a mathematical science, check here ☐ and go to H9. Otherwise go to H1.

H1. Please report the total number of your departmental majors who received their bachelors degrees from your institution between 01 July 2004 and 30 June 2005. Include joint majors and double majors[1] .... ☐ (1)

H2. Of the undergraduate degrees described in H1, please report the number who majored in each of the following categories. Each student should be reported only once. Include all double and joint majors[1] in your totals. Use "Other" category for a major in your department who does not fit into one of the earlier categories.

Area of Major	Male (1)	Female (2)
a) Mathematics (including applied)		
b) Mathematics Education		
c) Statistics		
d) Computer Science		
e) Actuarial Mathematics		
f) Operations Research		
g) Joint[1] Mathematics and Computer Science		
h) Joint[1] Mathematics and Statistics		
i) Joint[1] Mathematics and (Business or Economics)		
j) Other		

H3. Does your department teach any upper division Computer Science courses?

Yes.......................... ☐ (1)

No............................ ☐ (2)

H4. Can a major in your department count some upper division Computer Science course(s) from some other department toward the upper division credit hour requirement for your departmental major?

Yes.......................... ☐ (1)

No............................ ☐ (2)

H5. Does your department offer any upper division Statistics courses?

Yes.......................... ☐ (1)

No............................ ☐ (2)

H6. Can a major in your department count some upper division Statistics course(s) from some other department toward the upper division credit hour requirement for your departmental major?

Yes.......................... ☐ (1)

No............................ ☐ (2)

[1] *A "double major" is a student who completes the degree requirements of two separate majors, one in mathematics and a second in another program or department. A "joint major" is a student who completes a single major in your department that integrates courses from mathematics and some other program or department and typically requires fewer credit hours than the sum of the credit hours required by the two separate majors.*

H. Undergraduate Program (Fall 2005) cont.

Mathematics Questionnaire

H7. To what extent must majors in your department complete the following? Check one box in each row.

	Required of all majors (1)	Required of some but not all majors (2)	Not required of any major (3)
a) Modern Algebra I			
b) Modern Algebra I plus some other upper division Algebra course			
c) Real Analysis I			
d) Real Analysis I plus some other upper division Analysis course			
e) at least one Computer Science course			
f) at least one Statistics course			
g) at least one applied mathematics course beyond course C-25 (in Section C)			
h) a capstone experience (e.g. a senior project, a senior thesis, a senior seminar, or an internship)			
i) an exit exam (written or oral)			

H8. Many departments today use a spectrum of program-assessment methods. Please check all that apply to your department's undergraduate program-assessment efforts during the last six years.

(a) We conducted a review of our undergraduate program that included one or more reviewers from outside of our institution .... ☐ (1)

(b) We asked graduates of our undergraduate program to comment on and suggest changes in our undergraduate program .... ☐ (2)

(c) Other departments at our institution were invited to comment on the preparation that their students received in our courses .... ☐ (3)

(d) Data on our students' progress in subsequent mathematics courses was gathered and analyzed .... ☐ (4)

(e) We have a placement system for first-year students and we gathered and analyzed data on its effectiveness .... ☐ (5)

(f) Our department's program assessment activities led to changes in our undergraduate program .... ☐ (6)

H. Undergraduate Program (Fall 2005) cont.

H9. General Education Courses: Does your institution require all bachelors graduates to have credit for a quantitative literacy course as part of their general education requirements? Choose one of the following.

(a) Yes, all bachelors graduates must have such credit ☐ (1) ⟶ if (a), go to H10.

(b) Not (a), but all students in the academic unit to which our department belongs must have such credit[1] ☐ (2) ⟶ if (b), go to H10.

(c) neither (a) nor (b) ☐ (3) ⟶ if (c), go to H13.

H10. If you chose (a) or (b) in H9, is it true that all students (to whom the quantitative requirement applies) must fulfill it by taking a course in your department?

Yes........................... ☐ (1)

No............................. ☐ (2)

H11. Which courses in your department can be used to fulfill the general education quantitative requirement in H9?

(a) Any freshman course in our department ☐ (1) ⟶ go to H13.

(b) Only certain courses in our department ☐ (2) ⟶ go to H12.

H12. If you chose H11(b), which of the following departmental courses can be used to fulfill the general education quantitative requirement? Check all that apply.

Course	Can be used
a) College Algebra and/or Pre-calculus	
b) Calculus	
c) Mathematical Modeling	
d) a basic Probability and/or Statistics course	
e) a special general education course in our department not listed above	
f) some other course(s) in our department not listed above	

H13. Does your department or institution operate a mathematics lab or tutoring center intended to give students out-of-class help with mathematics or statistics problems?

Yes........................... ☐ (1) ⟶ If "Yes", go to H14.

No............................. ☐ (2) ⟶ If "No", go to H15.

[1] *For example, you would check H9(b) if students in the College of Fine Arts do not have a quantitative literacy requirement, and yet all students in the College of Science (to which our department belongs) must complete a quantitative literacy requirement.*

H. Undergraduate Program (Fall 2005) cont.

Mathematics Questionnaire

H14. Please check all services available through the mathematics lab or tutoring center mentioned in H13.

(a) Computer-aided instruction .... ☐ (1)

(b) Computer software such as computer algebra systems or statistical packages .... ☐ (2)

(c) Media such as video tapes, CDs, or DVDs .... ☐ (3)

(d) Tutoring by students .... ☐ (4)

(e) Tutoring by paraprofessional staff .... ☐ (5)

(f) Tutoring by part-time mathematics faculty .... ☐ (6)

(g) Tutoring by full-time mathematics faculty .... ☐ (7)

(h) Internet resources .... ☐ (8)

H15. Please check all of the opportunities available to your undergraduate mathematics students.

(a) Honors sections of departmental courses .... ☐ (1)

(b) An undergraduate Mathematics Club .... ☐ (2)

(c) Special mathematics programs to encourage women .... ☐ (3)

(d) Special mathematics programs to encourage minorities .... ☐ (4)

(e) Opportunities to participate in mathematics contests .... ☐ (5)

(f) Special mathematics lectures/colloquia not part of a mathematics club .... ☐ (6)

(g) Mathematics outreach opportunities in local K-12 schools .... ☐ (7)

(h) Undergraduate research opportunities in mathematics .... ☐ (8)

(i) Independent study opportunities in mathematics .... ☐ (9)

(j) Assigned faculty advisers in mathematics .... ☐ (10)

(k) Opportunity to write a senior thesis in mathematics .... ☐ (11)

(l) A career day for mathematics majors .... ☐ (12)

(m) Special advising about graduate school opportunities in mathematical sciences .... ☐ (13)

(n) Opportunity for an internship experience .... ☐ (14)

(o) Opportunity to participate in a senior seminar .... ☐ (15)

H. Undergraduate Program (Fall 2005) cont.

H16. If you offer a major in some mathematical science, please give your best estimate of the percentage of your department's graduating majors from the previous academic year (reported in H1) in each of the following categories. If you do not offer any mathematical sciences major, go to Section I

(a) who went into pre-college teaching .... ___ % (1)

(b) who went to graduate school in the mathematical sciences .... ___ % (2)

(c) who went to professional school or to graduate school outside of the mathematical sciences ___ % (3)

(d) who took jobs in business, industry, government, etc .... ___ % (4)

(e) who had other post-graduation plans known to the department .... ___ % (5)

(f) whose plans are not known to the department .... ___ % (6)

I. Pre-service Teacher Education in Mathematics

Mathematics Questionnaire

I-1. Does your institution offer a program or major leading to certification in some or all of grades K-8?

Yes............................ ☐ (1) ⟶ If "Yes", go to I-2.

No.............................. ☐ (2) ⟶ If "No", go to I-14.

I-2. Do members of your department serve on a committee that determines what mathematics courses are part of that certification program?

Yes............................ ☐ (1)

No.............................. ☐ (2)

I-3. Does your department offer a course or course-sequence that is designed specifically for the pre-service K-8 teacher certification program?

Yes............................ ☐ (1) ⟶ If "Yes", go to I-4.

No.............................. ☐ (2) ⟶ If "No", go to I-9.

I-4. Are you offering more than one section of the special course for pre-service K-8 teachers in fall 2005?

Yes............................ ☐ (1) ⟶ If "Yes", go to I-5.

No.............................. ☐ (2) ⟶ If "No", go to I-8.

I-5. Is there a designated departmental coordinator for your multiple sections of the special course for pre-service K-8 teachers in fall 2005?

Yes............................ ☐ (1) ⟶ If "Yes", go to I-6.

No.............................. ☐ (2) ⟶ If "No", go to I-8.

I-6. Please choose the box that best describes the coordinator mentioned in I-5.

(a) tenured or tenure-eligible .... ☐ (1)

(b) a postdoc[1] .... ☐ (2)

(c) a full-time faculty member not in (b) who holds a *visiting* appointment in your department ... ☐ (3)

(d) a full-time faculty member *without* a doctorate who is not in (a), (b), or (c) .... ☐ (4)

(e) a full-time faculty member *with* a doctorate who is not in (a), (b), (c), or (d) .... ☐ (5)

(f) a part-time faculty member .... ☐ (6)

(g) a graduate teaching assistant .... ☐ (7)

---

[1] *A postdoctoral appointment is a temporary position primarily intended to provide an opportunity to extend graduate education or to further research.*

I. Pre-service Teacher Education in Mathematics cont.

I-7. Given that you offer multiple sections of the special course for pre-service K-8 teachers in fall 2005, is it true that all sections of that course use the same textbook?

Yes............................ ☐ (1)

No.............................. ☐ (2)

I-8. During which year of their college careers are your pre-service K-8 teachers most likely to take your department's special course for pre-service K-8 teachers? If you have two such courses, consider only the first in responding to this question. Please check just one box.

a) Freshman	
b) Sophomore	
c) Junior	
d) Senior	

I-9. Are there any sections of other courses in your department (i.e., other than the special course for K-8 teachers mentioned in I-3) that are restricted to or designated for pre-service K-8 teachers?

Yes............................ ☐ (1)

No.............................. ☐ (2)

Special instructions for questions I-10, I-11, I-12, and I-13: Many institutions have different certification requirements for pre-service elementary teachers preparing for early grades and those preparing for later grades. However, there is no nationwide agreement on which grades are "early grades" and which are "later grades" except that grades 1 and 2 are "early" and grades 6 and above are usually considered "later grades," and that is how we use the terms in the next four questions.

I-10. Does your K-8 pre-service program have different requirements for students preparing to teach early grades and for those planning to teach later grades?.

Yes............................ ☐ (1) ⟶ If "Yes", go to I-12.

No.............................. ☐ (2) ⟶ If "No", go to I-11.

I-11. Given that your pre-service K-8 teacher education program *does not* distinguish between preparing for certification in early and later grades, how many courses are all pre-service elementary teachers required to take in your department (including general education requirements, if any)?

☐ Now go to I-13 and put all of your answers into column (3).

I-12. Given that your pre-service K-8 teacher education program *does* distinguish between preparing for certification to teach early grades and later grades, how many courses are pre-service K-8 teachers required to take in your department (including general education requirements, if any)?

(a) Number of courses required for early grade certification .... ☐ (1)

(b) Number of courses required for later grade certification .... ☐ (2)

Now go to I-13 and put all of your answers into columns (1) and (2).

I. Pre-service Teacher Education in Mathematics cont.

Mathematics Questionnaire

I-13. In your judgement, which three of the following courses in your department are most likely to be taken by pre-service K-8 teachers? If your program does NOT distinguish between early and later grades, please use the column (3) for your answers and check a total of only three boxes. If your program DOES distinguish between early and later grades, check exactly three boxes in each of columns (1) and (2) and ignore column (3).

Courses	Three most likely for early grade certification (1)	Three most likely for later grade certification (2)	Three most likely given that we do not distinguish between early & later grade (3)
a) A multiple-term course designed for elementary teachers			
b) A single-term course designed for elementary teachers			
c) College Algebra			
d) Elementary Functions, Pre-calculus, Analytic Geometry			
e) Introduction to Mathematical Modeling			
f) Mathematics for Liberal Arts			
g) Finite Mathematics			
h) Mathematics History			
i) Calculus			
j) Geometry			
k) Statistics			

I-14. How do students at your institution who are seeking certification for teaching mathematics in <u>secondary</u> schools learn about the history of mathematics? Choose one of the following boxes.

(a) We have no secondary school mathematics certification program .... ☐ (1)

(b) Students in our secondary school mathematics program are required to take a course in mathematics history .... ☐ (2)

(c) There is no required mathematics history course for our secondary school mathematics certification students and our secondary school certification students learn mathematics history from other courses they are required to take .... ☐ (3)

(d) Students in our secondary school mathematics certification program are not required to learn about mathematics history .... ☐ (4)

I. Pre-service Teacher Education in Mathematics cont.

I-15. Does your department offer any courses that are part of a graduate degree in mathematics education?

(a) No .... ☐ (1)

(b) Yes, and the degree is granted through our department .... ☐ (2)

(c) Yes, and the degree is granted through some other department or unit in our institution .... ☐ (3)

Thank you for completing this questionnaire. We know it was a time-consuming process and we hope that the resulting survey report, which we hope to publish in spring 2007, will be of use to you and your department.

Please keep a copy of your responses to this questionnaire in case questions arise.

Appendix V

Two-Year Mathematics Questionnaire

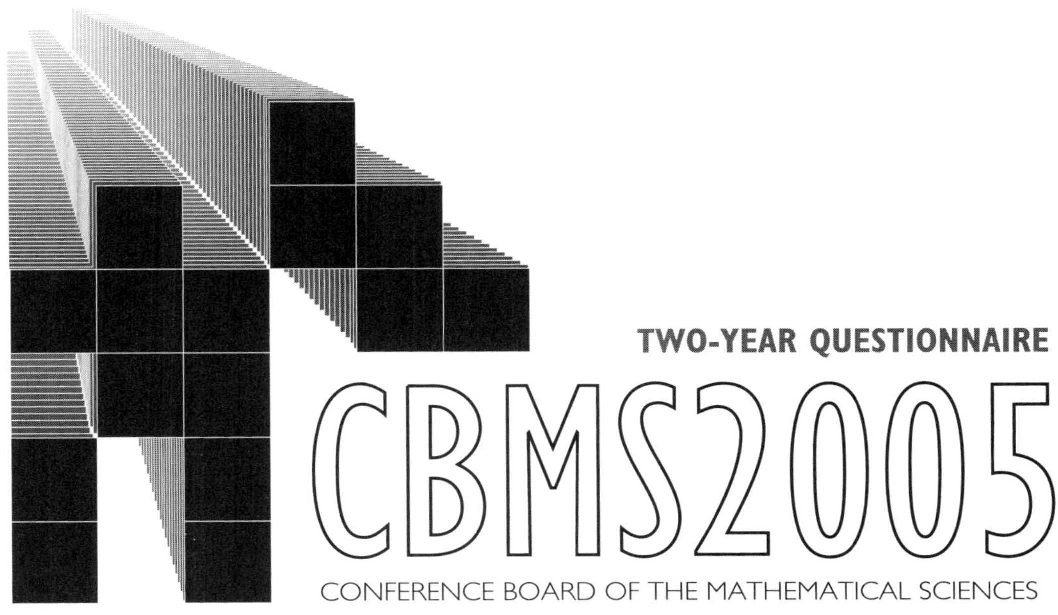

SURVEY OF UNDERGRADUATE PROGRAMS IN THE MATHEMATICAL SCIENCES

General Instructions

As part of a random sample, your department has been selected to participate in the CBMS2005 National Survey, the importance of which has been endorsed by all of our major professional societies. Please read the instructions in each section carefully and complete all of the pertinent items as indicated.

If your college does not have a departmental or divisional structure, consider the group of all mathematics instructors to be the "mathematics department" for the purpose of this survey.

Because some campuses are part of a multi-campus two-year college, special instructions may apply. Please consult the cover letter mailed with this questionnaire. If that letter asks you to report on the entire multi-campus system to which you may belong, please check this box ☐ and report data for the entire system. If you are NOT asked in that letter to report on your entire multi-campus system, then do not include data for branches or campuses of your college that are geographically or budgetarily separate from yours.

This questionnaire should be completed by the person who is directly in charge of the mathematics program or department on your campus.

Report on all of your courses and instructors that fall under the general heading of the mathematics program or department. Include all mathematics and statistics courses taught within your mathematics program or department.

We have classified your department as belonging to a two-year college, to a college or campus within a two-year system, or to a two-year branch of a university system. If this is not correct, please contact Stephen Rodi at the email address or telephone number given below.

If you have any questions, please contact Stephen Rodi, Associate Director for Two-Year Colleges, by email at srodi@austincc.edu or by phone at 512-223-3301.

Please return your completed questionnaire by October 15, 2005 in the enclosed envelope to:

CBMS Survey
UNC Survey Research Unit
730 Martin Luther King Boulevard, Suite 103
CB #2400, UNC-CH
Chapel Hill, NC 27599-2400

Please retain a copy of your responses to this questionnaire in case questions arise.

Mathematics Questionnaire

A. General Information

PLEASE PRINT CLEARLY

A1. Name of campus: ______________________________

A2. Name of your department: ______________________________

A3. Mailing address of the multi-campus organization to which your campus belongs (if any):

____

____

A4. We have classified your department as belonging to a two-year college or to a college campus within a two-year college system, or to a two-year branch of a university system. Do you agree?

Yes . . . . ☐ (1) ⟶ go to the next question.

No . . . . ☐ (2) ⟶ please contact Stephen Rodi, Survey Associate Director, by email (srodi@austincc.edu) or by phone (512-223-3301) before proceeding any further.

A5. What is the structural unit (= academic discipline group) that most directly administers the mathematics program on your campus or (if you checked the box in paragraph three on page one) for your system? (Check only <u>one</u> of the following boxes.)

	at my campus	at the district or multi-campus system level named in A3
a) Mathematics Department	☐ (1)	☐ (2)
b) Mathematics and Science Department or Division	☐ (3)	☐ (4)
c) Other Department or Division Structure	☐ (5)	☐ (6)
d) None of the above	☐ (7)	

A6. To help us project enrollment for the current academic year (2005–2006), please give the following enrollment figures for the <u>previous academic year</u> (2004–2005).

a) Fall 2004 total student enrollment in your mathematics program . . . . ☐ (1)

b) Entire academic year 2004–2005 enrollment in your mathematics program . . . . ☐ (2)

c) Calculus II in Winter/Spring 2005 total enrollment . . . . ☐ (3)

d) Calculus II in Winter/Spring 2005 total number of sections . . . . ☐ (4)

A. General Information (cont.)

A7. Are any of the developmental/remedial mathematics courses at your college administered separately from the mathematics department/program?

Yes .... ☐ (1)

No .... ☐ (2)

A8. Your name or contact person in your department: ☐

A9. Your email address or contact person's email address: ☐

A10. Your phone number or contact person's phone number, including area code: ☐

A11. Campus mailing address: ☐

B. Mathematics Faculty in the Mathematics Department/Program (Fall 2005)

- If you are part of a multi-campus college, please consult the third paragraph on page 1 before proceeding.
- Underlined faculty categories defined in this section will be used in later sections.

B1. For Fall 2005, what is the **total number of your full-time mathematics faculty**, both permanent and temporary, including those on leave or sabbatical?

Number of full-time mathematics faculty . . . . ☐

B2. Of the number in B1, how many are tenured, tenure-eligible, or on your permanent staff (including faculty who are on leave or sabbatical)? We will refer to these as "**permanent full-time faculty**".

Number tenured, tenure-eligible, or on permanent staff . . . . ☐

B3. Give the number of "**other full-time faculty**" by computing B1 minus B2 . . . . ☐

B4. For the **permanent full-time faculty** reported in B2,

a) give the required teaching assignment in weekly contact hours . . . . ☐ (1)

b) give the maximum percentage of the weekly teaching assignment in B4(a) that can be met by teaching distance-learning classes (= classes where at least half the students receive the majority of instruction by technological or other methods where the instructor is not physically present) . . . . ☐ (2)

c) give the number of office hours required weekly in association with the teaching assignment in B4(a) . . . . ☐ (3)

B5. Of the **permanent full-time faculty** reported in B2, how many teach extra hours for extra pay at your campus or within your organization or at other schools?

a) Number who teach extra hours for extra pay at your campus or within your organization . ☐ (1)

b) Number who teach extra hours for extra pay at other schools . . . . ☐ (2)

B6. Of the **permanent full-time faculty** reported in B5(a), how many extra hours per week do they teach?

a) Number who teach 1–3 hours extra weekly . . . . ☐ (1)

b) Number who teach 4–6 hours extra weekly . . . . ☐ (2)

c) Number who teach 7 or more hours extra weekly . . . . ☐ (3)

Mathematics Questionnaire

B. Mathematics Faculty in the Mathematics Department/Program (Fall 2005) cont.

B7. For Fall 2005, what is the number of your **part-time mathematics faculty**? (Note: None of these were reported above.)

a) Number of **part-time mathematics faculty paid by your college** . . . . ☐ (1)

b) Number of **part-time faculty paid by a third party**, such as a school district paying faculty who teach dual-enrollment couses (= courses taught in high school by high school teachers for which students may obtain high school credit and simultaneous college credit through your institution) . . . . . . ☐ (2)

c) **Total number of part-time faculty** (add B7(a) and B7(b) to get total). . . . . ☐ (3)

B8. How many **part-time faculty** in B7(a) (those paid by your college) teach six or more hours per week?

Number in B7(a) teaching six or more hours/week . . . . ☐

B9. Of the **part-time faculty** reported in B7(a) (those paid by your college), give the number who are:

a) employed full-time in a high school . . . . ☐ (1)

b) employed full-time in another two-year college . . . . ☐ (2)

c) employed full-time in another department of your campus or your larger organization . . . ☐ (3)

d) employed full-time in a four-year college or university. . . . . ☐ (4)

e) employed full-time in industry or other business . . . . ☐ (5)

f) graduate students . . . . ☐ (6)

g) not graduate students and not employed full-time anywhere . . . . ☐ (7)

B10. Are office hours required by college policy for the **part-time faculty** reported in B7(a) (those paid by your college)?

Yes . . . . ☐ (1)

No . . . . ☐ (2)

B11. Is the per contact hour or per course pay scale for the **part-time faculty** reported in B7(a) (those paid by your college) the same as the per contact hour or per course "extra hours" pay scale for **full-time faculty** reported in B5(a) who teach extra hours for extra pay?

Yes . . . . ☐ (1)

No, part-timers paid more . . . ☐ (2)

No, part-timers paid less . . . . ☐ (3)

C. Mathematics Courses (Fall 2005)

The following instructions apply throughout **Section C**. Read them carefully before you begin filling out the tables.

- If you are part of a multi-campus college, please consult the third paragraph on page 1 before proceeding.
- In this section, do **not** include courses taught in other departments, learning centers, or developmental/remedial programs separate from your mathematics program or department.
- Read the row and column labels carefully. If the titles of courses listed below do not coincide exactly with yours, use your best judgment about where to list your courses. List each course only **once**. Note that the **part-time faculty** in Column (6) are those reported in B7(a) (part-time faculty paid by your college). Column (6) should **not** include any of your full-time faculty who teach an overload section.
- If a course is **not** taught at your campus during the fall term or if it is never taught at your campus, leave the cell blank.
- Do not include dual-enrollment sections offered on a high school campus for simultaneous high school and college credit through your institution.

♦Cells left blank will be interpreted as zeros

				LIST THE NUMBER OF SECTIONS FROM COLUMN (4)								
Name of Course (or equivalent) (1)	Total number of students enrolled Fall 2005 via distance learning[a] (2)	Total number of on-campus students enrolled Fall 2005[b] (3)	Total number of on-campus sections Fall 2005[b] (4)	that have enrollment above 30 (5)	that are taught by part-time faculty[c] (6)	that use graphing calculators (7)	that include a writing component such as reports or projects (8)	that require computer assignments (9)	that assign group projects (10)	that use commercial or locally produced online-response homework or testing systems (11)	that are taught mostly by the standard lecture method (12)	if not offered in Fall 2005, was this course either offered in 2004–2005 or scheduled for Winter/Spring 2006? Y(es)/N(o) (13)
C1. Arithmetic/Basic Mathematics												
C2. Pre-Algebra												
C3. Elementary Algebra (high school level)												
C4. Intermediate Algebra (high school level)												
C5. Geometry (high school level)												

a At least half of the students in the section receive the majority of their instruction via Internet, TV, computer, programmed instruction, correspondence courses, or other method where the instructor is **not** physically present.

b These students or sections are **not** included in column (2).

c Do **not** include full-time mathematics faculty teaching an overload section in this column. Include only part-time faculty, reported in B7(a), those paid by your college.

C. Mathematics Courses (Fall 2005) cont.

◆Cells left blank will be interpreted as zeros

				LIST THE NUMBER OF SECTIONS FROM COLUMN (4)								
Name of Course (or equivalent) (1)	Total number of students enrolled Fall 2005 via distance learning[a] (2)	Total number of on-campus students enrolled Fall 2005[b] (3)	Total number of on-campus sections Fall 2005[b] (4)	that have enrollment above 30 (5)	that are taught by part-time faculty[c] (6)	that use graphing calculators (7)	that include a writing component such as reports or projects (8)	that require computer assignments (9)	that assign group projects (10)	that use commercial or locally produced online-response homework or testing systems (11)	that are taught mostly by the standard lecture method (12)	if not offered in Fall 2005, was this course either offered in 2004–2005 or scheduled for Winter/Spring 2006? Y(es)/N(o) (13)
C6. College Algebra (level beyond Intermediate Algebra)												
C7. Trigonometry												
C8. College Algebra and Trigonometry, combined												
C9. Introduction to Mathematical Modeling												
C10. Precalculus/Elementary Functions/Analytic Geometry												

a At least half of the students in the section receive the majority of their instruction via Internet, TV, computer, programmed instruction, correspondence courses, or other method where the instructor is **not** physically present.

b These students or sections are **not** included in column (2).

c Do **not** include full-time mathematics faculty teaching an overload section in this column. Include only part-time faculty, reported in B7(a), those paid by your college.

C. Mathematics Courses (Fall 2005) cont.

♦**Cells left blank will be interpreted as zeros**

				LIST THE NUMBER OF SECTIONS FROM COLUMN (4)								
Name of Course (or equivalent) (1)	Total number of students enrolled Fall 2005 via distance learning[a] (2)	Total number of on-campus students enrolled Fall 2005[b] (3)	Total number of on-campus sections Fall 2005[b] (4)	that have enrollment above 30 (5)	that are taught by part-time faculty[c] (6)	that use graphing calculators (7)	that include a writing component such as reports or projects (8)	that require computer assignments (9)	that assign group projects (10)	that use commercial or locally produced online-response homework or testing systems (11)	that are taught mostly by the standard lecture method (12)	if not offered in Fall 2005, was this course either offered in 2004–2005 or scheduled for Winter/Spring 2006? Y(es)/N(o) (13)
C11. Calculus I (typically for mathematics, physics, engineering majors)												
C12. Calculus II (typically for mathematics, physics, engineering majors												
C13. Calculus III												
C14. Non-Mainstream Calculus I[d]												
C15. Non-Mainstream Calculus II[d]												
C16. Differential Equations												
C17. Linear Algebra												
C18. Discrete Mathematics												

a At least half of the students in the section receive the majority of their instruction via Internet, TV, computer, programmed instruction, correspondence courses, or other method where the instructor is **not** physically present.

b These students or sections are **not** included in column (2).

c Do **not** include full-time mathematics faculty teaching an overload section in this column. Include only part-time faculty, reported in B7(a), those paid by your college.

d Typically for business, life sciences, and social science majors.

C. Mathematics Courses (Fall 2005) cont.

◆Cells left blank will be interpreted as zeros

Name of Course (or equivalent) (1)	Total number of students enrolled Fall 2005 via distance learning[a] (2)	Total number of on-campus students enrolled Fall 2005[b] (3)	Total number of on-campus sections Fall 2005[b] (4)	LIST THE NUMBER OF SECTIONS FROM COLUMN (4)								if not offered in Fall 2005, was this course either offered in 2004–2005 or scheduled for Winter/Spring 2006? Y(es)/N(o) (13)
				that have enrollment above 30 (5)	that are taught by part-time faculty[c] (6)	that use graphing calculators (7)	that include a writing component such as reports or projects (8)	that require computer assignments (9)	that assign group projects (10)	that use commercial or locally produced online-response homework or testing systems (11)	that are taught mostly by the standard lecture method (12)	
C19. Elementary Statistics (with or without probability)[d]												
C20. Probability (with or without statistics)[d]												
C21. Finite Mathematics												
C22. Mathematics for Liberal Arts/ Math Appreciation												
C23. Mathematics for Elementary School Teachers												

a At least half of the students in the section receive the majority of their instruction via Internet, TV, computer, programmed instruction, correspondence courses, or other method where the instructor is **not** physically present.

b These students or sections are **not** included in column (2).

c Do **not** include full-time mathematics faculty teaching an overload section in this column. Include only part-time faculty, reported in B7(a), those paid by your college.

d Do **not** count the same course in both lines C19 and C20.

C. Mathematics Courses (Fall 2005) cont.

◆Cells left blank will be interpreted as zeros

				LIST THE NUMBER OF SECTIONS FROM COLUMN (4)								
Name of Course (or equivalent) (1)	Total number of students enrolled Fall 2005 via distance learning[a] (2)	Total number of on-campus students enrolled Fall 2005[b] (3)	Total number of on-campus sections Fall 2005[b] (4)	that have enrollment above 30 (5)	that are taught by part-time faculty[c] (6)	that use graphing calculators (7)	that include a writing component such as reports or projects (8)	that require computer assignments (9)	that assign group projects (10)	that use commercial or locally produced online-response homework or testing systems (11)	that are taught mostly by the standard lecture method (12)	if not offered in Fall 2005, was this course either offered in 2004–2005 or scheduled for Winter/Spring 2006? Y(es)/N(o) (13)
C24. Business Mathematics (not a transfer course to four-year colleges)												
C25. Business Mathematics (transfer course)												
C26. Non-Calculus-Based Technical Mathematics (not a transfer course)												
C27. Calculus-Based Technical Mathematics (transfer course)												
C28. Other Mathematics Courses												

a At least half of the students in the section receive the majority of their instruction via Internet, TV, computer, programmed instruction, correspondence courses, or other method where the instructor is **not** physically present.

b These students or sections are **not** included in column (2).

c Do **not** include full-time mathematics faculty teaching an overload section in this column. Include only part-time faculty, reported in B7(a), those paid by your college.

D. Faculty Educational Level, by Subject Field

D1. For the **permanent full-time faculty** (including those on leave) reported in B2, complete the following table showing the area of each faculty member's highest earned degree. The total of all faculty listed in this table should equal the number reported in B2.

- If you are part of a multi-campus college, please consult the third paragraph on page 1 before proceeding.

HIGHEST DEGREE	MAJOR FIELD OF HIGHEST DEGREE			
	MATHEMATICS (1)	STATISTICS (2)	MATHEMATICS EDUCATION (3)	OTHER (4)
DOCTORATE (1)				
MASTER'S (2)				
BACHELOR'S (3)				
LESS THAN BACHELOR'S (4)				

D. Faculty Educational Level, by Subject Field cont.

D2. For the **part-time faculty** reported in B7(c) (including those paid by your college and those paid by a third party), complete the following table showing the area of each faculty member's highest earned degree. The total of all faculty listed in this table should equal the number reported in B7(c).

• If you are part of a multi-campus college, please consult the third paragraph on page 1 before proceeding.

	MAJOR FIELD OF HIGHEST DEGREE			
HIGHEST DEGREE	MATHEMATICS (1)	STATISTICS (2)	MATHEMATICS EDUCATION (3)	OTHER (4)
DOCTORATE (1)				
MASTER'S (2)				
BACHELOR'S (3)				
LESS THAN BACHELOR'S (4)				

E. Faculty by Gender and Ethnicity/Race

Instructions:

- If you are part of a multi-campus college, please consult the third paragraph on page 1 before proceeding.
- For the **permanent full-time faculty** (including those on leave) reported in B2 and for the **part-time faculty** reported in B7(a) (those paid by your college), complete the following table giving data about gender and ethnicity/race.
- The total of full-time faculty should equal the figure given in B2. The total of part-time faculty should equal the figure reported in B7(a).

ETHNIC/RACIAL STATUS AND GENDER			Permanent Full-Time Faculty From B2		Part-Time Faculty from B7(a)
			Age < 40 (1)	Age ≥ 40 (2)	(3)
American Indian, Eskimo, Aleut	Male	(1)			
	Female	(2)			
Asian, Pacific Islander	Male	(3)			
	Female	(4)			
Black or African American (non-Hispanic)	Male	(5)			
	Female	(6)			
Mexican American, Puerto Rican, or Other Hispanic	Male	(7)			
	Female	(8)			
White (non-Hispanic)	Male	(9)			
	Female	(10)			
Status Not Known or Other	Male	(11)			
	Female	(12)			

F. Faculty Age Profile

Complete the following table showing the number of faculty who belong in each of the age categories below.

- Consider only **permanent full-time faculty** (including those on leave) as reported in B2.
- If you are part of a multi-campus college, please consult the third paragraph on page 1 before proceeding.
- The total faculty listed should equal the number reported in B2.

FACULTY AGE	Under 30 (1)	30–34 (2)	35–39 (3)	40–44 (4)	45–49 (5)	50–54 (6)	55–59 (7)	60–64 (8)	65–69 (9)	70 & over (10)
MEN (1)										
WOMEN (2)										

Mathematics Questionnaire

G. Faculty Employment and Mobility

• If you are part of a multi-campus college, please consult the third paragraph on page 1 before proceeding.

G1. How many of the **permanent full-time faculty** members in B2 were newly appointed to a permanent full-time position this year (2005–2006)?

Number of faculty newly appointed on a permanent full-time basis . . . . ☐

if "zero" ⟶ go to G5.

if "1 or more" ⟶ go to G2.

G2. Of the faculty members counted in G1, how many had the following as their main activity in the academic year preceding their appointment? Report only **one** main activity per person. The total in G2 should equal the number reported in G1.

a) Attending graduate school . . . . ☐ (1)

b) Teaching in a four-year college or university . . . . ☐ (2)

c) Teaching in another two-year college . . . . ☐ (3)

d) Teaching in a secondary school . . . . ☐ (4)

e) Part-time or full-time temporary employment by your college . . . . ☐ (5)

f) Nonacademic employment . . . . ☐ (6)

g) Unemployed . . . . ☐ (7)

h) Status unknown . . . . ☐ (8)

G3. How many of the faculty reported in G1 had ever taught at your campus or in your larger organization either part-time or full-time? . . . . ☐

Mathematics Questionnaire

G. Faculty Employment and Mobility cont.

G4. For each **permanent full-time faculty** member reported in G1, give the following data. Add more lines at the bottom of the table if necessary. For each new hire complete an entire row.

	Age (1)	Gender (2)	Ethnicity/Race (3)	Highest Degree Earned (Bachelor's, Master's, or Doctorate) (4)
New Hire #1 (1)				
New Hire #2 (2)				
New Hire #3 (3)				
New Hire #4 (4)				
New Hire #5 (5)				
New Hire #6 (6)				

G5. How many of your faculty who were **permanent full-time faculty** in the previous year (2004–2005) are no longer part of your **permanent full-time faculty**? . . . . ☐

G6. Give the number of **permanent full-time faculty** (total for G6 should equal number reported in G5) who:

a) died while in full-time service . . . . ☐ (1)

b) left full-time service due to retirement . . . . ☐ (2)

c) left to teach at a four-year college or university . . . . ☐ (3)

d) left to teach at another two-year college . . . . ☐ (4)

e) left to teach at a secondary school . . . . ☐ (5)

f) left for a nonacademic position . . . . ☐ (6)

g) left to attend graduate school . . . . ☐ (7)

h) other (specify) ____________________ ☐ (8)

i) unknown . . . . ☐ (9)

H. Professional Activities of Permanent Full-Time Faculty

- If you are part of a multi-campus college, please consult the third paragraph on page 1 before proceeding.

H1. Is some form of continuing education or professional development required of your **permanent full-time faculty** reported in B2?

Yes . . . . ☐ (1) ⟶ go to H2.

No . . . . ☐ (2) ⟶ go to Section I.

H2. Estimate the number of **permanent full-time faculty** reported in B2 who fulfill the requirement in H1 in one or more of the following ways:

a) Activities provided by your college or organization at one of its locations . . . . ☐ (1)

b) Participation in professional association meetings and minicourses or other professional association activities. . . . . ☐ (2)

c) Publishing expository or research articles or textbooks . . . . ☐ (3)

d) Continuing graduate education . . . . ☐ (4)

e) Unknown . . . . ☐ (5)

I. Resources Available to Part-Time Mathematics Faculty

• If you are part of a multi-campus college, please consult the third paragraph on page 1 before proceeding.

I-1. How many of the **part-time faculty** paid by your college (reported in B7(a)) have campus office space that contains:

a) their own individual desk? . . . . ☐ (1)

b) a desk shared with one other person? . . . . ☐ (2)

c) a desk shared with more than one other person? . . . . ☐ (3)

I-2. How many of the **part-time faculty** paid by your college (reported in B7(a)) have no campus office space at all? . . . . ☐

• Note: The sum of all entries in I-1 and I-2 should equal the number reported in B7(a).

I-3. How many of the **part-time faculty** paid by your college (reported in B7(a)) have:

a) a computer in their campus office? . . . . ☐ (1)

b) no computer in their campus office but shared computers nearby? . . . . ☐ (2)

c) no convenient access, or no access at all, to a computer at your college? . . . . ☐ (3)

I-4. For which mathematics faculty do you periodically evaluate teaching? Check all that apply.

a) All **permanent full-time faculty** (reported in B2) . . . . ☐ (1)

b) All **part-time faculty** paid by your college (reported in B7(a)) . . . . ☐ (2)

If you checked <u>either</u> I-4(a) or I-4(b), then ⟶ go to I-5.

If you checked <u>neither</u> I-4(a) nor I-4(b), then ⟶ go to J.

I. Resources Available to Part-Time Mathematics Faculty cont.

I-5. Check all evaluation methods that are used for **part-time faculty** paid by your college (reported in B7(a)) or for **permanent full-time faculty** (reported in B2).

EVALUATION METHOD	Part-Time Faculty in B7(a) (1)	Full-Time Faculty in B2 (2)
a) Observation of classes by other faculty members or department chair		
b) Observation of classes by division head (if different from chair) or other administrator		
c) Evaluation forms completed by students		
d) Evaluation of written course material such as lesson plans, syllabi, or exams		
e) Self-evaluation such as teaching portfolios		
f) Other (specify) ______________		

J. Academic Support and Enrichment Opportunities for Students

• If you are part of a multi-campus college, please consult the third paragraph on page 1 before proceeding.

J1. Does your department or college offer a mathematics placement program for entering students?

Yes . . . . ☐ (1) ⟶ go to J2.

No . . . . ☐ (2) ⟶ go to J7.

J2. What is the source of the placement test(s)? (Check all that apply.)

a) Test written by your department . . . . ☐ (1)

b) Test provided by Educational Testing Service (ETS) . . . . ☐ (2)

c) Test provided by American College Testing Program (ACT) . . . . ☐ (3)

d) Test provided by professional association . . . . ☐ (4)

Name of professional association ______________________________

e) Test provided by other external source . . . . ☐ (5)

Name of external source ______________________________

J3. Is the placement examination <u>usually required</u> for first-time enrollees?

Yes . . . . ☐ (1) ⟶ go to J4.

No . . . . ☐ (2) ⟶ go to J7.

J4. Is it <u>usually required</u> that first-time enrollees discuss the results of the placement test with an advisor or a counselor before registering for their first mathematics course?

Yes . . . . ☐ (1)

No . . . . ☐ (2)

J5. Is placement in the student's first mathematics course mandatory based on:

Placement test score alone . . . . ☐ (1)

Placement test score and other information. . . . ☐ (2)

Not mandatory . . . . ☐ (3)

J. Academic Support and Enrichment Opportunities for Students cont.

J6. Does your department periodically assess the effectiveness of the mathematics placement test?

Yes .... ☐ (1)

No .... ☐ (2)

J7. Does your department or college operate a mathematics lab or tutoring center?

Yes .... ☐ (1) → go to J8.

No .... ☐ (2) → go to J9.

J8. Check all services available to students through your mathematics lab or tutoring center.

a) Computer-aided instruction .... ☐ (1)

b) Computer software such as computer algebra packages or statistical packages .... ☐ (2)

c) Internet resources .... ☐ (3)

d) Media such as CDs or DVDs .... ☐ (4)

e) Organized small group tutoring or study sessions .... ☐ (5)

f) Tutoring by students .... ☐ (6)

g) Tutoring by paraprofessional staff .... ☐ (7)

h) Tutoring by part-time mathematics faculty .... ☐ (8)

i) Tutoring by full-time mathematics faculty .... ☐ (9)

j) Other mathematics lab or tutoring center services (specify) ______________________ ☐ (10)

J. Academic Support and Enrichment Opportunities for Students cont.

J9. Check all opportunities available to your mathematics students.

a) Honors sections of mathematics courses .... ☐ (1)

b) Mathematics club .... ☐ (2)

c) Special mathematics programs to encourage women .... ☐ (3)

d) Special mathematics programs to encourage minorities .... ☐ (4)

e) Opportunities to compete in mathematics contests .... ☐ (5)

f) Special mathematics lectures/colloquia not part of a mathematics club .... ☐ (6)

g) Mathematics outreach opportunities in local K–12 schools .... ☐ (7)

h) Opportunities to participate in undergraduate research in mathematics .... ☐ (8)

i) Independent study opportunities in mathematics .... ☐ (9)

j) Assigned faculty advisors in mathematics .... ☐ (10)

k) Other (specify) ______________________ ☐ (11)

K. Dual-Enrollment Courses

- If you are part of a multi-campus college, please consult the third paragraph on page 1 before proceeding.
- In this questionnaire we use the term "dual-enrollment courses" to mean courses taught **in high school by high school teachers** for which students may obtain high school credit and simultaneous college credit through your institution.

K1. Does your department participate in any dual-enrollment program of the type defined above?

Yes . . . . ☐ (1) ⟶ go to K2.

No . . . . ☐ (2) ⟶ go to K6.

K2. Please complete the following table concerning your dual-enrollment program (as defined above) for the spring term of 2005 and for the current fall term of 2005.

Course	Total Dual Enrollments Last Term = Spring 2005 (1)	Number of Dual-Enrollment Sections Last Term = Spring 2005 (2)	Total Dual Enrollments This Term = Fall 2005 (3)	Number of Dual-Enrollment Sections This Term = Fall 2005 (4)
a) College Algebra				
b) Precalculus				
c) Calculus I				
d) Statistics				
e) Other				

K3. For the dual-enrollment courses in K2, which of the following are the responsibility of your department?

	Never Our Responsibility (1)	Sometimes Our Responsibility (2)	Always Our Responsibility (3)
a) Choice of textbook			
b) Design/approval of syllabus			
c) Design of final exam			
d) Choice of instructor			

K. Dual-Enrollment Courses cont.

K4. Does your department have a teaching evaluation program in which its own part-time department faculty (see B7(a)) are required to participate?

Yes . . . . ☐ (1) ⟶ go to K5.

No . . . . ☐ (2) ⟶ go to K6.

K5. Are instructors in the dual-enrollment courses reported in K2 required to participate in the teaching evaluation program for part-time departmental faculty?

Yes . . . . ☐ (1)

No . . . . ☐ (2)

K6. Does your department assign any of <u>its own</u> full-time or part-time faculty (faculty paid by your college as reported in either B1 or B7(a)) to teach courses on a high school campus for which high school students may receive both high school and college credit through your institution?

Yes . . . . ☐ (1) ⟶ go to K7.

No . . . . ☐ (2) ⟶ go to Section L.

K7. Please complete the following table describing high school student enrollments as taught by <u>your</u> faculty on a high school campus. See K6.

Course	Total Dual Enrollments Last Term = Spring 2005 (1)	Number of Dual-Enrollment Sections Last Term = Spring 2005 (2)	Total Dual Enrollments This Term = Fall 2005 (3)	Number of Dual-Enrollment Sections This Term = Fall 2005 (4)
a) College Algebra				
b) Precalculus				
c) Calculus I				
d) Statistics				
e) Other				

K8. For the courses described in K6 taught by <u>your</u> faculty, which of the following are the responsibility of your department?

	Never Our Responsibility (1)	Sometimes Our Responsibility (2)	Always Our Responsibility (3)
a) Choice of textbook			
b) Design/approval of syllabus			
c) Design of final exam			

L. Mathematics Preparation of K–12 Teachers

• If you are part of a multi-campus college, please consult the third paragraph on page 1 before proceeding.

L1. Does your department have a faculty member assigned to coordinate mathematics program courses for pre-service elementary school teachers?

Yes . . . . ☐ (1)

No . . . . ☐ (2)

L2. Other than the course "Mathematics for Elementary School Teachers" reported on line C23, do you designate any sections of your other mathematics program courses as "especially designed for pre-service elementary school teachers"?

Yes . . . . ☐ (1)

No . . . . ☐ (2)

L3. Which of the following groups can meet their entire mathematics course or licensure requirement for teaching via an organized program in your department? Consider "pre-service" and "career switchers" as distinct categories. "Career switchers" usually are post-baccalaureate older adults returning for teaching licensure after a non-teaching career and often under state-approved special licensure rules.

a) Pre-service elementary school teachers . . . . ☐ (1)

b) Pre-service middle school teachers . . . . ☐ (2)

c) Pre-service secondary school teachers . . . . ☐ (3)

d) In-service elementary school teachers . . . . ☐ (4)

e) In-service middle school teachers . . . . ☐ (5)

f) In-service secondary school teachers . . . . ☐ (6)

g) Career switchers moving to elementary school teaching . . . . ☐ (7)

h) Career switchers moving to middle school teaching . . . . ☐ (8)

i) Career switchers moving to secondary school teaching . . . . ☐ (9)

L4. Does your institution offer pedagogical courses in mathematics for teacher licensure?

Yes, in our mathematics department . . . . ☐ (1)

Yes, elsewhere in the institution . . . . ☐ (2)

No . . . . ☐ (3)

L. Mathematics Preparation of K–12 Teachers cont.

L5. How many mathematics courses (including general education requirements, if any) are required of students seeking their entire elementary school teacher licensure at your institution?

a) We have no students seeking elementary school teaching licensure entirely from us . . . . ☐ (1)

b) Number of mathematics courses required for early elementary grade licensure. . . . . ☐ (2)

c) Number of mathematics courses required for later elementary grade licensure . . . . ☐ (3)

L6. How do students seeking their entire secondary school teaching licensure at your institution learn about the history of mathematics?

a) We have no students seeking secondary school teaching licensure entirely from us . . . . ☐ (1)

b) We offer a course in the history of mathematics which students seeking secondary school teaching licensure are required to take . . . . ☐ (2)

c) There is no required mathematics history course for students seeking secondary school teaching licensure but these students learn mathematics history from other courses they are required to take . . . . ☐ (2)

d) Students in our secondary licensure program are not required to learn about mathematics history . . . . ☐ (4)

Mathematics Questionnaire

M. Issues of Professional Concern

M1. Below are problems often cited by two-year college mathematics departments. Please read each item carefully and check the box in each row that best reflects your view. (Check only **one box per row**.)

	Not a problem for us (1)	Minor problem for us (2)	Moderate problem for us (3)	Major problem for us (4)
a) Maintaining vitality of faculty	☐ (1)	☐ (2)	☐ (3)	☐ (4)
b) Dual-enrollment (high school and college credit) courses[a]	☐ (5)	☐ (6)	☐ (7)	☐ (8)
c) Staffing statistics courses	☐ (9)	☐ (10)	☐ (11)	☐ (12)
d) Unrealistic student understanding of the demands of college work	☐ (13)	☐ (14)	☐ (15)	☐ (16)
e) Need to use part-time faculty for too many courses	☐ (17)	☐ (18)	☐ (19)	☐ (20)
f) Faculty salaries too low	☐ (21)	☐ (22)	☐ (23)	☐ (24)
g) Class sizes too large	☐ (25)	☐ (26)	☐ (27)	☐ (28)
h) Low student motivation	☐ (29)	☐ (30)	☐ (31)	☐ (32)
i) Too many students needing remediation	☐ (33)	☐ (34)	☐ (35)	☐ (36)
j) Successful progress of students through developmental courses to more advanced mathematics courses	☐ (37)	☐ (38)	☐ (39)	☐ (40)
k) Low success rate in transfer-level courses	☐ (41)	☐ (42)	☐ (43)	☐ (44)
l) Too few students who intend to transfer actually do transfer	☐ (45)	☐ (46)	☐ (47)	☐ (48)
m) Inadequate travel funds for faculty	☐ (49)	☐ (50)	☐ (51)	☐ (52)
n) Inadequate classroom facilities for teaching with technology	☐ (53)	☐ (54)	☐ (55)	☐ (56)
o) Inadequate computer facilities for part-time faculty use	☐ (57)	☐ (58)	☐ (59)	☐ (60)
p) Inadequate computer facilities for student use.	☐ (61)	☐ (62)	☐ (63)	☐ (64)

[a] Courses taught in high school by high school teachers for which students may obtain high school credit and simultaneous college credit through your institution.

Mathematics Questionnaire

M. Issues of Professional Concern cont.

M1. Continued

	Not a problem for us (1)	Minor problem for us (2)	Moderate problem for us (3)	Major problem for us (4)
q) Outsourcing instruction to commerical companies	☐ (65)	☐ (66)	☐ (67)	☐ (68)
r) Heavy classroom and other duties prevent personal and teaching enrichment by faculty	☐ (69)	☐ (70)	☐ (71)	☐ (72)
s) Curriculum alignment between high schools and college	☐ (73)	☐ (74)	☐ (75)	☐ (76)
t) Lack of curricular flexibility because of transfer requirements	☐ (77)	☐ (78)	☐ (79)	☐ (80)
u) Use of distance education[b]	☐ (81)	☐ (82)	☐ (83)	☐ (84)
v) Other (specify) ____________	☐ (85)	☐ (86)	☐ (87)	☐ (88)

[b] At least half of the students in the section receive the majority of their instruction via Internet, TV, computer, programmed instruction, correspondence courses, or other method where the instructor is **not** physically present.

M2. Many departments today use a spectrum of program assessment methods. Please check all that apply to your department's program assessment efforts during the last six years.

a) We conducted a review of our mathematics program that included one or more reviewers from outside our institution . . . ☐ (1)

b) We asked students in our mathematics program to comment on and suggest changes in our program . . . ☐ (2)

c) Other departments at our institution were invited to comment on the preparation that their students received in our courses . . . ☐ (3)

d) Data on students' progress in subsequent mathematics courses were gathered and analyzed . . . ☐ (4)

e) We have a placement system for first-year students, and we gathered and analyzed data on its effectiveness . . . ☐ (5)

f) Our department's program assessment activities led to changes in our mathematics program . . . ☐ (6)

M. Issues of Professional Concern cont.

The next four questions deal with general education requirements at your institution.

M3. Does your institution require all associate degree graduates to have a quantitative course as part of their general education requirements? Choose one of the following.

a) Yes, all associate degree graduates must have such credit . . . . ☐ (1) ⟶ go to M4.

b) Not (a), but all Associate of Arts or Associate of Science graduates must have such credit . . . . ☐ (2) ⟶ go to M4.

c) Neither (a) nor (b) . . . . ☐ (3) ⟶ go to Section N.

M4. If you chose (a) or (b) in M3, is it true that all students (to whom the quantitative requirement applies) must fulfill it by taking a course in your mathematics department?

Yes . . . . ☐ (1)

No . . . . ☐ (2)

M5. Which courses in your department can be used to fulfill the general education quantitative requirement in M3?

a) Any course in the department, including all high school-level courses . . . . ☐ (1)

b) Intermediate Algebra (see C4) or any course beyond Intermediate Algebra . . . . ☐ (2)

c) Not Intermediate Algebra, but any course beyond Intermediate Algebra . . . . ☐ (3)

d) Only certain courses beyond Intermediate Algebra . . . . ☐ (4)

M6. If you chose M5(d), which of the following departmental courses can be used to fulfill the general education quantitative requirement? Check all that apply. If you did not choose M5(d), omit this question and go to Section N.

Course	Can be used
a) College Algebra and/or Precalculus	
b) Calculus (any course)	
c) Introduction to Mathematical Modeling	
d) A basic Probability and/or Statistics course	
e) A special general education course in our department not listed above	
f) Some other course(s) in our department not listed above	

N. Mathematics Enrollments Outside Your Mathematics Department/Program (Fall 2005)

Data to answer the following questions often are beyond the information normally available to a mathematics department chair. Please invest the extra effort needed to give an accurate account of all enrollments in the following courses that are **not** taught in the mathematics department/program. (Give enrollments, not the number of sections taught.)

Instructions:

- Please consult the third paragraph on page 1 before proceeding to determine whether to report on your campus or on your entire multi-campus system.
- Report all enrollments at your campus or in your multi-campus system that are **not** taught in the mathematics department/program (and so are not listed in Section C).
- Please consult appropriate sources outside the mathematics program such as schedules, registrar's data, or the heads of these programs to get accurate data on enrollments.

COURSE (1)	Occupational Programs (2)	Business (3)	Learning Center (4)	Other Dept/Division[a] (5)
N1. Arithmetic/Pre-Algebra				
N2. Elementary Algebra (high school level)				
N3. Intermediate Algebra (high school level)				
N4. Business Mathematics				
N5. Statistics/Probability				
N6. Technical Mathematics				

[a] Such as a Developmental Studies Division separate from the mathematics department/program.

Mathematics Questionnaire

O. Comments and Suggestions

O1. If you have found some question(s) difficult to interpret or answer, please let us know. We welcome comments or suggestions to improve future surveys (e.g., CBMS2010).

Thank you for completing this questionnaire. We know it was a time-consuming process. We hope the final survey report, which should be published and online in spring 2007, will be useful to you and your department.

Please retain a copy of this questionnaire in case questions arise.

Appendix VI

Four-Year Statistics Questionnaire

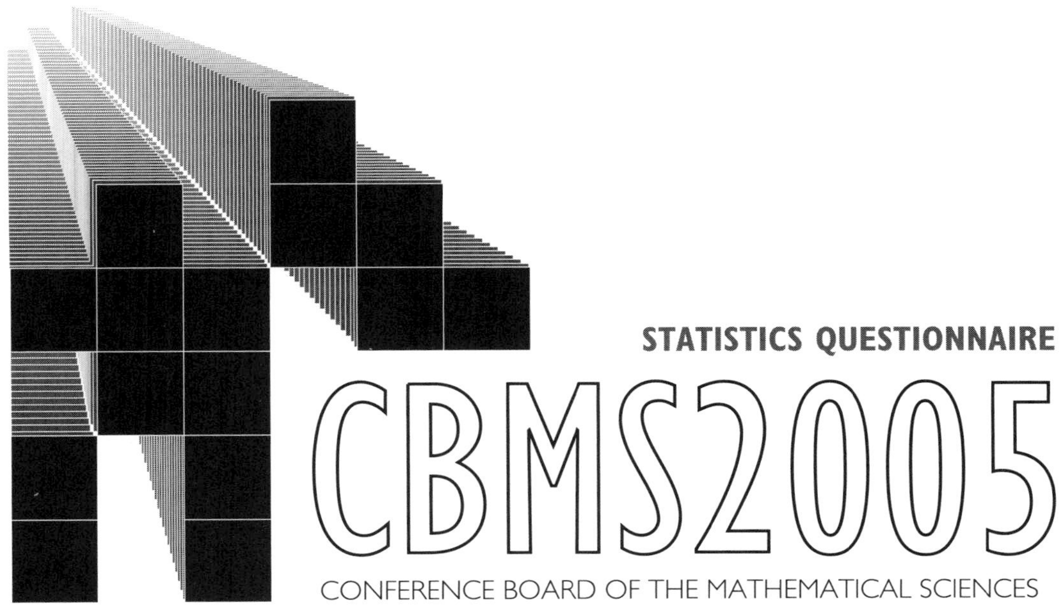

SURVEY OF UNDERGRADUATE PROGRAMS IN THE MATHEMATICAL SCIENCES

General Information

Statistics Questionnaire

As part of a random sample, your department has been chosen to participate in the NSF-funded CBMS2005 National Survey of Undergraduate Mathematical and Statistical Sciences. Even though it is a very complicated survey, the presidents of all U.S. mathematical and statistical sciences organizations have endorsed it and ask for your cooperation.

We assure you that no individual departmental data, except the names of responding departments, will be released.

This survey provides data about the nation's undergraduate statistical effort that is available from no other source. You can see the results of a similar survey five years ago by going to www.ams.org/cbms where the CBMS 2000 report is available on-line.

This survey studies the undergraduate programs in universities and colleges that offer at least a bachelors degree. Many of the departments in our random sample also offer higher degrees in the statistical sciences.

We have classified your department as belonging to a university or four-year college. If this is not correct, please contact David Lutzer, Survey Director, at 757-221-4006 or at Lutzer@math.wm.edu.

If you have any questions while filling out this survey form, please call the Survey Director, David Lutzer, at 757-221-4006 or contact him by e-mail at Lutzer@math.wm.edu.

Please report on undergraduate programs in the broadly defined mathematical and statistical sciences including applied mathematics, statistics, operations research, and computer science that are under the direction of your department. Do not include data for other departments or for branches or campuses of your institution that are budgetarily separate from your own.

Please return your completed questionnaire by October 15, 2005 in the enclosed envelope to:

CBMS Survey
UNC-CH Survey Research Unit
730 Martin Luther King, Jr. Blvd
Suite 103, CB#2400, UNC-CH
Chapel Hill, NC 27599-2400

Please retain a copy of your responses to this questionnaire in case questions arise.

A. General Information

Statistics Questionnaire

PLEASE PRINT CLEARLY

A1. Name of your institution: ______________________________

A2. Name of your department: ______________________________

A3. We have classified your department as being part of a university or four-year college. Do you agree?

Yes............................ ☐ (1) ⟶ If "Yes", go to A4 below.

No.............................. ☐ (2) ⟶ If "No", please call David Lutzer, Survey Director, at 757-221-4006 before proceeding any further.

A4. Your institution is ....public ☐ (1) ; ....private ☐ (2)

A5. Which programs leading to the following degrees does your department offer? Please check at least one box in each row.

Program	None (1)	Baccalaureate Degree (2)	Masters Degree (3)	Doctoral Degree (4)
a) Mathematics				
b) Statistics				
c) Biostatistics				
d) Computer Science				
e) Other (please specify below)				

If you offer bachelors, masters, or doctoral degrees in a mathematical or statistical science other than those in A5-a, b, c, and d, please enter the name(s) of the field(s) here: ______________________________

A6. Responses to this question will be used to project total enrollment in the current (2005-2006) academic year based on the pattern of your departmental enrollments in 2004-2005. Do NOT include any numbers from dual-enrollment courses[1] in answering question A6.

a) Previous fall (2004) total student enrollment in your department's undergraduate courses (remember: do not include dual-enrollment courses[1]): .... ☐ (1)

b) Previous academic year (2004-2005) total enrollment in your department's undergraduate courses, excluding dual enrollments[1] and excluding enrollments in summer school 2005: .... ☐ (2)

c) Total enrollment in your department's undergraduate courses in summer school 2005: .... ☐ (3)

[1] *In this question, the term "dual-enrollment courses" is used to mean courses taught on a high school campus, by high school teachers, for which high school students may obtain high school credit and simultaneously college credit through your institution.*

A. General Information cont.

Statistics Questionnaire

A7. Which of the following best describes your institution's academic calendar? Check only one box.

a) Semester	
b) Trimester	
c) Quarter	
d) Other (please specify below)	

Academic calendar description if not a), b), or c): ______________________

A8. If your college or university does not recognize tenure, check the following box ☐ and follow the special instructions in subsequent sections for counting departmental faculty of various types.

A9. Contact person in your department: ☐

A10. Contact person's e-mail address: ☐

A11. Contact person's phone number including area code: ☐

A12. Contact person's mailing address: ☐

B. Dual Enrollment Courses

Statistics Questionnaire

In this questionnaire the term dual enrollment courses refers to courses conducted on a high school campus and taught by high school teachers, for which high school students may obtain high school credit and simultaneously college credit through your institution.

B1. Does your department participate in any dual enrollment programs of the type defined above?

Yes............................ ☐ (1) ⟶ If "Yes", go to B2.

No.............................. ☐ (2) ⟶ If "No", go to B6.

B2. Please complete the following table concerning your dual enrollment program (as defined above) for the previous term (spring 2005) and the current fall term of 2005.

Course	Total Dual Enrollments Last Term =Spring 2005 (1)	Number of Dual-Enrollment Sections Last Term =Spring 2005 (2)	Total Dual Enrollments This Term =Fall 2005 (3)	Number of Dual-Enrollment Sections This Term =Fall 2005 (4)
a) Statistics				
b) Other				

B3. For the dual enrollment courses in B2, to what extent are the following the responsibility of your department?

	Never Our Responsibility (1)	Sometimes Our Responsibility (2)	Always Our Responsibility (3)
a) Choice of textbook			
b) Design/approval of syllabus			
c) Design of final exam			
d) Choice of instructor			

B4. Does your department have a teaching evaluation program in which your part-time department faculty are required to participate?

Yes............................ ☐ (1) ⟶ If "Yes", go to B5.

No.............................. ☐ (2) ⟶ If "No", go to B6.

B5. Are instructors in the dual-enrollment courses reported in B2 required to participate in the teaching evaluation program for part-time departmental faculty described in B4?

Yes............................ ☐ (1)

No.............................. ☐ (2)

B. Dual Enrollment Courses cont.

B6. Does your department assign any of its own full-time or part-time faculty to teach courses conducted on a high school campus for which high school students may receive both high school and college credit (through your institution)?

Yes............................ ☐ (1) ⟶ If "Yes", go to B7.

No............................. ☐ (2) ⟶ If "No", go to Section C.

B7. How many students are enrolled in the courses conducted on a high school campus and taught by your full-time or part-time faculty and through which high school students may receive both high school and college credit (through your institution) in fall 2005? .... ☐

C. Probability and Statistics Courses (Fall 2005)

The following instructions apply throughout sections C and D (pages 6-12).

- If your departmental course titles do not match exactly with the ones that we suggest, please use your best judgment to match them.
- Report distance-learning enrollments separately from other enrollments. A *distance-learning* section is one in which a majority of students receive the majority of their instruction by Internet, TV, correspondence courses, or other methods where the instructor is NOT physically present.
- Do NOT include any dual-enrollment sections or enrollments in these tables. (In this questionnaire, a *dual-enrollment* section is one that is conducted on a high-school campus, taught by a high-school teacher, and which allows students to receive high-school credit and simultaneously college credit from your institution for the course. These courses were reported in Section B.)
- Except in C1-2, please count any lecture course along with its associated recitation/problem/laboratory sessions as one section of the course. (Special instructions for C1-2 are given in a footnote.)
- In course C-1 below, we ask you to list those lecture sections with several recitation/problem/laboratory sessions separately from other sections of the course that do not have such recitation/problem/laboratory sessions.
- Report a section of a course as being taught by a *graduate teaching assistant (GTA)* if and only if that section is taught *independently* by the GTA, i.e., when it is the GTA's own course and the GTA is the instructor of record.
- If your institution does not recognize tenure, report sections taught by your permanent full-time faculty in column (5) and sections taught by other full-time faculty in columns (6) or (7) as appropriate.
- Full-time faculty teaching in your department and holding joint appointments with other departments should be counted in column (5) if they are tenured, tenure-eligible, or permanent in your department. Faculty who are not tenured, tenure-eligible, or permanent in your department and who teach more than 50% of their fall term teaching assignment in your department should be counted in column (6) or (7) depending upon their highest degree. Faculty who are not tenured, tenure-eligible, or permanent in your department and who teach in your department for at most half of their fall-term teaching assignment should be counted in column (8). (Example: If a tenured psychology professor with a joint appointment in your department teaches a total of two courses in fall 2005, with exactly one being in your department, then that person would be counted as part-time in your department.)
- Do not fill in any shaded rectangles.
- Any unshaded rectangle that is left blank will be interpreted as reporting a count of zero.
- Except where specifically stated to the contrary, the tables in Sections C and D deal with enrollments in fall term 2005.

C. Probability & Statistics Courses (Fall 2005)

◆**Cells left blank will be interpreted as zeros**

				Of the number in Column 4, how many sections are taught by:					Of the number in Column 4, how many sections:				
Name of Course (or equivalent) (1)	Total distance-education enrollment[a] (2)	Total enrollment NOT in Col (2) and NOT dual enrollments[b] (3)	Number of sections corres-ponding to Column (3) (4)	Tenured or Tenure-eligible Faculty (5)	Other Full-time Faculty with Ph.D. (6)	Other Full-time Faculty without Ph.D. (7)	Part-time Faculty (8)	Graduate Teaching Assist.[c] (9)	Use graphing calculators (10)	Include writing components such as reports or projects (11)	Require computer assign-ments (12)	Use on-line homework generating and grading packages (13)	Assign group projects (14)
PROBABILITY & STATISTICS													
ELEMENTARY LEVEL													
C1. Elementary Statistics (no calculus prerequisite):													
C1-1. Lecture with separately scheduled recitation/problem/laboratory sessions[d]													
C1-2. Number of recitation/problem/laboratory sessions associated with courses reported in C1-1[e]													
C1-3. Other sections with enrollment of 30 or less													
C1-4. Other sections with enrollment above 30													

[a] A majority of students receive the majority of their instructor via Internet, TV, correspondence courses, or other methods where the instructor is NOT physically present.

[b] Do not include any dual-enrollments courses, i.e., courses taught on a high school campus by a high school instructor, for which high school students may obtain both high school credit and simultaneously college credit through your institution.

[c] Sections taught independently by GTAs.

[d] A class along with its recitation/problem/laboratory sessions is to be counted as one section in C1-1.

[e] Example: suppose your department offers four 100-student sections of a course and that each is divided into five 20-student discussion sessions that meet separately from the lectures. Report 4•5=20 recitation/problem/laboratory sessions associated with the course, even if each discussion meets several times per week.

C. Probability & Statistics Courses (Fall 2005) cont.

Statistics Questionnaire

♦Cells left blank will be interpreted as zeros

Name of Course (or equivalent) (1)	Total distance-education enrollment[a] (2)	Total enrollment NOT in Col (2) and NOT dual enrollments[b] (3)	Number of sections corresponding to Column (3) (4)	Of the number in Column 4, how many sections are taught by: Tenured or Tenure-eligible Faculty (5)	Other Full-time Faculty with Ph.D. (6)	Other Full-time Faculty without Ph.D. (7)	Part-time Faculty (8)	Graduate Teaching Assist.[c] (9)
PROBABILITY & STATISTCS								
ELEMENTARY LEVEL CONT.								
C2. Probability and statistics (no calculus prerequisite)								
C3. Statistical Literacy/Statistics and Society								
C4. Statistics for pre-service elementary or middle grades teachers								
C5. Statistics for pre-service high school teachers								
C6. All other elementary-level statistics courses								

[a] A majority of students receive the majority of their instruction via Internet, TV, correspondence courses, or other method where the instructor is NOT physically present.

[b] Do not include any dual-enrollments courses, i.e., courses taught on a high school campus by a high school instructor, for which high school students may obtain both high school credit and simultaneously college credit through your institution.

[c] Sections taught independently by GTAs.

C. Probability & Statistics Courses (Fall 2005) cont.

◆Cells left blank will be interpreted as zeros

Name of Course (or equivalent) (1)	Total enrollment Fall 2005 (2)	Number of sections corresponding to Column (2) (3)	Number of sections corresponding to Column (3) taught by Tenured or Tenure-eligible Faculty (4)	Was this course taught in ANY term of the previous academic year? Y(es) / N(o) (5)	Will this course be offered in the next term (Spring 2006)? Y(es) / N(o) (6)
PROBABILITY & STATISTICS					
INTERMEDIATE AND ADVANCED LEVEL					
C7. Mathematical Statistics (calculus prerequisite)					
C8. Probability (calculus prerequisite)					
C9. Combined Probability & Statistics (calculus prerequisite)					
C10. Stochastic Processes					
C11. Applied Statistical Analysis					
C12. Design & Analysis of Experiments					
C13. Regression (and Correlation)					
C14. Biostatistics					
C15. Nonparametric Statistics					
C16. Categorical Data Analysis					
C17. Sample Survey Design & Analysis					
C18. Statistical Software & Computing					
C19. Data Management					
C20. Senior Seminar/ Independent Studies					
C21. All other upper level Probability & Statistics courses					

D. Computer Science Courses (Fall 2005)

- Please refer to the course reporting instructions at the beginning of Section C.
- In December 2001, a joint IEEE Computer Society/ACM Task Force issued its recommendations on "Model Curricula for Computing." That report replaced the curricular recommendations published by ACM in 1991 and is available from http://www.computer.org/education/cc2001/. Course numbers and, to the degree possible, course names in the table below are taken from the detailed course outlines in the appendices of that CC2001 report.

D. Does your department offer any Computer Sciences courses?

Yes........................... ☐ (1) ⟶ If "Yes", go to D1, below.

No............................. ☐ (2) ⟶ If "No", go to Section E

◆Cells left blank will be interpreted as zeros

				Of the number in Column 4, how many sections are taught by:				
Name of Course (or equivalent) (1)	Total distance-education enrollment[a] (2)	Total enrollment NOT in Col (2) and NOT dual enrollments[b] (3)	Number of sections corresponding to Column (3) (4)	Tenured or Tenure-eligible Faculty (5)	Other Full-time Faculty with Ph.D. (6)	Other Full-time Faculty without Ph.D. (7)	Part-time Faculty (8)	Graduate Teaching Assist.[c] (9)
COMPUTER SCIENCE								
GENERAL EDUCATION COURSES								
D1. Computers and Society, Issues in CS								
D2. Intro. to Software Packages								
D3. Other CS General Education Courses								

[a] A majority of students receive the majority of their instruction via Internet, TV, correspondence courses, or other method where the instructor is NOT physically present.
[b] Do not include any dual-enrollments (see Section B).
[c] Sections taught independently by GTAs.

D. Computer Science Courses (Fall 2005) cont.

◆Cells left blank will be interpreted as zeros

Name of Course (or equivalent) (1)	Total distance-education enrollment[a] (2)	Total enrollment NOT in Col (2) and NOT dual enrollments[b] (3)	Number of sections corresponding to Column (3) (4)	Of the number in Column 4, how many sections are taught by: Tenured or Tenure-eligible Faculty (5)	Other Full-time Faculty with Ph.D. (6)	Other Full-time Faculty without Ph.D. (7)	Part-time Faculty (8)	Graduate Teaching Assist.[c] (9)
COMPUTER SCIENCE								
INTRODUCTORY CS COURSES								
D4. Computer Programming I (CS101 or 111)[d]								
D5. Computer Programming II (CS102 or 112 and 113)[d]								
D6. Discrete Structures for CS (CS105, 106, or 115)[d],								
D7. All other introductory Level CS courses								
INTERMEDIATE LEVEL								
D8. Algorithm Design and Analysis (CS210)[d]								
D9. Computer Architecture (CS220, 221, or 222)[d]								
D10. Operating Systems (CS225, 226)[d]								

[a] A majority of students receive the majority of their instruction via Internet, TV, correspondence courses, or other method where the instructor is NOT physically present.
[b] Do not include any dual-enrollments (see Section B).
[c] Sections taught independently by GTAs.
[d] Course numbers from CC2001.

D. Computer Science Courses (Fall 2005) cont.

◆**Cells left blank will be interpreted as zeros**

Name of Course (or equivalent) (1)	Total distance-education enrollment[a] (2)	Total enrollment NOT in Col (2) and NOT dual enrollments[b] (3)	Number of sections corresponding to Column (3) (4)	Of the number in Column 4, how many sections are taught by: Tenured or Tenure-eligible Faculty (5)	Other Full-time Faculty with Ph.D. (6)	Other Full-time Faculty without Ph.D. (7)	Part-time Faculty (8)	Graduate Teaching Assist.[c] (9)
COMPUTER SCIENCE								
INTERMEDIATE LEVEL CONT.								
D11. Net-centric Computing (CS230)[d]								
D12. Programming Language Translation (CS240)[d]								
D13. Human-Computer Interaction (CS250)[d]								
D14. Artificial Intelligence (CS260, 261, 262)[d]								
D15. Databases (CS270, 271)[d]								
D16. Social and Professional Issues in Computing (CS280)[d]								
D17. Software Development (CS290, 291, 292)[d]								
D18. All other intermediate Level CS courses								
UPPER LEVEL								
D19. All upper level CS Courses (numbered 300 or above in CC2001)								

[a] A majority of students receive the majority of their instruction via Internet, TV, correspondence courses, or other method where the instructor is NOT physically present.
[b] Do not include any dual-enrollments (see Section B).
[c] Sections taught independently by GTAs.
[d] Course numbers from CC2001.

E. Faculty Profile (Fall 2005)

Statistics Questionnaire

E1. Number of faculty in your department in fall 2005

NOTES for E1:

- In responding to questions in this section, use the same rules for distinguishing between full-time and part-time faculty that you used in sections C and D. Often, one easy way to distinguish between full-time and part-time faculty is to ask whether a given faculty member participates in the same kind of insurance and retirement programs as does your department chair. Part-time faculty are often paid by the course and do not receive the same insurance and retirement benefits as does the department chair.

- If your institution does not recognize tenure, please report departmental faculty who are permanent on line E1-(a) and report all other faculty on lines E1-(c), (d), or (e) as appropriate.

(a) Number of full-time tenured faculty (not including visitors or those on leave) in fall 2005 .... ☐ (1)

(b) Number of full-time tenure-eligible-but-not-tenured faculty (not including visitors or those on leave) in fall 2005 .... ☐ (2)

(c) Number of tenured or tenure-eligible faculty on leave in fall 2005 .... ☐ (3)

(d) Number of post-docs in your department in fall 2005 (where a postdoctoral appointment is a temporary position primarily intended to provide an opportunity to extend graduate training or to further research) .... ☐ (4)

(e) Number of full-time faculty in your department in fall 2005 not included in (a), (b), (c), or (d) and who hold visiting appointments .... ☐ (5)

(f) Number of full-time faculty in your department in fall 2005 who are not in (a), (b), (c), (d), or (e) ☐ (6)

(g) Number of part-time faculty in your department in fall 2005 .... ☐ (7)

E2. What is the expected (or average) teaching assignment for the tenured and tenure-eligible faculty reported in E1-(a), (b)? (If your institution does not recognize tenure, report on those faculty who are "permanent full-time.")

(a) Expected classroom contact hours per week for tenured and tenure-eligible faculty in fall 2005 .... ☐ (1)

(b) Expected classroom contact hours per week for tenured and tenure-eligible faculty last year in winter/spring term 2005 .... ☐ (2)

E. Faculty Profile (Fall 2005) cont.

E3. During fall 2005, how many faculty members are teaching the undergraduate statistics courses that you reported in Section C, above? .... ☐ (1)

E4. Of the faculty members reported in E3, how many had a masters degree or a doctoral degree in statistics or biostatistics as of 01 September, 2005?

Number with a doctoral degree in statistics/biostatistics.......... ☐ (1)

Number with a master's degree, but not a doctoral degree, in statistics/biostatistics .... ☐ (2)

E5. For the faculty members teaching statistics courses (number given in E3), what are the major fields of study for their highest earned degree? Complete the following table by showing the number of faculty belonging to each box.

HIGHEST DEGREE	Statistics (1)	Biostatistics (2)	Mathematics (3)	Mathematics Education (4)	Computer Science (5)	Social Science (6)	Education (7)	Other (8)
Doctorate (1)								
Masters (2)								
Other (3)								

F. Undergraduate Program (Fall 2005)

Statistics Questionnaire

F1. Please report the total number of your departmental majors who received their bachelors degrees from your institution between 01 July 2004 and 30 June 2005. Include joint majors and double majors[1] .... ☐ (1)

F2. Of the undergraduate degrees described in F1, please report the number who majored in each of the following categories. Each student should be reported only once. Include all double and joint majors[1] in your totals. Use "Other" category for a major in your department who does not fit into one of the earlier categories.

Area of Major	Male (1)	Female (2)
a) Statistics		
b) Biostatistics		
c) Actuarial Science		
d) Computer Science		
e) Joint[1] Statistics and Mathematics		
f) Joint[1] Statistics and (Business or Economics)		
g) Statistics Education		
h) Other		

F3. Does your department teach any upper division Computer Science courses?

Yes.......................... ☐ (1)

No............................ ☐ (2)

F4. Can a major in your department count some upper division Computer Science course(s) from some other department toward the upper division credit hour requirement for your departmental major?

Yes.......................... ☐ (1)

No............................ ☐ (2)

F5. Can a major in your department count some upper division Mathematics course(s) from some other department toward the upper division credit hour requirement for your departmental major?

Yes.......................... ☐ (1)

No............................ ☐ (2)

[1] A "double major" a student who completes the degree requirements of two separate majors, one in statistics and a second in another program or department. A "joint major" is a student who completes a single major in your department that integrates courses from statistics and some other program or department and typically requires fewer credit hours than the sum of the credit hours required by the two separate majors.

F. Undergraduate Program (Fall 2005) cont.

Statistics Questionnaire

F6. To what extent must majors in your department complete the following? Check one box in each row.

	Required of all majors (1)	**Required of some but not all majors** (2)	**Not required of any major** (3)
a) Calculus I			
b) Calculus II			
c) Multivariable Calculus			
d) Linear Algebra/Matrix Theory			
e) at least one Computer Science course			
f) at least one applied mathematics course (not including a, b, c, d above)			
g) a capstone experience (e.g., a senior project, a senior thesis, a senior seminar, or an internship)			
h) an exit exam (written or oral)			

F7. Many departments today use a spectrum of program-assessment methods. Please check all that apply to your department's undergraduate program-assessment efforts during the last six years.

(a) We conducted a review of our undergraduate program that included one or more reviewers from outside of our institution .... ☐ (1)

(b) We asked graduates of our undergraduate program to comment on and suggest changes in our undergraduate program .... ☐ (2)

(c) Other departments at our institution were invited to comment on the preparation that their students received in our courses .... ☐ (3)

(d) Data on our students' progress in subsequent statistics courses were gathered and analyzed .... ☐ (4)

(e) We have a placement system for first-year students and we gathered and analyzed data on its effectiveness .... ☐ (5)

(f) Our department's program assessment activities led to changes in our undergraduate program .... ☐ (6)

F. Undergraduate Program (Fall 2005) cont.

Statistics Questionnaire

F8. General Education Courses: Does your institution require all bachelors graduates to have credit for a quantitative literacy course as part of their general education requirements? Choose one of the following.

(a) Yes, all bachelors graduates must have such credit ☐ (1) → if (a), go to F9.

(b) Not (a), but all students in the academic unit to which our department belongs must have such credit[1] ☐ (2) → if (b), go to F9.

(c) neither (a) nor (b) ☐ (3) → if (c), go to F12.

F9. If you chose (a) or (b) in F8, is it true that all students (to whom the quantitative requirement applies) must fulfill it by taking a course in your department?

Yes........................... ☐ (1)

No............................. ☐ (2)

F10. Which courses in your department can be used to fulfill the general education quantitative requirement in F8?

(a) Any freshman course in our department ☐ (1) → go to F12.

(b) Only certain courses in our department ☐ (2) → go to F11.

F11. If you chose F10(b), which of the following departmental courses can be used to fulfill the general education quantitative requirement? Check all that apply.

Course	Can be used
a) Elementary Statistics (no calculus prerequisite)	
b) Probability and Statistics (no calculus prerequisite)	
c) Statistical Literacy/Statistics and Society	
d) a special general education course in our department not listed above	
e) some other course(s) in our department not listed above	

F12. Does your department or institution operate a statistics lab or tutoring center intended to give students out-of-class help with statistics problems?

Yes........................... ☐ (1) → If "Yes", go to F13.

No............................. ☐ (2) → If "No", go to F14.

[1] *For example, you would check F8(b) if students in the College of Fine Arts do not have a quantitative literacy requirement, and yet all students in the College of Science (to which our department belongs) must complete a quantitative literacy requirement.*

F. Undergraduate Program (Fall 2005) cont.

Statistics Questionnaire

F13. Please check all services available through the statistics lab or tutoring center mentioned in F12.

(a) Computer-aided instruction ☐ (1)

(b) Computer software such as computer algebra systems or statistical packages ☐ (2)

(c) Media such as video tapes, CDs, or DVDs ☐ (3)

(d) Tutoring by students ☐ (4)

(e) Tutoring by paraprofessional staff ☐ (5)

(f) Tutoring by part-time statistics faculty ☐ (6)

(g) Tutoring by full-time statistics faculty ☐ (7)

(h) Internet resources ☐ (8)

F14. Please check all of the opportunities available to your undergraduate statistics students.

(a) Honors sections of departmental courses ☐ (1)

(b) An undergraduate Statistics Club ☐ (2)

(c) Special statistics programs to encourage women ☐ (3)

(d) Special statistics programs to encourage minorities ☐ (4)

(e) Opportunities to participate in statistics contests ☐ (5)

(f) Special statistics lectures/colloquia not part of a statistics club ☐ (6)

(g) Outreach opportunities in local K-12 schools ☐ (7)

(h) Undergraduate research opportunities in statistics ☐ (8)

(i) Independent study opportunities in statistics ☐ (9)

(j) Assigned faculty advisers in statistics ☐ (10)

(k) Opportunity to write a senior thesis in statistics ☐ (11)

(l) A career day for statistics majors ☐ (12)

(m) Special advising about graduate school opportunities in statistical sciences ☐ (13)

(n) Opportunity for an internship experience ☐ (14)

(o) Opportunity to participate in a senior seminar ☐ (15)

F. Undergraduate Program (Fall 2005) cont.

Statistics Questionnaire

F15. Please give your best estimate of the percentage of your department's graduating majors from the previous academic year (2004-2005) in each of the following categories:

(a) who went into pre-college teaching .... [%] (1)

(b) who went to graduate school in the statistical sciences .... [%] (2)

(c) who went to professional school or to graduate school outside of the statistical sciences .... [%] (3)

(d) who took jobs in business, industry, government, etc. .... [%] (4)

(e) who had other post-graduation plans known to the department .... [%] (5)

(f) whose plans are not known to the department .... [%] (6)

F16. For fall 2005, how many students received credit for an introductory course in your department as a result of their score on the AP statistics examination?

Number receiving credit based on AP statistics exam .... [ ]

F17. During the last five years, has your department introduced any new courses or course options as a result of the statistics AP examination?

Yes........................... [ ] (1)

No............................. [ ] (2)

G. Pre-service Teacher Education in Statistics and Mathematics

G1. Does your institution offer a program or major leading to certification in some or all of grades K-8?

Yes............................ ☐ (1) ⟶ If "Yes", go to G2.

No.............................. ☐ (2) ⟶ If "No", go to G14.

G2. Do members of your department serve on a committee that determines what statistics and mathematics courses are part of that certification program?

Yes............................ ☐ (1)

No.............................. ☐ (2)

G3. Does your department offer a course or course-sequence that is designed specifically for the pre-service K-8 teacher certification program?

Yes............................ ☐ (1) ⟶ If "Yes", go to G4.

No.............................. ☐ (2) ⟶ If "No", go to G9.

G4. Are you offering more than one section of the special course for pre-service K-8 teachers in fall 2005?

Yes............................ ☐ (1) ⟶ If "Yes", go to G5.

No.............................. ☐ (2) ⟶ If "No", go to G8.

G5. Is there a designated departmental coordinator for your multiple sections of the special course for pre-service K-8 teachers in fall 2005?

Yes............................ ☐ (1) ⟶ If "Yes", go to G6.

No.............................. ☐ (2) ⟶ If "No", go to G8.

G6. Please choose the box that best describes the coordinator mentioned in G5.

(a) tenured or tenure-eligible .... ☐ (1)

(b) a postdoc[1] .... ☐ (2)

(c) a full-time faculty member not in (b) who holds a *visiting* appointment in your department ... ☐ (3)

(d) a full-time faculty member *without* a doctorate who is not in (a), (b), or (c) .... ☐ (4)

(e) a full-time faculty member *with* a doctorate who is not in (a), (b), (c), or (d) .... ☐ (5)

(f) a part-time faculty member .... ☐ (6)

(g) a graduate teaching assistant .... ☐ (7)

---

[1] *A postdoctoral appointment is a temporary position primarily intended to provide an opportunity to extend graduate education or to further research.*

G. Pre-service Teacher Education in Statistics and Mathematics cont.

Statistics Questionnaire

G7. Given that you offer multiple sections of the special course for pre-service K-8 teachers in fall 2005, is it true that all sections of that course use the same textbook?

Yes.......................... ☐ (1)

No............................ ☐ (2)

G8. During which year of their college careers are your pre-service K-8 teachers most likely to take your department's special course for pre-service K-8 teachers? If you have two such courses, consider only the first in responding to this question. Please check just one box.

a) Freshman	
b) Sophomore	
c) Junior	
d) Senior	

G9. Are there any sections of other courses in your department (i.e., other than the special course for K-8 teachers mentioned in G3) that are restricted to or designated for pre-service K-8 teachers?

Yes.......................... ☐ (1)

No............................ ☐ (2)

Special instructions for questions G10, G11, G12, and G13: Many institutions have different certification requirements for pre-service elementary teachers preparing for early grades and those preparing for later grades. However, there is no nationwide agreement on which grades are "early grades" and which are "later grades" except that grades 1 and 2 are "early" and grades 6 and above are usually considered "later grades", and that is how we use the terms in the next four questions.

G10. Does your K-8 pre-service program have different requirements for students preparing to teach early grades and for those planning to teach later grades?

Yes.......................... ☐ (1) ⟶ If "Yes", go to G12.

No............................ ☐ (2) ⟶ If "No", go to G11.

G11. Given that your pre-service K-8 teacher education program *does not* distinguish between preparing for certification in early and later grades, how many courses are all pre-service elementary teachers required to take in your department (including general education requirements, if any)?

☐ Now go to G13 and put all of your answers into column (3).

G12. Given that your pre-service K-8 teacher education program *does* distinguish between preparing for certification to teach early grades and later grades, how many courses are pre-service K-8 teachers required to take in your department (including general education requirements, if any)?

(a) Number of courses required for early grade certification .... ☐ (1)

(b) Number of courses required for later grade certification .... ☐ (2)

Now go to G13 and put all of your answers into columns (1) and (2).

G. Pre-service Teacher Education in Statistics and Mathematics cont.

G13. In your judgement, which three of the following courses in your department are most likely to be taken by pre-service K-8 teachers? If your program does NOT distinguish between early and later grades, please use the column (3) for your answers and check a total of only three boxes. If your program DOES distinguish between early and later grades, check exactly three boxes in each of columns (1) and (2) and ignore column (3).

Courses	Three most likely for early grade certification (1)	Three most likely for later grade certification (2)	Three most likely given that we do not distinguish between early & later grade (3)
a) A multiple-term course designed for K-8 teachers			
b) A single-term course designed for K-8 teachers			
c) Introductory Statistics (in line C1, above)			
d) Probability and Statistics (in line C2, above)			
e) Statistical Literacy/Statistics and Society (in line C3, above)			

G14. Does your department offer any courses that are part of a graduate degree in mathematics/statistics education?

(a) No .... ☐ (1)

(b) Yes, and the degree is granted through our department .... ☐ (2)

(c) Yes, and the degree is granted through some other department or unit in our institution .... ☐ (3)

Thank you for completing this questionnaire. We know it was a time-consuming process and we hope that the resulting survey report, which we hope to publish in spring 2007, will be of use to you and your department. Please retain a copy of this questionnaire in case questions arise.

Appendix VII

Tables of Standard Errors

Standard error tables for S.1 to S.4.

Table S.1	Four-year	SE	Two-year	SE
Mathematics	1607	45	1580	72
Statistics	260	15	117	9
Computer Science	59	10	na	na
Total	1925	51	1697	75

Table S.3	Four-year	SE
Semester	91%	3
Trimester	1%	1
Quarter	6%	2
Other	2%	2

Table S.2	Mathematics	SE	Statistics	SE	TYC	SE
Precollege	201	19	na	na	965	51
Introductory	706	29	na	na	321	24
Calculus level	587	24	na	na	108	6
Advanced	112	6	na	na	0	
Total Mathematics	1607	45	na	na	1580	72
Statistics						
Elementary	148	14	54	4	117	9
Upper level	34	3	24	2	0	
Total Stat	182	15	78	5	117	9
CS						
Lower	44	8	2	1		
Middle	8	1	0			
Upper	5	1	0			
Total	57	10	2	1	0	
Grand Total	1845	51	80	5	1697	75

Table S.4		SE
Math	12316	786
Math Ed	3369	476
Statistics	527	68
Actuarial	499	91
OR	31	26
Math & CS	719	185
Math & Stat	203	97
Math & Econ	214	65
Other	954	200
Total M & S degrees	18833	1065
Women	8192	541
CS degrees	2603	573
Women	465	110
Total degrees	21437	1280
Women	8656	560

Standatd Error Table for S.5 and S.6

Table S.5	TTE	SE	OFT	SE	PT	SE	GTA	SE	Unkn	SE	Enroll	SE
Math Depts												
Math courses	46	2	21	1	20	1	8	1	5	1	1607	45
Stat courses	52	3	24	4	19	3	2	1	2	1	182	15
CS courses	70	5	11	4	11	3	0	0.3	7	4	57	10
All Math Dept	48	2	21	1	19	1	7	0.6	5	0.7	1845	51
Stat Depts												
All Stat courses	47	3	23	2	7	1.4	11	3	13	6	80	5
TYC												
All courses	56	1			44	1					1697	75
Table S.6												
Math, Precollege	9	2	25	3	46	4	14	3	5	1.8	199	19
Math, Intro	31	2	25	2	28	2	10	1	6	1	695	29
Math, Calculus	61	2	17	1	9	1	7	1	6	1	583	24
Math, Upper	84	2							16	2	112	6
Math, Elem Stat	49	4	16	3	28	3	3	1	3	1	145	14
Math Adv Stat	59	8							41	8	34	3
Math, CS Lower	63	6	12	4	17	5	1	0.4	8	4	43	8
Stat Dept Elem	25	4	21	3	13	3	20	5	21	10	53	4
Stat Dept Upper	74	3							26	3	23	2
TYC, All	56	1.5			44	1.5					1739	77

Standard Error Table for S.7

Table S.7	TTE	SE	OFT	SE	PT	SE	GTA	SE	Unkn	SE	Enroll	SE	Avg Sect	SE
MS Calc 1														
Lect/Recit	52	4	27	4	9	3	5	2	7	3	80	8	46	4
Reg < 31	77	3	10	2	5	1	5	1	3	2	63	7	22	1
Reg > 30	49	4	17	3	10	2	16	3	8	3	58	6	36	1
MS Calc I Total	63	2	17	2	7	1	8	1	5	1	201	10	32	1
MS Calc II														
Lect/Recit	58	5	24	4	5	2	5	3	8	3	36	4	50	5
Reg < 31	80	3	8	2	3	1	7	2	2	1	25	3	22	1
Reg > 30	51	5	19	3	11	2	11	3	7	3	24	3	36	1
MS Calc II Total	66	3	15	2	6	1	8	1	5	1	85	5	33	1
Total I & II	64	2	16	1	7	1	8	1	5	1	286	13	32	1
TYC	Full-time				Part-time									
MS Calc I	88	2			12	2					49	3	22	1
MS Calc II	87	2			13	2					19	1	18	1
Total I&II	87	1			13	1					68	4	21	1

Standard Error Table for S.8

Table S.8	TTE	SE	OFT	SE	PT	SE	GTA	SE	Unkn	SE	Enroll	SE	Avg Sect	SE
NMS Calc 1														
Lect/Recit	19	5	33	7	9	4	9	3	30	9	28	4	64	7
Reg < 31	40	6	18	4	20	6	14	4	8	4	30	7	23	2
Reg > 30	36	5	24	4	26	4	13	4	2	1	50	6	44	3
NMS Calc I Total	35	4	23	3	21	3	13	3	9	3	107	9	37	2
NMS Calc II	33	7	26	6	23	5	17	4	1	1	10	2	46	5
NMSC I & II	35	3	23	3	21	3	13	3	8	3	118	9	38	2
TYC	Full-time				Part-time									
NMS Calc I	73	4			27	4					20	2	23	1
NMS Calc II	66	9			34	9					1	0.2	21	2
Total I & II	72	4			28	4					21	2	23	1

Standard Error Table for S.9 and S.10

S.9 Math Dept	T/TE	SE	OFT	SE	PT	SE	GTA	SE	Ukn	SE	Enroll	SE	Av Sect	SE
Elem Stat														
Lecture/recitation	30	7	27	8	34	8	2	1	7	3	12	4	32	6
Regular <31	56	5	12	3	28	5	2	1	2	1	54	11	24	1
Regular >30	49	4	18	4	22	4	6	2	5	3	56	9	40	1
Course total	51	4	16	3	27	3	3	1	4	1	122	13	31	1
Prob & Stat														
Course total	29	8	24	8	44	12	1	1	2	1	18	5	30	2
Total All Elem. P & S	48	4	17	3	29	3	3	1	3	1	140	13	31	1
TYC														
ElemStat	65	2			35	2					101	8	26	1
S.10 Stat Depts														
Elem Stat														
Lecture/recitation	19	4	27	4	16	4	17	4	21	11	28	3	82	13
Regular <31	33	6	18	6	7	3	23	9	20	6	1	0.3	12	5
Regular >30	33	8	14	3	18	4	30	13	5	2	13	3	50	4
Course total	26	4	21	3	16	3	22	6	15	6	42	3	63	7
Prob & Stat														
Course total	34	8	38	7	0	0	16	5	13	5	2	0.6	68	12
Total All Elem. P & S	26	4	22	3	15	3	22	5	15	6	44	3	64	6

Standard Error Table for S.11 and S.12.

Table S.11	Calculators	SE	Writing	SE	Computer	SE	On-line	SE	Group	SE	Enroll	SE	Avg Sect	SE
MS Calc I														
Lecture/recitation	48	7	13	5	24	6	6	2	12	5	80	8	46	4
Regular <31	58	7	16	4	20	5	2	1	8	2	63	7	22	1
Regular >30	43	6	10	4	20	5	6	2	13	4	58	6	35	1
Course total	51	4	13	3	21	3	4	1	10	2	201	10	32	1
MS Calc II														
Lecture/recitation	38	6	9	4	20	5	4	2	7	4	36	4	50	5
Regular<31	47	8	13	5	24	5	2	1	5	2	25	3	21	1
Regular >30	42	7	5	3	18	5	5	2	5	3	24	3	36	1
Course total	43	5	9	3	21	3	3	1	6	2	85	5	33	1
Total MS Calc I & II	49	4	12	2	21	3	4	1	9	2	285	13	32	1
TYC														
MS Calc I	79	4	19	3	20	3	5	1	19	3	49	3	22	1
MS Calc II	81	4	18	4	30	4	25	4	7	2	19	1	18	1
Total MS Calc I & II	80	4	18	3	23	3	5	1	21	3	68	4	21	1
Table S.12														
NMS Calc I														
Lecture/recitation	60	8	7	4	8	3	7	3	4	2	28	4	64	7
Regular <31	63	9	1	0.4	5	2	4	2	1	1	30	7	23	2
Regular >30	37	7	7	3	4	2	5	2	6	2	50	6	44	3
Course total	53	6	4	1	5	1	5	2	3	1	108	9	37	2
TYC														
NMS Calc I	77	5	14	4	9	3	3	1	14	3	20	2	23	1

Standard Error Tables for S.13, S.14, S.15, and S.16.

Table S.13	Calculators	SE	Writing	SE	Computers	SE	On-line	SE	Groups	SE	Enroll	SE	Avg sect	SE
Math Depts														
Lecture/recitation	42	18	48	17	83	6	0	0	38	19	12	4	32	6
Regular <31	30	9	30	10	56	10	4	2	19	10	54	11	24	1
Regular >30	44	9	21	8	46	9	2	2	5	2	56	9	40	1
Course total	36	7	28	6	55	7	3	1	16	6	122	13	31	1
Stat Depts														
Lecture/recitation	9	4	42	6	59	6	26	9	30	7	28	3	82	13
Regular <31	0	0	19	11	85	7	30	14	16	10	1	0.2	12	5
Regular >30	1	1	57	10	52	11	1	1	22	10	13	3	50	4
Course total	5	2	46	6	58	6	16	6	26	6	42	3	63	7
TYC Course total	73	5	44	5	45	5	10	3	24	4	101	8	26	1

Table S.14		
Math Depts	2005	
FT faculty	21885	595
PT faculty	6536	338
Stat Depts		
FT faculty	946	8
PT faculty	112	3
TYC		
FT faculty	9403	425
PT faculty	18227	900

Table S.15	Total	SE	T&TE	SE	OFT	SE	Posdoc	SE
Full-time	21885	595	17256	464	4629	177	819	25
with PhD	18071	400	15906	363	2165	79	813	24
Doctoral Stat								
FT faculty	946	8	783	7	163	3	51	2
with PhD	915	8	781	7	133	3	51	2
Total M & S	22831	595	18039	464	4792	177	870	25
TYC	Total FT	SE	FT perm	SE	FT temp	SE		
FT faculty	9402	425	8793	398	610	163		
Grand Total	32251	na	26837	na	5415	na	874	na

Table S.16	PhD = 16%	SE=2	MA = 82%	SE=2	BA= 2%	SE=1

Standard Error Tables for S.17, S.18, S.19, S.20, and S.21.

Table S.17	Total	SE	T	SE	TE	SE	OFT	SE	PD	SE
FT faculty	21885	595	12874	320	4382	193	4629	177	819	25
#women	5641	239	2332	111	1250	72	2059	111	191	24
Doc Stat Depts										
FT faculty	946	8	604	5	179	3	163	3	51	2
#women	212	3	79	1	66	1	66	2	16	1
TYC	All	SE	FT < 40	SE						
FTfaculty	8793	398	2326	169						
#women	4387	256	1148	102						
Table S.18	<30	30-34	35-39	40-44	45-49	50-54	55-59	60-64	65-69	>69
Ages, Math total %	2	9	13	14	13	14	14	13	6	2
SE	0	0	0	1	0	0	1	0	0	0
TYC	<30	30-34	35-39	40-44	45-49	50-54	55-59	>59	Avg	
Perm fac ages %	5	8	12	13	15	18	17	11	47.8	
SE	1	1	1	1	1	1	1	1	0.4	
Table S.19										
Ages Stat total %	5	15	15	12	12	12	12	9	6	2
SE	1	2	1	2	2	2	2	2	1	1
Table S.20	Asian	Black	Hisp.	White	Other					
FT men %	9	2	2	59	2					
SE men	0	0	0	1	0					
FT women %	3	1	1	21	1					
SE women	1	0	0	1	0					
Table S.21										
FT men %	18	1	1	55	2					
SE men	1	0	0	1	0					
FT women %	7	1	0	16	0					
SE women	2	1	0	2	0					

Standard Error Tables for S.22 and S.23.

Table S.22	D&Ret	SE	Number	SE
PhD Math	139	5	5652	0
MA Math	140	23	3563	92
BA Math	219	51	8041	455
Total Math	499	56	17256	464
Total Doc Stat	14	2	783	7
TYC total	292	56	8793	398

Table S.23	< 6	6	7-8	9-11	12	>12	Avg
Math Doc Fall	24	42	25	5	2	2	6.3
SE Math Doc	4	5	4	2	2	2	0.2
Math MA Fall	0	4	5	44	48	0	10.3
SE Math MA	0	2	3	8	8	0	0.3
Math BA Fall	0	0	3	30	53	14	11.3
SE Math BA	0	0	2	6	7	5	0.3
Stat Doc Fall	48	45	4	0	4	0	5.3
SE Stat Doc	6	6	3	0	3	0	0.3

Standard Error Tables for SP.1, SP.2, SP.3, and SP.4.

Table SP.1			Table SP.3					
	Have K-8	SE	Committee	SE	Special course	SE	Designate	SE
Math PhD	78	4	58	5	81	4	31	5
Math MA	92	4	86	6	96	3	45	9
Math BA	88	4	82	6	85	5	21	5
Math total	87	3	80	4	86	4	25	4
Stat PhD	40	5	29	9	11	7	0	0
Stat MA	59	14	25	17	33	21	0	0
Stat total	44	5	28	8	16	7	0	0

Table SP.2								
PS elem = 30	PS mid = 19	PS sec = 3	IS elem = 16	IS mid = 15	IS sec = 2	CSw elem = 19	CSw mid = 14	CSw sec = 6
S E = 5	S E = 4	S E= 1	S E=4	S E=4	S E=1	S E= 4	S E=3	S E= 2

Table SP.4			
Coord = 38	Special = 11	In dept = 9	Out dept = 10
S E=5	S E=3	S E=3	S E=3

Standard Error Table for SP.5

Table SP.5			
Several tracks	SE	Unified track	SE
44	5	56	5

	Early	SE	Later	SE	Unified	SE
0 req	11	6	16	7	4	3
1 req	17	6	7	4	26	7
2 req	31	7	5	4	37	7
3 req	17	5	2	1	22	6
4 req	17	7	11	6	11	5
5 or more	8	4	58	8	0	0
	Avg req		Avg req		Avg req	
Math PhD	3.3	0.5	5.5	0.8	2.4	0.2
Math MA	3.3	0.6	6.9	0.8	2.5	0.2
Math BA	2.5	0.4	5.3	0.9	2	0.2
All Math	2.7	0.3	5.6	0.7	2.1	0.2

Standard Error Table for SP.6

Table SP.6	PhD, early	SE	MA, early	SE	BA, early	SE	PhD, later	SE	MA, later	SE	BA, later	SE
Multi-term	59	7	70	11	64	10	28	7	47	12	38	9
Single term	21	6	37	12	33	10	16	6	10	7	12	7
College algebra	41	8	40	12	56	11	21	7	40	12	23	9
Precalculus	15	5	6	6	46	11	13	4	13	8	15	8
Math Mod	5	2	0	0	0	0	8	5	0	0	0	0
Lib Arts	28	7	30	11	25	10	8	5	7	6	2	1
Finite	23	7	7	6	15	8	10	4	7	6	8	6
History	5	4	0	0	0	0	31	7	23	10	18	9
Calculus	21	6	6	6	12	7	64	8	50	12	77	9
Geometry	10	5	24	10	0	0	43	8	47	12	53	11
Elem Stat	31	8	26	11	27	10	41	8	44	12	55	11

Standard Error Tables for SP.7, SP.8, SP.9, SP.10, and SP.11.

Table SP.7	PhD Math	SE	MA, Math	SE	BA, Math	SE
Same text	97	2	91	5	100	0
Coordinator	90	3	82	7	69	10
a) T/TE	65	7	81	7	68	12
b) Postdoc	0	0	0	0	0	0
c) Visitor	2	2	9	6	0	0
d) FT PhD not a,b,c	28	7	9	6	32	12
e) FT, not a,b,c,d	2	2	0	0	0	0
f) PT	3	2	0	0	0	0
g) GTA	0	0	0	0	0	0

Table SP.10	No program	In dept	Other dept
Math, PhD	43	29	28
SE	5	4	5
Math, MA	21	35	44
SE	7	8	8
Math, BA	89	2	9
SE	3	1	3
Stat, PhD	58	23	19
SE	6	6	5
Stat, MA	56	29	15
SE	14	13	10

Table SP.8	Fresh	SE	Soph	SE	Junior	SE	Senior	SE
Math, PhD	23	5	45	6	27	6	5	2
Math, MA	43	9	36	9	17	7	4	3
Math BA	23	7	64	8	13	5	0	0

Table SP.9		SE		SE		SE
Req course	58	5	69	8	41	7
Included in other	22	5	25	7	43	7
No history req	19	4	7	4	16	6

Table SP.11	Placement	Required	Discuss	Mandatory	Assess	Department	ETS	ACT	Professional	Other
Percent	96	97	90	88	81	11	22	51	12	25
SE	3	2	2	4	4	4	4	5	3	4

Standard Error Tables for SP.12 and SP.13.

Table SP.12	Math	SE	Stat	SE	TYC	SE
PhD, Math	96	2	79	5	xx	xx
MA, Math	91	4	85	10	xx	xx
BA, Math	88	4	na		xx	xx
All, Math	89	3	80	5	95	3

Table SP.13	CAI	SE	Software	SE	Media	SE	Students	SE	Paraprof	SE	PT fac	SE	FT fac	SE	Internet	SE
PhD, Math	33	5	48	5	20	4	98	1	29	4	22	4	27	5	38	5
MA, Math	33	8	55	8	40	8	96	3	43	9	23	7	28	7	37	8
BA, Math	25	6	33	7	27	6	99	1	20	5	9	3	19	5	21	5
All, Math	27	4	38	5	27	5	98	1	24	4	13	3	21	4	25	4
PhD Stat	44	6	68	6	13	3	96	3	13	5	9	4	17	4	27	6
MA Stat	51	15	83	12	17	12	100	0	17	12	0	0	17	12	69	13
All Stat	46	6	71	5	14	4	97	2	14	5	7	3	17	4	37	5
All TYC	75	4	72	5	68	5	94	2	67	5	48	5	51	5	77	4

Standard Error Tables for SP.14 and SP.15.

Table SP.14 / Table SP.15	Honors	Club	Women	Minorities	Contests	Colloquia	Outreach	REU	Ind. Stud.	Advisor	Thesis	Career	Grad Sch	Intern	Sen Sem
PhD, Math	70%	88%	15%	10%	92%	70%	51%	90	95	85	62	24	49	47	39
SE	5	3	3	2	3	4	5	3	2	3	5	4	5	5	5
MA, Math	44	92	21	23	68	71	63	74	91	97	53	15	61	55	46
SE	8	4	7	7	8	7	8	7	4	3	8	5	8	8	8
BA, Math	18	66	4	6	62	37	26	54	79	88	48	10	45	35	38
SE	5	6	2	3	7	7	6	7	6	4	7	3	6	6	7
All Math	28	72	8	8	67	46	34	62	83	89	50	12	47	39	39
SE	4	5	2	2	5	5	4	5	4	3	5	3	5	5	5
PhD, Stat	27	27	0	7	22	47	11	60	62	73	27	15	56	47	15
SE	6	5	0	3	5	6	4	6	6	5	5	5	6	6	5
MA, Stat	41	29	0	0	29	44	15	59	100	85	44	15	59	71	29
SE	14	13	0	0	13	14	10	14	0	10	14	10	14	13	13
All Stat	30	27	0	6	23	46	12	60	70	76	31	15	57	52	18
SE	5	5	0	3	5	6	4	6	5	5	5	4	6	5	5
All TYC	24	22	7	15	37	6	25	9	38	40	na	na	na	na	na
SE	4	4	2	3	4	2	4	3	5	5	na	na	na	na	na

Standard Error Tables for SP.16 and SP.17.

Table SP.16	Math, 4-yr			TYC			Stat, 4-yr		
Depts with D.En.	14%			50%			8%		
SE	3			5			3		
Dual Enrollments	Math spring 05	Math fall 05	Math Other, fall 05	TYC spring 05	TYC fall 05	TYC Other, fall 05	Stat spring 05	Stat, fall 05	Stat Other fall 05
College algebra	2673	8046	201000	9913	11362	206000	na	na	na
SE	1424	6138	17200	3941	3176	20000			na
Precalculus	2944	597	93000	14650	13801	58000	na	na	na
SE	1650	367	8900	5678	5636	7000			na
Calculus I	5540	8490	201000	8218	11188	51000	na	na	na
SE	2374	3320	9600	2290	4143	3000			na
Statistics	340	981	124000	3648	2440	111000	1563	1295	43000
SE	330	900	13800	1047	937	8000	844	812	3700
Other	3470	723	na	5452	3045	na	0	0	na
SE	1374	405		1988	866	na	0	0	na
Dept. Control	Never	Sometimes	Always	Never	Sometimes	Always	Never	Sometimes	Always
Textbook	41%	15%	44%	14%	12%	74%	36%	30%	34%
SE	12	11	11	5	4	6	19	16	18
Syllabus	2%	6%	92%	4%	7%	89%	36%	0%	64%
SE	1	4	4	2	4	4	19	0	19
Final exam	40%	30%	30%	36%	28%	37%	100%	0%	0%
SE	12	12	10	7	6	7	0	0	0
Instructor	32%	20%	48%	35%	13%	52%	36%	0%	64%
SE	12	10	12	7	5	7	19	0	19
Teaching evals			16%			64%			0%
SE			7			9			0

Table SP.17						
Assign own	4-yrMath = 4%	S E=2	TYC = 12%	S E= 3	4-yr Stat = 0%	S E=0
Number enrolled	2874	SE = 1929	2008	SE = 736	0	SE = 0

Standard Error Table for SP.18.

Table SP.18	PhD, Math	SE	MA, Math	SE	BA, Math	SE	PhD, Stat	SE	MA, Stat	SE
Quant. Requirement	87%	4	98%	2	91%	4	86%	4	88%	7
In department only	51	5	68	7	61	7	8	3	0	0
Any freshman course	26	5	28	7	32	7	27	6	17	12
Only certain courses	74	5	72	7	69	7	73	6	83	12
Departmental courses										
Coll. alg./Precalculus	56	6	61	9	62	9	na		na	
Calculus	97	3	87	6	86	6	na		na	
Math models	23	5	11	6	13	6	na		na	
Prob/Stat	55	6	60	10	66	8	94	3	60	17
Stat literacy	na		na		na		27	7	20	14
Special gen ed	52	6	73	9	55	9	0	0	0	0
Other courses	50	6	71	9	57	9	33	7	20	14

Standard Error Tables for SP.19 and SP.20.

Table SP.19	PhD, all	SE	MA, all	SE	BA, all	SE	PhD, some	SE	MA, some	SE	BA, some	SE	PhD, none	SE	MA, none	SE	BA, none	SE
Mod Alg I	24	4	48	8	56	7	59	5	42	8	36	7	18	4	10	5	8	4
> 1 Algebra	5	2	8	4	8	4	40	5	28	7	17	5	55	5	63	7	75	6
Analysis I	36	4	39	8	46	7	49	5	54	8	29	7	15	4	7	4	25	6
> 1 Analysis	10	2	4	3	8	4	49	5	36	8	20	6	41	5	60	8	71	7
CS	55	5	76	7	64	7	27	5	16	6	14	5	18	4	8	4	22	6
Stat	32	5	56	8	32	6	40	5	32	8	32	6	28	5	11	5	35	7
Appl Math	16	4	23	7	21	6	52	5	41	8	25	7	32	5	36	8	54	8
Capstone	27	5	52	8	59	7	23	4	13	5	8	4	50	5	35	7	33	7
Exit exam	8	3	8	4	29	7	4	2	16	6	3	3	88	3	76	7	68	7

Table SP.20	PhD, Stat, all	SE	MA, Stat, all	SE			PhD, Stat, some	SE	MA, Stat, some	SE			PhD, Stat, none	SE	MA, Stat, none	SE		
Calc I	92	4	86	8			4	3	0	0			4	3	14	8		
Calc II	87	4	86	8			4	3	0	0			8	4	14	8		
Multivar Calc	78	5	51	15			9	4	17	12			13	4	31	13		
Lin Alg	84	4	69	13			3	2	0	0			13	4	31	13		
CS	72	5	86	8			16	4	0	0			12	4	14	8		
Appl Math	24	6	14	8			12	4	17	12			64	7	69	13		
Capstone	34	6	51	15			9	4	17	12			57	7	31	13		
Exit exam	0	0	0	0			0	0	17	12			100	0	83	12		

Standard Error Table for SP.21.

Table SP.21	PhD, Math	MA, Math	BA, Math	PhD, Stat	MA, Stat
Teach own CS	17%	25%	42%	4%	29%
	4	5	7	3	13
Allow other CS	69	31	22	55	100
	5	7	6	6	0
Teach own stat	64	94	87	na	na
	4	4	5		
Allow other stat	55	12	15	na	na
	5	5	5	na	na
Allow udiv math	na	na	na	66	86
	na	na	na	6	8

Standard Error Tables for SP.22 and SP.22 contd.

Table SP.22	All Math 04-5&05-6	SE	PhD, Math	SE	MA, Math	SE	BA, Math	SE
Modern Algebra I	61%	5	86%	4	87%	6	52%	7
Modern Algebra II	21	3	40	5	40	8	15	4
Number Theory	37	4	61	5	61	8	29	5
Combinatorics	22	4	55	5	38	8	14	5
Actuarial Mathematics	11	2	24	4	23	7	6	3
Foundations/Logic	11	2	27	4	16	6	7	3
Discrete Structures	14	3	27	4	22	7	10	4
History of Mathematics	35	4	43	5	68	8	28	5
Geometry	55	5	81	4	89	5	44	6
Math for secondary teachers	37	5	41	5	50	8	35	6
Adv Calculus/ Real Analysis I	66	5	95	3	86	6	57	6
Adv Calculus/Real Analysis II	26	4	62	5	44	8	17	5
Adv Math Engin/Phys	16	3	50	5	28	7	7	4
Advanced Linear Algebra	19	3	52	5	42	8	9	3
SP.22 contd								
Vector Analysis	9%	3	21%	4	6%	4	7%	3
Advanced Differential Equations	13	2	45	5	28	7	5	3
Partial Differential Equations	19	3	57	5	29	7	11	4
Numerical Analysis I and II	47	5	83	4	76	7	36	6
Applied Math/Modeling	26	4	48	5	47	8	18	5
Complex Variables	37	4	80	4	53	8	26	5
Topology	32	4	61	5	33	8	26	5
Mathematics of Finance	8	2	24	4	8	4	5	3
Codes & Cryptology	8	2	17	3	8	4	7	3
Biomathematics	8	2	24	4	9	5	4	3
Intro to Operations Research	12	3	17	4	20	6	10	4
Intro to Linear Programming	6	1	19	4	21	7	1	1
Math senior seminar/Ind study	45	5	61	5	48	8	42	6

Standard Error Table for SP.23

Table SP.23	All Math Depts	PhD Math	MA Math	BA Math	All Stat Depts	PhD Stat	MA Stat
Math Stat	38%	52%	63%	31%	76%	73%	88%
SE	5	4	8	6	4	5	7
Probability	51	72	69	43	86	90	73
SE	5	4	8	6	4	4	12
Stochastic Pr	6	21	13	2	43	42	44
SE	2	4	6	2	5	6	14
App stat analysis	13	26	32	7	65	63	73
SE	2	4	8	3	5	6	12
Exp design	6	14	23	2	54	49	73
SE	1	3	7	1	5	6	12
Reg & Corr	6	20	12	3	62	55	88
SE	2	4	5	2	5	6	7
Biostatistics	4	11	13	2	25	28	15
SE	1	3	6	2	5	5	10
Nonparametric	2	6	8	0	38	33	59
SE	1	2	4	0	5	6	14
Cat data analysis	1	5	3	1	21	19	29
SE	0.6	2	3	0.5	4	4	13
Survey design	4	13	8	1	49	43	73
SE	1	3	4	0.7	5	6	12
Stat software	3	11	7	0.5	43	35	73
SE	0.7	3	4	0.5	5	6	12
Data mgmt	0	0	0	0	5	6	0
SE	0	0	0	0	2	3	0
Sen sem/Ind study	3	8	8	0.5	41	36	59
SE	1	3	4	0.5	5	5	14

Standard Error Tables for SP.24 and SP.25.

Table SP.24	PhD, Math	SE	MA, Math	SE	BA, Math	SE	PhD, Stat	SE	MA, Stat	SE
Pre-college teaching	16%	2	44%	4	32%	4	1%	0.5	0%	0
Grad or profess school	21	2	16	2	19	2	18	4	29	2
Jobs in bus, gov, etc.	19	2	21	3	29	3	16	4	36	6
Other known plans	4	1	1	0.6	2	0.7	0	0	6	2
Plans unknown	39	4	18	4	17	4	65	8	28	5
Table SP.25										
Outside reviewers	47%	5	45%	8	29%	6	37%	7	59%	14
Survey graduates	62	5	81	7	74	7	54	7	71	13
Other departments	51	5	41	8	35	7	29	7	56	14
Student progress	45	5	52	8	38	7	30	7	56	14
Eval placement	72	5	72	8	51	7	5	3	15	10
Change program	76	4	72	7	76	6	69	7	29	13

Standard Error Table for E.1

Table E.1	PhD,Math	MA,Math	BA,Math	Total, Math	PhD, Stat	MA, Stat	Total, Stat	Total M&S
Men, Math	4112	1350	3358	8820				8820
SE	337	213	374	547				547
Women, Math	2282(36%)	1027(43%)	2482(43%)	5791(40%)				5791(40%)
SE	233(1.4%)	183(2.9%)	336(2.5%)	448(1.3%)				448(1.3%)
Total Math	6393	2377	5839	14610				14610
SE	540	371	653	925				925
Men, Math Ed	296	401	645	1341				1341
SE	48	91	187	213				213
Women, Math Ed	470(61%)	628(61%)	930(59%)	2028(60%)				2028(60%)
SE	70(2.5%)	161(3.9%)	231(4.3%)	290(2.4%)				290(2.4%)
Total Math Ed	766	1029	1575	3369				3369
SE	111	239	396	476				476
Men, Stat	64	44	17	125	237	120	357	482
SE	16	22	11	30	44	44	62	69
Women, Stat	69(52%)	41(48%)	6(25%)	116(48%)	184(44%)	73(38%)	257(42%)	373(44%)
SE	21(5%)	24(5%)	5(16%)	32(4%)	35(2.2%)	24(3.6%)	43(2.0%)	53(2.0%)
Total Stat	133	85	23	241	421	193	614	855
SE	34	45	14	58	77	67	102	117
Men, CS	413	314	1412	2139				2139
SE	183	158	423	487				487
Women, CS	58(12%)	72(19%)	335(19%)	465(18%)				465(18%)
SE	27(1%)	35(2.9%)	101(3.6%)	110(2.5%)				110(2.5%)
Total CS	471	386	1747	2603				2603
SE	209	191	499	573				573
Total,Men	4884	2109	5431	12424	237	120	357	12780
SE	384	294	672	827	44	44	62	830
Total, Women	2879(37%)	1768(46%)	3752(41%)	8399(40%)	184(44%)	73(38%)	257(42%)	8656(40%)
SE	242(1.3%)	273(2.4%)	422(2.4%)	558(1.3%)	35(2.2%)	24(3.6%)	43(2.0%)	560(1.2%)
Total	7763	3877	9183	20823	421	193	614	21437
SE	589	535	998	1276	77	67	102	1280

Standard Error Table for E.2

Table E.2	PhD,Math	MA, Math	BA, Math	Total, Math	PhD, Stat	MA, Stat	Total, Stat
Precollege	55	60	87	201			
SE	7	10	14	19			
Intro	269	190	248	706			
SE	17	11	21	29			
Calculus	345	88	154	587			
SE	17	8	14	24			
Adv Math	52	24	36	112			
SE	3	3	4	6			
Total Math	720	362	525	1607			
SE	26	18	32	45			
Elem Stat	30	32	86	148	42	13	54
SE	4	6	12	14	4	3	4
Upper Stat	15	9	10	34	20	3	23
SE	2	2	2	3	2	0.5	2
Total Stat	44	42	96	182	62	16	78
SE	4	7	12	15	4	3	5
Lower CS	3	11	30	44	0	1	2
SE	1	4	7	8	0.1	0.2	1
Middle CS	1	1	6	8	0	0	0
SE	0.6	0.3	1	1	0	0	0.2
Upper CS	1	1	3	5	0	0	0
SE	0.5	0,3	1	1	0	0.2	0
Total CS	5	13	39	57	0	2	2
SE	2	4	9	10	0.2	1	1
Total all	769	417	659	1845	62	18	80
SE	26	20	39	51	4	3	5

Standard Error Table for E.3

Table E.3	PhD, Math	MA, Math	BA, Math	Total, Math	PhD, Stat	MA, Stat	Total, Stat
Precoll, Math	1363	1902	3862	7126			
SE	174	305	581	679			
Intro, Math	5518	5543	9895	20955			
SE	454	391	812	1009			
Calculus	7696	3237	7388	18321			
SE	356	275	584	738			
Adv Math	2625	1622	3507	7754			
SE	119	150	369	416			
Total Math	17202	12303	24652	54157			
SE	719	724	1341	1685			
Elem Stat	629	924	3191	4744	696	186	882
SE	104	158	437	476	123	34	127
Upper Stat	869	714	771	2354	499	156	654
SE	241	206	141	347	38	28	47
Total Stat	1498	1638	3962	7098	1195	342	1537
SE	261	261	455	586	143	46	149
Lower CS	114	512	1629	2254	11	22	33
SE	42	157	373	407	8	12	15
Middle CS	61	121	739	921	2	14	16
SE	31	37	149	157	1	10	10
Upper CS	61	83	444	587	0	0	0
SE	30	34	142	149	0	0	0
Total CS	236	715	2811	3762	13	36	49
SE	96	199	558	600	9	22	24
Total, All	18935	14656	31425	65017	1208	378	1586
SE	752	821	1634	1978	141	44	147

SE Tables for E.4 and E.5

Table E.4	Dist-Lrn, 4-Yr	SE	Other , 4-Yr	SE	Dist-Lrn , TYC	SE	Other, TYC	SE
Pre-coll Math	2489	955	198760	18648	37036	5619	927697	50844
Coll Alg	5856	2268	352591	20962	15721	2423	298081	22492
Calculus I	593	173	308518	13511	3620	944	68919	4452
Calculus II	577	310	94858	5342	270	97	20003	1334
D Eq & Lin Alg	238	146	82034	4504	83	46	7423	731
El Stat in Math	3075	1577	140077	13336	9894	1331	107304	7922
El Stat in Stat	990	485	44303	3400				

Table E.5	T/TE	OFT	PT	GTAs	Ukn	# Math	T/TE	OFT	PT	GTAs	Ukn	# Stat	T/TE	OFT	PT	GTAs	Ukn	# CS
PhD, Math	35	24	14	21	6	17202	39	44	7	9	2	1498	39	38	9	7	6	236
SE	1	1	1	2	1	719	6	9	2	3	1	261	9	9	3	2	4	96
MA, Math	45	20	22	8	6	12303	49	33	15	1	2	1639	43	8	18	0	30	715
SE	3	4	3	2	2	724	7	9	4	1	1	261	11	6	8	0	15	199
BA, Math	54	20	23	1	3	24652	59	13	25	0	3	3962	80	9	9	0	1	2811
SE	3	2	3	0.3	1	1341	5	3	4	0	1	455	6	5	4	0	1	558
Total, Math	46	21	20	9	5	54157	52	24	19	2	2	7099	70	11	11	0	7	3762
SE	2	1	1	1	1	1685	3	4	3	1	1	586	5	4	3	0	4	600
PhD, Stat							41	22	7	14	15	1195						13
SE							4	3	2	4	7	143						9
MA, Stat							64	27	7	0	2	342						36
SE							4	4	2	0	1	46						22
Total, Stat							46	23	7	11	12	1537						49
SE							3	2	1	3	6	149						24

Standard Error Tables for E.6, E.7, and E.8

Table E.6	TTE	SE	OFT tot	SE	OFT doc	SE	PT	SE	GTA	SE	Ukn	SE	Total Sect	SE
PhD, Math	29	11	312	58	34	10	579	112	376	81	66	46	1363	174
MA, Math	55	33	491	177	43	18	616	161	641	167	99	66	1902	305
BA, Math	576	161	980	247	209	118	2091	377	23	17	192	108	3862	581
Total	660	165	1783	309	286	119	3286	425	1040	187	357	134	7126	679
Table E.7														
PhD, Math	588	82	1457	171	341	46	1176	129	1902	235	394	96	5517	454
MA, Math	1849	232	1373	312	197	85	1657	252	295	104	369	129	5543	391
BA, Math	4079	388	2385	413	423	111	2998	469	0	0	432	136	9895	812
Total	6517	460	5215	545	960	147	5831	548	2196	257	1196	211	20955	1009
Table E.8														
PhD, Math	3199	175	1860	141	1155	107	726	82	1261	153	650	159	7696	356
MA, Math	2196	192	375	114	159	69	402	109	16	14	249	101	3237	275
BA, Math	5754	483	900	168	526	126	520	120	107	75	108	48	7388	584
Total	11149	549	3135	247	1841	179	1648	182	1384	171	1006	194	18321	738

Standard Error Tables for E.9, E.10. E.11, and E.12

Table E.9	T/TE	OFT tot	OFT-doc	PT	GTA	Ukn	# Sect
PhD, Math	145	219	73	104	136	25	629
SE	33	68	37	23	35	12	104
MA, Math	441	185	34	250	15	34	924
SE	93	53	21	66	10	17	158
BA, Math	1738	366	90	987	0	100	3191
SE	308	82	38	206	0	47	437
Total Math	2324	770	197	1341	151	159	4744
SE	323	119	57	218	37	52	476
PhD, Stat	144	111	60	88	172	180	696
SE	26	16	10	17	42	106	123
MA, Stat	80	75	22	24	0	7	186
SE	27	12	12	9	0	3	34
Total Stat	224	186	82	112	172	187	882
SE	38	20	16	20	42	106	127

Table E.10							
PhD, Math	31	44	24	10	14	15	114
SE	14	20	12	6	9	11	42
MA, Math	187	50	0	127	0	149	512
SE	65	45	0	63	0	95	157
BA, Math	1199	168	55	256	0	6	1629
SE	276	95	32	119	0	5	373
Total Math	1416	262	79	393	14	169	2254
SE	284	107	34	135	9	96	407

Table E.11	T/TE	OFT tot	OFT-doc	PT	GTA	Ukn	# Sect
PhD, Math	19	36	19	3	3	0	61
SE	12	26	14	3	3	0	32
MA, Math	72	11	0	6	0	33	121
SE	28	10	0	5	0	30	37
BA, Math	613	98	70	6	0	22	739
SE	139	52	47	5	0	22	149
Total Math Depts	703	145	89	15	3	55	921
SE	142	59	49	8	3	37	157

Table E.12	T/TE	Total	Stat Dept	T/TE	Total
PhD, Math	2184	2625			
SE	98	119			
MA, Math	1382	1622			
SE	136	150			
BA, Math	2941	3507			
SE	309	369			
Tot Adv Math	6506	7754			
SE	352	416			
PhD, Math	434	869	PhD, Stat	343	499
SE	38	241	SE	33	38
MA, Math	359	714	MA, Stat	140	156
SE	51	206	SE	31	28
BA, Math	604	771			
SE	90	141			
Tot Adv Stat	1398	2354	Total, Adv Stat	483	654
SE	111	347	SE	44	47
Tot, All Adv	7904	10108	Total, All Adv	483	654
SE	378	545	SE	44	47

Standard Error Tables for E.13 and E.14

Table E.13	PhD,Math	MA, Math	BA, Math	PhD, Stat	MA, Stat	All Depts 05
Precoll, Math	40	31	22			28
SE	3	2	1			1
Intro, Math	48	34	25			33
SE	2	1	1			1
Calculus	45	27	21			32
SE	2	1	1			1
Adv Math	20	15	10			14
SE	1	1	1			1
Elem Stat	47	34	26	60	63	35
SE	4	2	1	8	15	1
Adv Stat	17	13	13	40	22	19
SE	5	5	2	3	4	2
Lower CS	25	22	18	16	66	19
SE	3	2	1	0	7	1
Middle CS	19	8	8	48	16	9
SE	2	0.4	1	0	0	1
Upper CS	15	8	7	0	0	8
SE	2	1	1	0	0	1

Table E.14	Univ (PhD)	Univ (MA)	College (BA)
MS Calc I	28	19	21
SE	1	1	3
MS Calc II	26	20	15
SE	1	0	4
Other Calc	29	na	na
SE	2	na	na
El Stat in Math	30	32	22
SE	2	0	3
El Stat in Stat	32	19	na
SE	2	0.4	na

Standard Error Table for F.1

Table F.1	T	TE	OFT	PD	PT	T	TE	OFT	PD	PT	T	TE	OFT	PD	PT
Mathematics	PhD Depts					MA Depts					BA Depts				
Doc Fac	4699	930	1381	760	412	2412	990	268	5	383	4697	2179	516	48	837
SE	0	0	0	0	0	68	57	26	3	47	233	163	74	24	80
Doc(F)	420	218	336	147	95	480	319	97	2	102	1080	614	166	41	210
SE	0	0	0	0	0	25	23	14	2	18	82	61	41	24	33
Tot Math	4719	933	2049	764	1046	2544	1019	1027	7	1860	5612	2429	1553	48	3630
SE	0	0	0	0	0	78	59	73	3	199	310	184	162	24	273
Tot Math(F)	427	220	735	148	386	532	337	532	2	689	1373	693	792	41	1503
SE	0	0	0	0	0	32	24	47	2	78	106	68	100	24	132
PhD Stat Depts															
Doc Fac	603	178	133	51	76										
SE	5	3	3	2	3										
Doc(F)	79	66	46	16	16										
SE	1	1	1	1	1										
Tot PhD Stat	604	179	163	51	112										
SE	5	3	3	2	3										
Tot PhD Stat(F)	79	66	66	16	33										
SE	1	1	2	1	1										

See the Appendix on statistical methods for a discussion of SE values for mathematics Ph.D. department estimates.

Standard Error Tables for F.2, F.3, and F.4.

Table F.2	T	TE	OFT	PD	T	TE	OFT	PD	T	TE	OFT	PD	T	TE	OFT	PD
Men, 2005	4292	713	1314	616	2011	682	495	4	4239	1737	761	8	10542	3132	2570	628
SE	0	0	0	0	62	45	38	2	231	134	87	5	239	142	95	5
Women, 2005	427	220	735	148	532	337	532	2	1373	693	792	41	2332	1250	2059	191
SE	0	0	0	0	32	24	47	2	106	68	100	24	111	72	111	24
Total, 2005	4719	933	2049	764	2544	1019	1027	7	5612	2429	1553	48	12874	4382	4629	819
SE	0	0	0	0	78	59	73	3	310	184	162	24	320	193	177	25

Table F.3	T	TE	OFT	PD
Men, 2005	525	113	97	35
SE	5	2	2	1
Women, 2005	79	66	66	16
SE	1	1	2	1
Total, 2005	604	179	163	51
SE	5	3	3	2

Table F.4	<30	30-34	35-39	40-44	45-49	50-54	55-59	60-64	65-69	>69
Total PhD Math	1	8	10	13	14	15	14	13	8	4
SE	0	0	0	0	0	0	0	0	0	0
Total MA Math	3	9	16	12	15	13	12	13	5	3
SE	0	1	1	1	1	1	1	1	1	0
Total BA Math	2	10	13	16	13	13	15	12	4	1
SE	0	1	1	1	1	1	1	1	1	0

Standard Error Tables for F.5 and F.6.

Table F.5	Asian	Black	Mex Am	White	Oth/Ukn
PhD Math					
FT Men	12%	1%	2%	66%	1%
SE	0	0	0	0	0
FT Women	3	0	1	14	0
SE	0	0	0	1	0
MA Math					
FT Men	10	3	2	54	2
SE	1	1	0	1	0
FT Women	4	1	2	22	1
SE	2	1	1	2	0
BA Math					
FT Men	6	2	2	57	2
SE	1	1	1	1	0
FT Women	3	1	1	25	1
SE	1	1	0	2	1
PhD Stat					
FT Men	18	1	1	55	2
SE	1	0	0	1	0
FT Women	7	1	0	16	0
SE	2	1	0	2	0

Table F.6	Asian	Black	Mex Am	White	Oth/Ukn
PhD Math					
PT Men	4%	2%	0%	50%	6%
SE	1	0	0	1	1
PT Women	3	0	0	31	2
SE	1	0	0	1	0
MA Math					
PT Men	3	2	2	46	7
SE	1	1	1	2	1
PT Women	2	3	1	33	3
SE	2	1	1	2	0
BA Math					
PT Men	3	3	2	44	8
SE	2	1	1	2	1
PT Women	1	2	1	31	6
SE	1	1	1	2	0
PhD Stat					
PT Men	11	2	1	44	12
SE	2	1	1	3	3
PT Women	1	0	0	23	5
SE	3	0	0	3	0

Standard Error Table for FY.1

Table FY.1	PhD	MA	BA	PhD	MA	BA	PhD	MA	BA	PhD	MA	BA	PhD	MA	BA	PhD	MA	BA	PhD	MA	BA
Math Lib Arts	18	36	43	19	13	16	5	4	4	28	38	32	25	3	0	11	10	9	46	34	25
SE	3	6	6	4	4	4	2	2	2	4	6	7	5	2	0	5	5	4	3	2	1
Fin Math	17	49	31	32	28	14	7	4	4	12	17	55	23	0	0	16	6	0	74	34	23
SE	4	8	5	4	7	8	2	3	3	3	6	8	5	0	0	5	6	0	6	2	2
Bus Math (N-C)	14	30	36	20	23	30	9	5	11	21	41	32	43	2	0	2	3	3	47	34	26
SE	2	4	11	6	9	10	3	4	7	4	10	10	6	1	0	1	2	2	6	3	2
Math Elem Tch	19	45	59	38	24	24	10	2	3	22	24	12	14	1	0	6	6	6	29	27	22
SE	2	5	6	4	7	6	2	1	1	3	6	3	3	1	0	2	3	3	1	1	1
College Algebra	4	24	34	25	36	31	3	5	3	21	26	29	44	6	0	6	7	5	46	41	27
SE	1	6	6	3	10	8	1	3	1	3	6	5	4	4	0	2	3	2	3	3	2
Trigonometry	10	31	30	26	36	32	3	0	2	19	19	39	43	0	0	2	14	0	37	31	27
SE	3	10	9	8	11	14	1	0	2	4	9	15	7	0	0	1	10	0	2	2	2
Col A&T (comb)	6	26	61	45	8	29	10	2	8	19	36	11	29	30	0	1	0	0	57	28	25
SE	2	10	20	7	6	20	3	2	8	5	7	3	5	11	0	1	0	0	8	2	1
El Fnctns, Precal	7	32	43	22	21	22	8	3	0	24	33	35	40	10	0	7	4	0	48	31	25
SE	2	7	10	4	8	7	2	2	0	3	7	8	5	5	0	3	3	0	3	3	1
Int Math Mod	25	36	11	75	14	78	38	0	22	0	50	11	0	0	0	0	0	0	81	31	20
SE	16	20	9	16	11	11	8	0	16	0	9	8	0	0	0	0	0	0	11	4	2
Total Intro Lev	11	33	41	26	25	24	6	4	4	21	30	30	34	5	0	7	7	4	48	34	25
SE	1	4	4	3	5	3	1	2	1	2	4	4	3	2	0	2	2	1	2	1	1

Standard Error Table for FY.2.

Table FY.2	PhD	MA	BA	PhD	MA	BA	PhD	MA	BA	PhD	MA	BA	PhD	MA	BA	PhD	MA	BA	PhD	MA	BA
Math El Tchr	14	38	14	36	58	55	10	13	20	3	2	2	25	31	43	15	20	37	29	27	22
SE	4	10	5	7	10	9	3	5	9	2	1	1	6	7	9	2	3	5	1	1	1
College Algebra	47	41	47	4	13	3	18	3	5	18	6	7	4	3	3	71	63	62	46	41	27
SE	8	10	9	2	8	2	6	1	4	5	3	4	2	1	2	9	10	10	3	3	2
Trigonometry	31	51	70	1	18	5	12	0	5	15	0	5	1	7	5	17	6	7	37	31	27
SE	8	13	12	1	11	4	5	0	4	5	0	4	1	5	4	2	1	2	2	2	2
Coll A&T(comb)	32	57	19	4	4	0	2	0	0	12	0	0	0	4	0	18	7	9	57	28	25
SE	9	15	16	4	3	0	1	0	0	5	0	0	0	3	0	4	2	6	8	2	1
El Fnctns, Precal	47	50	77	2	6	13	6	2	11	17	2	4	2	7	9	47	20	25	48	31	25
SE	8	12	8	1	5	7	3	2	8	6	2	2	1	5	7	6	3	6	3	3	1
Intro Math Mod	25	59	48	25	59	44	0	0	59	0	0	4	13	0	56	1	4	3	81	31	20
SE	16	32	32	15	32	32	0	0	24	0	0	4	10	0	25	1	2	2	11	4	2
All in FY.2	39	44	42	7	23	21	12	4	12	10	3	4	6	10	17	169	120	143	44	34	25
SE	5	7	6	1	6	5	3	1	4	2	1	2	1	2	4	13	11	15	2	2	1

Standard Error Table for FY.3

Table FY.3	PhD	MA	BA	PhD	MA	BA	PhD	MA	BA	PhD	MA	BA	PhD	MA	BA	PhD	MA	BA	PhD	MA	BA
Lect/rec	42	72	62	31	16	24	19	3	17	6	2	14	9	0	0	11	11	0	65	29	23
SE	6	9	9	3	8	8	3	2	8	2	1	6	3	0	0	4	8	0	6	3	1
Regular <31	42	78	83	19	5	9	10	1	5	5	5	5	32	4	0	2	7	2	25	24	21
SE	6	7	3	4	4	3	3	1	2	1	2	2	6	4	0	2	5	2	1	1	1
Regular >30	28	71	94	21	16	0	14	6	0	12	8	6	29	0	0	11	5	0	37	34	33
SE	4	6	5	4	6	0	3	3	0	3	3	5	5	0	0	5	2	0	1	2	1
Total MS Calc I	36	73	79	25	12	12	15	4	7	8	6	7	22	1	0	9	7	2	46	29	22
SE	3	5	3	2	4	3	2	1	2	1	2	2	3	1	0	3	3	1	3	1	1
Lect/rec	51	63	79	29	0	18	20	0	4	4	21	0	7	0	0	8	16	4	64	23	19
SE	5	11	13	4	0	13	4	0	3	2	10	0	4	0	0	4	11	3	6	7	2
Regular<31	38	70	96	20	7	4	14	4	4	6	13	0	36	0	0	1	9	0	26	22	20
SE	5	9	3	4	4	3	3	3	3	2	5	0	5	0	0	0.4	6	0	1	1	1
Regular>30	34	78	100	25	12	0	13	12	0	14	4	0	18	0	0	9	6	0	38	31	35
SE	5	9	0	4	7	0	3	7	0	3	4	0	4	0	0	4	3	0	1	2	1
Total MS Calc II	42	73	94	26	8	6	16	7	3	8	10	0	17	0	0	7	9	1	47	25	20
SE	3	6	3	3	3	3	2	3	2	1	4	0	3	0	0	2	3	1	3	2	1
Total MS Calc I&II	38	73	83	25	11	10	15	5	6	8	7	5	20	1	0	9	7	1	46	28	22
SE	2	5	3	2	3	2	2	2	2	1	2	1	3	1	0	3	3	1	2	1	1

Standard Error Table for FY.4

Table FY.4	PhD	MA	BA	PhD	MA	BA	PhD	MA	BA	PhD	MA	BA	PhD	MA	BA	PhD	MA	BA	PhD	MA	BA
Lect/rec	37	69	57	5	9	25	14	39	33	10	6	0	4	0	27	60	5	14	65	29	23
SE	7	19	14	2	6	13	6	15	13	3	6	0	2	0	13	7	2	4	6	3	1
Regular <31	44	66	59	2	27	16	9	10	25	4	1	2	5	19	6	11	8	44	25	24	21
SE	8	12	9	1	10	5	4	8	6	2	1	1	3	7	3	2	2	7	1	1	1
Regular >30	42	36	65	5	18	14	26	4	32	11	0	0	11	10	32	34	17	7	37	34	33
SE	9	11	15	2	10	9	7	3	17	4	0	0	5	5	15	5	3	2	1	2	1
Total MS Calc I	40	52	59	5	20	18	18	12	27	9	2	2	7	11	12	105	30	65	46	29	22
SE	6	9	7	1	7	5	4	5	5	2	1	1	3	4	4	6	3	6	3	1	1
Lect/rec	23	75	64	4	0	25	8	46	43	6	0	0	1	0	28	31	2	3	64	23	19
SE	6	14	17	2	0	16	3	14	17	3	0	0	1	0	16	4	1	1	6	7	2
Regular<31	42	54	47	6	12	15	17	6	31	3	0	2	3	12	4	6	4	15	26	22	20
SE	8	13	11	2	6	8	6	5	8	2	0	2	2	6	3	1	1	3	1	1	1
Regular>30	37	44	86	1	8	28	15	16	57	8	0	0	2	8	28	16	6	1	38	31	35
SE	9	14	10	1	7	20	5	8	21	3	0	0	1	7	20	2	1	1	1	2	1
Total MS Calc II	32	53	52	3	8	17	13	16	34	6	0	2	2	8	9	54	12	19	47	25	20
SE	5	10	10	1	4	7	3	6	8	2	0	2	1	4	4	4	1	3	3	2	1
Total MS Calc I&II	38	52	57	4	16	18	16	14	29	8	1	2	5	10	11	159	42	84	46	28	22
SE	5	9	7	1	6	4	4	5	5	2	1	1	2	4	4	9	4	9	2	1	1

Standard Error Tables for FY.5 and FY.6

Table FY.5	PhD	MA	BA	PhD	MA	BA	PhD	MA	BA	PhD	MA	BA	PhD	MA	BA	PhD	MA	BA	PhD	MA	BA
Lect/recit	16	27	40	33	9	60	13	9	0	11	0	0	11	0	0	29	64	0	72	28	22
SE	4	22	32	8	7	32	4	7	0	4	0	0	3	0	0	10	29	0	9	2	0.3
Reg. <31	7	46	47	24	7	20	4	1	5	12	27	20	36	0	13	20	20	0	26	23	24
SE	3	14	8	7	3	6	2	1	4	4	14	8	9	0	6	12	15	0	1	1	2
Reg. >30	21	40	75	27	27	6	11	8	3	24	31	19	27	0	0	1	2	0	53	39	28
SE	4	8	7	5	9	3	2	6	1	4	8	7	7	0	0	1	1	0	4	3	3
Total NMS Calc I	17	42	51	28	18	19	10	5	4	17	28	19	24	0	10	14	12	0	52	33	25
SE	3	7	8	4	6	5	2	4	3	3	7	7	5	0	6	5	7	0	4	2	2
Total NMS Calc II	25	47	100	31	13	0	9	0	0	20	40	0	22	0	0	1	0	0	56	18	14
SE	6	15	0	7	8	0	3	0	0	5	14	0	5	0	0	1	0	0	5	4	0
Total NMS Calc I&II	18	42	52	29	18	19	10	5	4	18	28	19	23	0	10	12	12	0	53	32	25
SE	3	7	8	4	5	5	2	3	3	3	7	7	4	0	5	5	7	0	4	2	2

Table FY.6	PhD	MA	BA	PhD	MA	BA	PhD	MA	BA	PhD	MA	BA	PhD	MA	BA	PhD	MA	BA	PhD	MA	BA
Lect/recit	60	36	80	4	0	60	10	0	0	8	0	0	5	0	0	26	1	1	72	28	22
SE	9	29	16	2	0	32	4	0	0	3	0	0	3	0	0	3	1	1	9	2	0.3
Reg. <31	45	44	75	1	2	0	1	0	7	1	0	5	1	0	1	5	5	20	26	23	24
SE	13	16	11	1	2	0	1	0	4	1	0	3	1	0	1	1	2	6	1	1	2
Reg. >30	31	47	35	6	9	6	7	0	0	6	0	13	4	7	6	30	15	5	53	39	28
SE	9	13	20	2	8	6	3	0	0	3	0	11	2	5	6	4	3	2	5	3	3
Total NMS Calc I	43	45	68	4	6	3	7	0	6	6	0	6	4	4	2	61	21	26	52	33	25
SE	6	10	11	1	5	2	2	0	3	2	0	4	1	3	1	5	3	6	4	2	2

Standard Error Tables for FY.7 and FY.8

Table FY.7	PhD	MA	BA	PhD	MA	BA	PhD	MA	BA	PhD	MA	BA	PhD	MA	BA	PhD	MA	BA	PhD	MA	BA
Lec/recit	15	13	41	58	14	17	32	9	0	14	63	34	9	0	0	4	9	8	70	37	22
SE	8	0.4	8	15	12	7	10	8	0	6	20	8	5	0	0	3	8	4	21	0	3
Reg. < 31	1	35	61	51	28	8	22	4	3	14	31	29	33	6	0	0	0	2	24	26	24
SE	1	13	6	14	15	3	11	3	2	4	11	6	13	5	0	0	0	1	0.4	2	1
Reg. > 30	31	53	54	25	20	13	5	2	5	12	22	27	26	1	0	6	3	6	48	41	36
SE	5	6	8	6	6	6	2	2	4	5	8	7	2	1	0	4	3	6	3	1	1
Tot. El Stat	21	45	57	38	21	10	14	3	3	13	28	29	24	2	0	4	3	4	46	37	27
SE	5	5	4	8	6	3	6	2	1	3	7	4	6	1	0	2	2	2	5	1	1
Tot P&S (N-C)	25	53	15	29	17	27	2	5	4	37	25	58	10	0	0	0	6	0	49	33	23
SE	13	9	7	13	4	15	1	3	4	8	9	16	6	0	0	0	3	0	6	1	2
Tot both	21	47	53	37	20	12	13	4	3	17	27	32	22	2	0	3	4	3	47	36	26
SE	5	5	5	8	5	3	5	2	1	4	6	4	5	1	0	2	2	2	4	1	1

Table FY.8	PhD	MA	BA	PhD	MA	BA	PhD	MA	BA	PhD	MA	BA	PhD	MA	BA	PhD	MA	BA	PhD	MA	BA
Lec/recit	0	33	62	0	67	62	69	67	92	0	0	0	0	0	65	7	1	5	70	37	22
SE	0	28	23	0	28	23	14	28	4	0	0	0	0	0	22	2	1	4	21	0	3
Reg. < 31	0	59	29	3	27	31	57	35	58	0	7	4	0	25	20	3	4	47	24	26	24
SE	0	15	10	3	12	11	14	12	11	0	5	3	0	12	12	1	1	11	1	2	1
Reg. > 30	36	39	52	24	12	26	17	40	64	0	1	3	12	6	2	14	20	23	48	41	36
SE	13	18	15	10	6	15	8	15	12	0	1	3	7	4	2	3	6	6	3	1	1
Tot. El Stat	21	43	37	14	21	33	36	41	62	0	2	4	7	10	20	23	24	74	46	35	27
SE	9	14	8	6	6	8	10	12	8	0	2	2	4	4	9	4	6	11	5	2	1
Tot P&S (N-C)	19	3	35	8	0	79	85	13	61	0	0	0	19	0	53	4	7	7	49	33	23
SE	12	3	14	6	0	15	9	11	15	0	0	0	12	0	18	1	2	5	6	1	2
Tot both	21	34	37	13	16	37	43	35	62	0	2	3	8	7	23	27	31	81	43	32	26
SE	8	13	8	5	5	8	9	10	8	0	2	2	4	3	8	4	6	11	4	3	1

Standard Error Tables for FY.9 and FY.10

Table FY.9	PhD	MA	PhD	MA	PhD	MA	PhD	MA	PhD	MA	PhD	MA	PhD	MA
Lect/Recit	18	26	21	63	8	0	16	11	20	0	25	0	75	121
SE	5	10	3	15	2	0	5	7	5	0	12	0	12	39
Reg <31	31	40	8	60	8	60	8	0	28	0	24	0	21	29
SE	6	0	2	0	2	0	2	0	11	0	5	0	2	0
Reg > 30	18	58	11	20	10	4	18	17	48	0	5	5	58	38
SE	6	5	4	4	4	3	7	2	15	0	3	4	5	1
Tot El Stat	19	46	17	37	9	6	16	14	30	0	18	3	67	66
SE	4	8	2	10	2	4	4	3	7	0	8	2	7	18
Prob&Stat	41	25	19	63	13	63	0	0	28	0	13	13	95	30
SE	14	7	8	3	4	3	0	0	7	0	4	10	9	0.03
Tot ElStat& P&S	20	44	17	39	9	11	16	13	29	0	18	4	64	62
SE	4	8	2	10	2	7	4	4	7	0	8	2	6	16
Stat Lit	13	0	22	67	12	33	10	33	20	0	35	0	61	94
SE	6	0	5	12	5	12	5	12	9	0	18	0	12	14
Total, FY.9	19	43	17	40	9	12	14	13	27	0	23	4	68	63
SE	4	8	2	10	2	7	3	3	7	0	10	2	7	15

Table FY.10	PhD	MA	PhD	MA	PhD	MA	PhD	MA	PhD	MA	PhD	MA	PhD	MA
Lect/Recit	10	0	37	74	56	74	28	15	29	41	22	7	75	121
SE	5	0	6	18	6	18	11	9	8	23	2	3	12	39
Reg <31	2	0	24	0	82	100	20	80	20	0	0	0	21	29
SE	1	0	14	0	10	0	13	0	13	0	0	0	2	0
Reg > 30	2	0	62	48	43	67	0	2	6	48	9	4	58	38
SE	1	0	14	13	16	6	0	2	4	13	2	2	5	1
Tot El Stat	7	0	44	54	54	71	18	11	20	43	31	11	67	66
SE	3	0	7	10	7	7	8	6	6	12	3	2	7	18

Standard Error Tables for TYE.1, TYE.2, TYE.3, TYE.4, TYE.5, and TYE.6.

Table TYE.1	See NCES

Table TYE.3	2005	SE
Arith	104	13
Pre-alg	137	16
El alg (HS)	380	22
Int alg(HS)	336	20
Geom(HS)	7	1
Intro lev		
Col alg	206	20
Trig	36	3
Col A&T	14	4
In Math Mod	7	3
Precalc/El F	58	7
Calc level		
MS Calc I	51	3
MS Calc II	19	1
MS Calc III	11	1

Table TYE.2	1739	SE = 77
	2005	SE
NMS-Calc I	21	2
NMS-Calc II	1	0.2
DEq	4	0.4
Lin Alg	3	0.5
Disc Math	2	0.4
El Stat(no Pr)	111	8
Pr (w/wo St)	7	3
Fin Math	22	4
Math Lib Arts	59	7
Math El Tchrs	29	3
Bus Math(cmb)	26	5
Tech Math (NC)	16	4
Tech Math (C)	1	0.4
Other	28	7
Tot TYC Math	1696	75

Table TYE.4		SE
Precoll	964	51
Precalc	321	24
Calc	107	6
Stat	118	9
Other	186	14
Total	1696	75

Table TYE.6		
MS Calc I	82	4
DEq	25	3
Lin Alg	19	3
Disc Math	12	3
El Stat	78	4
Fin Math	28	4
Math Lib Arts	56	5
Math El Tchrs	59	5
Tech Math (NC)	35	5
Tech Math (C)	5	2

Table TYE.5		SE			SE
Arith	48	5	Disc Math	22	4
Pre-alg	47	5	El Stat	80	4
El Alg	80	4	Prob	8	3
Int Alg	88	3	Fin Math	35	4
Geom	24	4	Math Lib Arts	65	5
Coll Alg	79	4	Math El Tchrs	66	5
Trig	63	5	Bus Math (NTr)	22	4
Coll A&T	17	3	Bus Math (Tr)	17	3
In Math Mod	7	3	Tech Math (NC)	36	5
Precalc/El F	60	5	Tech Math (C)	7	2
MS Calc I	87	3			
MS Calc II	78	4			
MS Calc III	70	4			
NMS-Calc I	46	5			
NMS-Calc II	6	2			
DEq	58	5			
Lin Alg	41	5			

Standard Error Tables for TYE.7, TYE.8, TYE.9, and TYE.11. (See next table for TYE.10.)

Table TYE.7	2005	SE	>30	SE
Precoll	23.9	0.8	21%	2
Precalc	23.6	0.9	23%	2
Calculus	20	0.6	16%	2
Stat	25.9	0.6	33%	3
All courses	23	0.6	21%	2

Table TYE.8	Size	SE		Size	SE
Arith	22.7	2	DEq	14.2	1
Pre-alg	22.3	1	Lin Alg	16.3	1
El Alg	24	0.9	Misc Math	14.3	2
Int Alg	25.1	0.8	El Stat	26.1	0.6
Geom (HS)	17.8	3	Prob	22.6	1
Coll Alg	24.7	0.9	Fin Math	25.3	0.9
Trig	22.5	0.9	Math Lib Arts	24	0.7
Coll A&T	21.7	1	Math El Tchrs	15.4	3
Math Mod	24.6	2	Bus Math(NT)	21.1	1
Precalc	21.2	2	Bus Math(T)	8.6	5
MS Calc I	21.9	0.6	Tech Math(NC)	18.7	1
MS Calc II	18.2	0.8	Tech Math (C)	18.1	2
MS Calc III	15.6	1	Other	22	2
NMS Calc I	22.9	0.9			
NMS-Calc II	20.8	2			

Table TYE.9		SE		SE	%PT	SE
Precoll	38814	2327	21696	1595	56%	2
Precalc	12898	972	3914	373	30	2
MS Calc	3973	231	493	58	12	1
NMS Calc	923	104	254	36	28	4
Adv Lv	617	53	58	20	9	3
Stat	4142	286	1452	131	35	2
Serv Crs	6710	1021	1913	196	29	5
Tech Math	927	171	339	85	37	6
Other	1193	249	552	126	46	7
Total	70197	3420	30671	1988	44	1

Table TYE.10 See next SE table.

Table TYE.11						
Type	Group	SE	Writing	SE	# Sect	SE
MS Calc I	19	3	19	3	2226	138
MS Calc II	25	4	18	3	1054	78
MS Calc III	20	4	16	4	693	55
NMS Calc I	14	3	14	4	883	103
NMS Calc II	27	16	21	16	40	11

Standard Error Table for TYE.10

Table TYE.10	Graph calc	SE	Writing	SE	Cmptr	SE	Group	SE	On-line	SE	Std Lect	SE	# Sect	SE
Arith	2	1	3	1	13	4	9	4	14	4	64	6	4400	544
Pre-alg	5	3	9	3	18	5	9	3	7	2	74	5	5954	715
El Alg	17	4	7	2	14	3	8	2	11	2	74	4	15331	1022
Int Alg	32	4	8	2	13	3	9	2	11	2	77	3	12773	771
Geom(HS)	33	19	25	12	23	16	15	6	0	0	68	12	356	74
Col Alg	60	6	17	5	8	2	14	3	14	4	74	4	7866	749
Trig	67	5	14	5	3	1	16	4	7	4	81	5	1529	137
Col A&T	53	14	8	3	25	13	10	3	13	5	78	7	654	174
Math mod	80	12	38	13	17	7	59	12	6	4	64	16	248	97
Precalc	75	8	14	4	9	4	21	5	6	2	76	8	2601	369
MS Calc I	79	4	19	3	20	3	19	3	5	1	81	4	2226	138
MS Calc II	81	4	18	3	30	4	25	4	7	2	86	3	1054	78
MS Calc III	74	5	16	4	28	5	20	4	4	2	83	5	693	55
NMS Calc I	77	5	14	4	9	3	14	3	3	1	76	5	883	103
NMS Calc II	40	17	21	16	0	0	27	16	0	0	89	8	40	11
DEq	81	7	11	5	27	6	21	7	5	3	93	6	290	33
Lin Alg	60	8	18	6	29	7	14	5	0	0	68	9	204	31
Disc Math	47	12	39	12	33	11	23	11	0	0	82	7	123	27
El Stat	73	5	44	5	45	5	24	4	10	3	85	3	3872	270
Prob	83	11	55	13	49	10	50	15	0	0	68	7	270	125
Fin Math	55	9	17	6	19	10	11	5	3	2	68	8	844	146
Math Lib Arts	33	6	36	5	7	2	25	5	6	2	79	5	2232	244
Math Elem Tchrs	21	7	52	12	13	5	48	11	3	2	48	11	1665	401
Bus Math(NT)	6	4	2	2	18	9	1	1	0	0	87	5	539	167
Bus Math(T)	18	10	7	5	7	3	6	5	2	2	24	14	1430	864
Tech Math(NC)	39	8	4	2	5	3	5	2	5	3	72	7	863	170
Tech Math(C)	63	16	17	12	21	12	30	15	0	0	83	12	64	20
Other	27	9	10	4	5	2	7	3	6	3	63	10	1193	249

Standard Error Tables for TYE.12, TYE.13. TYE.14, TYE.15, and TYE.16.

Table TYE.12					
Arithmetic	104	13	NMS Calc I	21	2
Pre-algebra	137	16	NMS Calc II	1	0.2
El Alg (HS)	380	22	D Eq	4	0.4
Int Alg (HS)	336	20	Lin Alg	3	0.5
Geom (HS)	7	1	Discr Math	2	0.4
Col Alg	206	20	El Stat	111	8
Trig	36	3	Prob	7	3
Coll A&T	14	4	Fin Math	22	4
Math Model	7	3	Math Lib Arts	59	6
Precalc	58	7	Math El Tchrs	29	3
MS Calc I	51	3	Bus Math (NT)	13	2
MS Calc II	19	1	Bus Math (T)	14	3
MS Calc III	11	1	Tech Math (NC)	16	4
			Tech Math (C)	1	0.4

Table TYE.13		
Diag Tests	96	3
Math Lab	95	3
Advising	40	5
Contests	37	4
Honors	24	4
Club	22	4
Minority Prog	15	3
Colloq	6	2
Women Prog	7	2
K-12 Outreach	25	4
REU	9	3
Indep Stud	38	5
Other	4	1

Table TYE.14	2005	SE
CAI	75	4
Software	72	5
Internet	77	4
Media	68	5
Study Sess	62	5
Tutor/students	94	2
Tutor/parapr	67	5
Tutor/PT	48	5
Tutor/FT	51	5

Table TYE.15	2005	SE
Arith/Pre Alg	60	15
El Alg (HS)	65	27
Int Alg (HS)	26	10
Bus Math	15	2
Stat & Prob	12	2
Tech Math	10	3
Total	188	44

Table TYE.16	OP	SE	Bus	SE	LC	SE	Other	SE
Arith/Pre Alg	0.9	0.5	0.7	0.4	9	4	50	15
El Alg (HS)	0.7	0.3	0.1	0.1	5	3	59	26
Int Alg (HS)	0	0	0	0	3	2	22	10
Bus Math	0.5	0.3	14	2	0	0	0.6	0.4
Stat & Prob	0.5	0.5	8	2	0	0	4	1
Tech Math	8	3	0.1	0.1	0	0	1	0.9
Total	11	3	23	3	17	8	137	43

Standard Error Tables for TYF.1, TYF.2, TYF.3, TYF.4, TYF.5, and TYF.6.

Table TYF.1	2005	SE
FT Perm	8793	398
FT Temp	610	163
PT(by TYC)	18227	900
PT(by other)	1915	509

Table TYF.2	<10	10 to 12	13 to 15	16 to 18	19 to 21	>21
% TYC	0	6	79	8	4	3
SE	0	2	4	2	2	2

Avg CH	SE	Extra	SE	Hrs	SE	Other	SE
15.3		52.5	2.9	3.6	0.1	7.6	1.2

Table TYF.3	2005	SE
HS	25%	3
Other TYC	2	0.4
Other dept	5	1
4-yr coll	2	0.3
Indust	14	2
Grad Sch	3	0.4
None above	49	4
# PT	18227	900

Table TYF.5	PhD	MA	BA
Math	8	61	1
SE	2	2	0.6
Stat	0.3	2	0
SE	0.2	0.5	0
Math Ed	4	14	0
SE	1	1.5	0
Other	3	5	1
SE	1	1	0.4
Total	16	82	2
SE	2	2	1

Table TYF.4	2005	SE
PhD	16	2
MA	82	2
BA	2	0.8
# FT	8793	398

Table TYF.6	2005	SE
PhD	6	1
MA	72	2
BA	22	2
# PT	20142	1066

Standard Error Tables for TYF.7, TYF.8, TYF.9, TYF.10, TYF.11, TYF.12, TYF.13 and TYF.14.

Table TYF.7	PhD	SE	MA	SE	BA	SE	Total	SE
Math	2	0.3	36	3	11	1.4	49%	3
Math Ed	0.6	0.2	20	3	7	1.5	27%	3
Stat	0	0	2.5	0.5	0	0	3%	0.4
Other	3	0.5	14	2	4	1	21%	2
Total	6	xxx	72	2	22	2		

Table TYF.10	2005	SE
%FTP Minority	14	1
#FTP Minority	1198	114
# FTP	8793	398

Table TYF.8	2005	SE
Men	4420	231
Women	4373	256
Total	8793	398

Table TYF.9	FT Perm	SE	PT	SE
Men	50	1.4	53	1.3
Women	50	1.4	47	1.3
Number	8793	398	18227	900

Table TYF.14		SE
% Minority PT	16	1
# PT	18227	900

Table TYF.11	2005	SE
Am In	0.3	0.2
Asian	6	1
Black	5	0.7
Mex Am	3	0.5
White	84	1.3
Ukn	2	0.7
# FT	8793	398

Table TYF.12	# FT	SE	% Ethnic	SE	% Women	SE
Am In	27	16	0.3	0.2	0	0
Asian	538	83	6	1	52	5
Black	413	63	5	0.7	47	5
Mex Am	280	41	3	0.5	43	7
White	7353	376	84	1	51	2
Ukn	182	59	2	0.7	34	7
# FT	8793	398				

Table TYF.13	% among FTP	% among <40
Minority	14	23
SE	1	3
White	84	76
SE	1	3
Unknown	2	1
SE	0.7	0.6

Standard Error Tables for TYF.15, TYF.16, TYF.17, TYF.18, TYF.19, TYF.20, TYF.21, and TYF.22.

Table TYF.15	% Ethnic	SE	% Women	SE
Am Ind	0.6	0.3	18	11
Asian	5.7	0.7	46	4
Black	6.5	1	47	4
Mex Am	2.9	0.6	45	7
White	81	1.7	48	1
Unknown	2.9	0.7	45	7
Total	18227	900		

Table TYF.16	2005	SE	2005	SE
<30	5	1	478	83
30–34	8	1	716	87
35–39	12	1	1037	89
40–44	13	1	1163	99
45–49	15	1	1298	114
50–54	18	1	1574	141
55–59	17	1	1528	126
>59	11	1	999	103
Total			8793	398

Table TYF.17	Women	SE	Men	SE	% Women	SE
<35	7	1	7	1	49	4
35–44	13	1	12	1	50	3
45–54	18	1	15	1	55	3
>54	12	1	16	1	43	3
Total	50%		50%			

Table TYF.18	2005%	SE
Grad Sch	23	5
4-yr Coll	18	5
TYC	11	3
Sec Sch	13	4
PT at TYC	29	5
Nonacademic	5	3
Unemployed	0	0
Unknown	1	0.5
# hired	605	66

Table TYF.19	2005–2006	SE
PhD	12	4
MA	84	5
BA	5	3
Ukn	0	0

Table TYF.20	2005–2006	SE	% Women	SE
Asian	7	3	49	23
Black	1	1	100	0
MexAm	11	4	62	19
White	80	5	52	6
Other	1	1	31	13

Table TYF.21	2005	SE
<30	22	6
30–34	20	5
35–39	17	5
40–44	15	5
45–49	15	5
50–54	5	2
55–59	0	0
>59	6	4

Table TYF.22		SE
Dies/Ret	292	56
To 4 yr Coll	9	4
To TYC	14	9
To Sec Sch	2	2
Nonacad	5	5
Grad Sch	3	2
Other	107	35
Unknown	7	5
Total	439	63

Standard Error Tables For TYF.23, TYF.24, TYF.25, TYF.26, TYF.27, and TYF.30. (Tables for TYF.28 amd TYF.29 are separate.)

Table TYF.23	2005	SE
Own Desk	5	2
Share with 1	7	2
Share more	65	4
None	23	4

Table TYF.24	% PT	SE
In Office	63	5
Nearby	35	4
None	2	0.6

Table TYF.25	% in 05	SE
All FT	89	3
All PT	89	3

Table TYF.26	PT	SE	FT	SE
Other Fac	64	5	52	5
Div Head	33	5	61	5
Students	94	3	96	2
Lesson Plans	49	5	55	5
Self-eval	19	4	46	5
Other	0	0	5	2

Table TYF.27	% PermFT	SE
Employer	53	1
Prof Assoc	38	1
Publish	6	0.7
Grad Study	7	1

TYF.28: See Later Std Error Table

TYF.29: See Later Std Error Table

Table TYF.30	Own Campus	SE	Multicampus	SE
Math Dept	39	4	2	1
Math & Sci	35	5	1	0.7
Other Str	15	4	2	1
None Above	6	3		

Standard Error Tables for TYF.28 and TYF.29.

Table TYF.28	Minor/None	SE	Somewhat	SE	Major	SE
Maintaining vitality	77	4	21	4	2	2
Dual-enrollment	74	5	21	4	5	3
Staffing statistics courses	88	3	9	3	3	2
Students misunderstand coll wk	10	3	35	5	55	5
PT faculty for too many courses	38	5	32	4	30	4
Faculty salaries too low	32	4	46	5	22	4
Class sizes too large	72	4	23	4	5	2
Low student motivation	20	4	31	5	50	5
Remediation	8	3	28	5	63	5
Lack of student progress	29	5	37	5	34	4
Low success rate	58	5	35	5	7	2
Too few transfers	73	4	23	4	4	2
Inadequate travel funds	56	5	22	4	22	4
Inadequate classroom technology	74	4	14	3	12	4
Inadequate computers for PT	72	4	18	4	9	3
Inadequate computers/students	89	3	10	3	1	1
Commercial outsourcing	98	2	2	2	0	0
Heavy classroom duties prevent	47	5	39	5	14	4
Coordinating with high schools	77	4	17	4	7	3
Lack of curricular flexibility	77	4	17	4	7	3
Use of distance education	83	4	11	3	6	2
Table TYF.29	cf TYF.28					